# Springer Series in Optical Sciences

Volume 178

AF775070



For further volumes:
http://www.springer.com/series/624

## Springer Series in Optical Sciences

The Springer Series in Optical Sciences, under the leadership of Editor-in-Chief William T. Rhodes, Georgia Institute of Technology, USA, provides an expanding selection of research monographs in all major areas of optics: lasers and quantum optics, ultrafast phenomena, optical spectroscopy techniques, optoelectronics, quantum information, information optics, applied laser technology, industrial applications, and other topics of contemporary interest.
With this broad coverage of topics, the series is of use to all research scientists and engineers who need up-to-date reference books.
The editors encourage prospective authors to correspond with them in advance of submitting a manuscript. Submission of manuscripts should be made to the Editor-in-Chief or one of the Editors. See also www.springer.com/series/624

PD Lin

# New Computation Methods for Geometrical Optics

Springer

PD Lin
Department of Mechanical Engineering
National Cheng Kung University
Tainan
Taiwan

ISSN 0342-4111 ISSN 1556-1534 (electronic)
ISBN 978-981-10-1341-6 ISBN 978-981-4451-79-6 (eBook)
DOI 10.1007/978-981-4451-79-6
Springer Singapore Heidelberg New York Dordrecht London



Printed on acid-free paper

Springer is part of Springer Science+Business Media (www.springer.com)

*To my former advisor*

Kornel F. Ehmann

*To my wife*

Chiung-Jung Huang

*and*

*In memory of my past wife*

Su-Chin Wang

# Preface

The traditional geometrical optics is based on raytracing only. It is very difficult, if possible, to compute the first- and second-order derivatives of a ray and optical path length with respect to system variables, since they are recursive functions. Consequently, current commercial software packages use a finite difference approximation methodology to estimate these derivatives for use in optical design and analysis. Furthermore, previous publications of geometrical optics use vector notation, which is comparatively awkward for computations for non-axially symmetrical systems. In order to circumvent these limitations, this book employs homogeneous coordinate notation to compute the first- and second-order derivative matrices of various optical quantities. It will be one of the important mathematical tools for automatic optical design.

This book is dedicated to Department of Mechanical Engineering, National Cheng Kung University in Taiwan. It is almost impossible to complete this book without its wonderful environment. My special thanks are also delivered to National Science Council of Taiwan for its financial supports every year.

Taiwan PD Lin

# Contents

# Chapter 1
# Homogeneous Coordinate Notation

Homogeneous coordinate notation is one of powerful mathematical tools for engineering field. It was used to study the motion of rigid bodies [1, 2], robotics [3], theory of gearing [4], and computer graphics [5]. Previous publications of geometrical optics use vector notation, which is comparatively awkward for computations for non-axially symmetrical systems. In order to circumvent these limitations, this book employs homogeneous coordinate notation to compute various optical quantities. The discussion in this chapter is designed mainly for those readers who have not so far encountered this mathematical tool. The treatment is elementary and covers only what will needed to understand the rest of the book.

## 1.1 Set, Sequence, and Matrix Notations

Sets and their theory are important foundational tools in mathematics and philosophy. Any collection of objects that can either be listed or described by some predicate constitutes a set. As shown in Table 1.1, the components of a set are surrounded by two curly brackets. To use set notation we need the short list of commonly used symbols. The second symbol, $\in$, indicates membership of an object in a particular set. The negation of this, or nonmembership is indicated by "$\notin$" ($x$ is not in G). Given two sets G and F, G $=$ F if and only if they have exactly the same components. Another operation on two sets is the difference "G minus F", written G $-$ F, which 'subtracts' from G all components which are in F.

In mathematics, a sequence is an ordered list of objects or events (called as components) with double rounded brackets around its components. The number of ordered components (possibly infinite) is called the length of the sequence. Unlike a set, exactly the same components can appear multiple times at different positions in the sequence. Therefore, (1,2,3) and (1,2,2,3) are two different sequences. $\{1,2,3\} = \{3,2,1\}$ is a set.

In mathematics, a 2D matrix is a rectangular array of numbers, symbols, or expressions, arranged in rows and columns surrounded by squared brackets. The

PD Lin, *New Computation Methods for Geometrical Optics*,
Springer Series in Optical Sciences 178, DOI: 10.1007/978-981-4451-79-6_1,

**Table 1.1** Five commonly used symbols of set theory

| Symbol | Symbol name | Meaning/definition | Example |
|---|---|---|---|
| {} | Set | A collection of elements | G = {3,7,9,14}, F = {9,14,28} |
| $a \in G$ | Element of | Set membership | G = {3,9,14}, $3 \in G$ |
| $X \notin G$ | Not element of | No set membership | G = {3,9,14}, $1 \notin G$ |
| G = F | Equality | Both sets have the same members | G = {3,9,14}, F = {3,9,14}, G = F |
| G − F | Relative complement | Objects that belong to G and not to F | G = {3,9,14}, F = {1,2,3} G − F = {9,14} |

individual items in a matrix are called its components. The component in the wth row and vth column of a matrix is referred as the (w,v)th component of the matrix. Sometimes a matrix is referred to by giving a formula for its (w,v)th component, often with double square brackets around the formula for the component, for example if the (w,v)th component of $\bar{A}_i$ were given by $a_{wv}$, $\bar{A}_i$ would be denoted as $\bar{A}_i = [a_{wv}]$.

One has to note that in this book a matrix is denoted by using an overbar above its symbol. Consequently, $\bar{R}_i$ represents a matrix and $R_i$ is a scalar. A symbol with an underbar (e.g., $\underline{\theta}_i$) or with a caret over it (e.g., $\hat{m}$) is also used to represent a scalar. One has to note that in this book a vector can be expressed as (1,2,3) in sequence notation or a column matrix (e.g., $[1 \quad 2 \quad 3 \quad 1]^T$) in homogeneous coordinate notation.

## 1.2 Vectors

In homogeneous coordinate notation of 3-Dimensional (3D) space, the ith position vector ${}^g\bar{P}_i = P_{ix}\bar{i} + P_{iy}\bar{j} + P_{iz}\bar{k}$, where $\bar{i}$, $\bar{j}$, and $\bar{k}$ are the unit vectors along the $x_g$, $y_g$, and $z_g$ axes, respectively, of a coordinate frame $(xyz)_g$, is written as a column matrix

$$ {}^g\bar{P}_i = \begin{bmatrix} P_{ix} \\ P_{iy} \\ P_{iz} \\ 1 \end{bmatrix}. \tag{1.1} $$

The pre-superscript "g" of the leading symbol ${}^g\bar{P}_i$ indicates that this ith vector is defined with respect to the coordinate frame $(xyz)_g$. For simplicity, we do not use $[{}^gP_{ix} \quad {}^gP_{iy} \quad {}^gP_{iz} \quad 1]^T$ to represent the components of ${}^g\bar{P}_i$, since we always know its reference coordinate frame is $(xyz)_g$ from the leading symbol. To save space while designating a vector, we also represent the position vector by the row matrix, ${}^g\bar{P}_i = [P_{ix} \quad P_{iy} \quad P_{iz} \quad 1]^T$, where the post-superscript T indicates the transpose of a row vector as a column vector. This notation is also applicable to the ith unit directional vector

$$
{}^{g}\overline{\ell}_{i} = \begin{bmatrix} \ell_{ix} \\ \ell_{iy} \\ \ell_{iz} \\ 0 \end{bmatrix}, \tag{1.2}
$$

where $\ell_{ix}^2 + \ell_{iy}^2 + \ell_{iz}^2 = 1$. The vector $[0 \quad 0 \quad 0 \quad 0]^T$ is undefined. Given two vectors, ${}^{g}\overline{P}_{i} = [P_{ix} \quad P_{iy} \quad P_{iz} \quad 1]^T$ and ${}^{g}\overline{P}_{j} = [P_{jx} \quad P_{jy} \quad P_{jz} \quad 1]^T$, the dot product is a scalar given by

$$
{}^{g}\overline{P}_{i} \cdot {}^{g}\overline{P}_{j} = P_{ix}P_{jx} + P_{iy}P_{jy} + P_{iz}P_{jz}. \tag{1.3}
$$

Note that both ${}^{g}\overline{P}_{i}$ and ${}^{g}\overline{P}_{j}$ have to be expressed with respect to the same coordinate frame $(xyz)_g$. The cross product of ${}^{g}\overline{P}_{i}$ and ${}^{g}\overline{P}_{j}$, indicated by ${}^{g}\overline{P}_{i} \times {}^{g}\overline{P}_{j}$, is another vector perpendicular to the plane formed by ${}^{g}\overline{P}_{i}$ and ${}^{g}\overline{P}_{j}$. It can be computed by

$$
{}^{g}\overline{P}_{i} \times {}^{g}\overline{P}_{j} = \begin{bmatrix} P_{iy}P_{jz} - P_{iz}P_{jy} \\ P_{iz}P_{jx} - P_{ix}P_{jz} \\ P_{ix}P_{jy} - P_{iy}P_{jx} \\ 1 \end{bmatrix}. \tag{1.4}
$$

Again, both ${}^{g}\overline{P}_{i}$ and ${}^{g}\overline{P}_{j}$ are expressed with respect to the same coordinate frame $(xyz)_g$.

*Example 1.1* The position vector of point $\overline{P}_1$ shown in Fig. 1.1 can be represented as ${}^{4}\overline{P}_{1} = [3 \quad 4 \quad 2 \quad 1]^T$ and ${}^{5}\overline{P}_{1} = [-4 \quad 3 \quad 2 \quad 1]^T$ with respect to the coordinate frames $(xyz)_4$ and $(xyz)_5$, respectively.

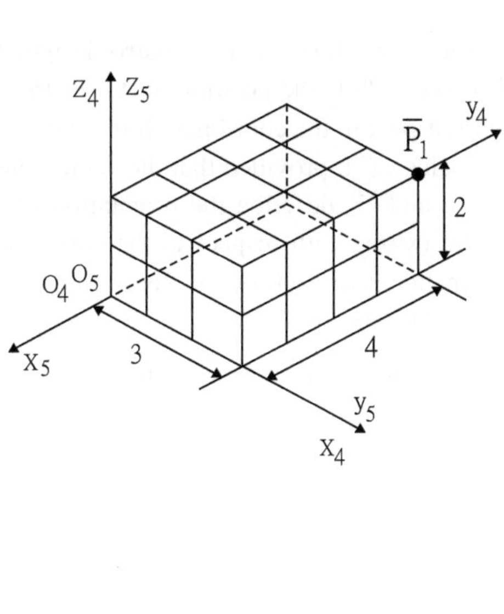


**Fig. 1.1** The position vector of a point

Fig. 1.2 An unit directional vector $\bar{\ell}_4$

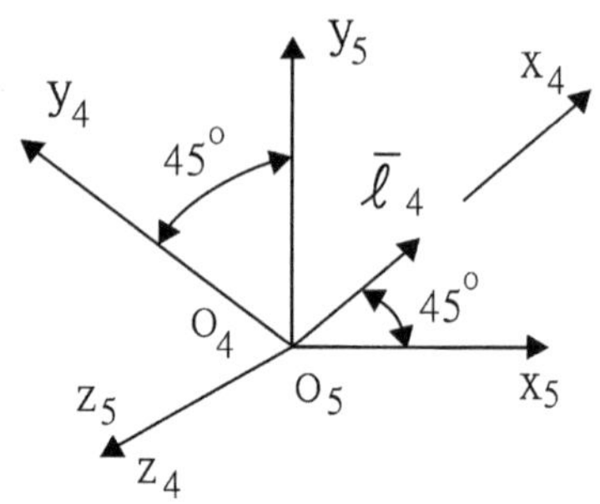


*Example 1.2* The unit directional vector $\bar{\ell}_4$ of Fig. 1.2 can be represented as ${}^4\bar{\ell}_4 = \begin{bmatrix} 1 & 0 & 0 & 0 \end{bmatrix}^T$ and ${}^5\bar{\ell}_4 = \begin{bmatrix} 1/\sqrt{2} & 1/\sqrt{2} & 0 & 0 \end{bmatrix}^T$ with respect to the coordinate frames $(xyz)_4$ and $(xyz)_5$, respectively.

## 1.3 Coordinate Transformation Matrix

In this book, the orientation of the three axes of a coordinate frame relative to each other should always comply with the right-hand rule, unless specifically stated otherwise. A homogeneous coordinate transformation matrix ${}^h\bar{A}_g$ describes the relative position and orientation (referred as pose hereafter) of $(xyz)_g$ with respect to another coordinate frame $(xyz)_h$ in 3D space. It has the following form:

$$ {}^h\bar{A}_g = \begin{bmatrix} {}^h\bar{I} & {}^h\bar{J} & {}^h\bar{K} & {}^h\bar{t} \end{bmatrix} = \begin{bmatrix} I_x & J_x & K_x & t_x \\ I_y & J_y & K_y & t_y \\ I_z & J_z & K_z & t_z \\ 0 & 0 & 0 & 1 \end{bmatrix}. \tag{1.5} $$

Equation (1.5) is a 4×4 matrix in which each column corresponds to a vector. The vector ${}^h\bar{t}$ is the position vector of the origin $o_g$ of the coordinate frame $(xyz)_g$ with respect to the coordinate frame $(xyz)_h$ (see Fig. 1.3). There is no restriction to the value of ${}^h\bar{t}$, provided that the vector can reach the desired position. The vectors ${}^h\bar{I}$, ${}^h\bar{J}$, and ${}^h\bar{K}$ describe the orientation of the three unit vectors of the coordinate frame $(xyz)_g$ with respect to the coordinate frame $(xyz)_h$. ${}^h\bar{I}$, ${}^h\bar{J}$, and ${}^h\bar{K}$ are not independent of each other. Their nine components are related with the following six equations:

(1) ${}^h\bar{I}$ is the vector cross product of ${}^h\bar{J}$ and ${}^h\bar{K}$, i.e.

$$ {}^h\bar{I} = {}^h\bar{J} \times {}^h\bar{K} = \begin{bmatrix} J_yK_z - J_zK_y \\ J_zK_x - J_xK_z \\ J_xK_y - J_yK_x \\ 0 \end{bmatrix}. \tag{1.6} $$

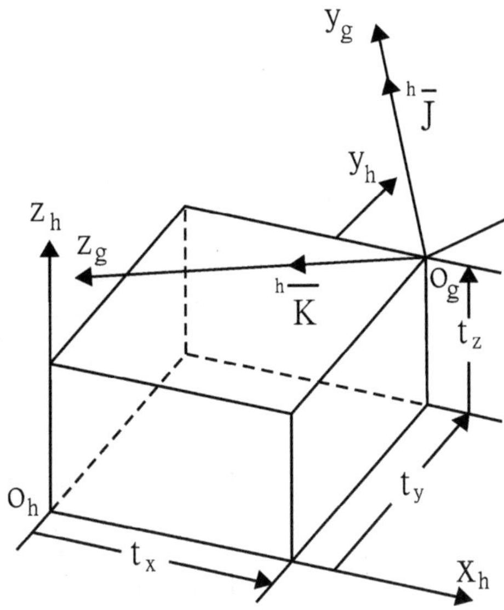


**Fig. 1.3** The interpretation of homogeneous coordinate transformation matrix ${}^h\bar{A}_g$

(2) The vectors ${}^h\bar{J}$ and ${}^h\bar{K}$ must be of unit magnitude and perpendicular to each other, i.e.

$$J_x^2 + J_y^2 + J_z^2 = 1, \tag{1.7}$$

$$K_x^2 + K_y^2 + K_z^2 = 1, \tag{1.8}$$

$$J_x K_x + J_y K_y + J_z K_z = 0. \tag{1.9}$$

Given a vector ${}^g\overline{P}_i$, its transformation ${}^h\overline{P}_i$ can be obtained by the matrix product ${}^h\overline{P}_i = {}^hA_g{}^g\overline{P}_i$.

To perform successive coordinate transformation, we need only to follow the concatenation property. For instance, the matrix equation

$${}^h\bar{A}_g = {}^h\bar{A}_a{}^a\bar{A}_b \ldots \ldots {}^d\bar{A}_e{}^e\bar{A}_g \tag{1.10}$$

represents successive coordinate transformation from $(xyz)_h$ to $(xyz)_a$, from $(xyz)_a$ to $(xyz)_b$,…., and from $(xyz)_e$ to $(xyz)_g$.

*Example 1.3* The pose matrix ${}^4\bar{A}_5$ of the two coordinate frames shown in Fig. 1.4 is given by

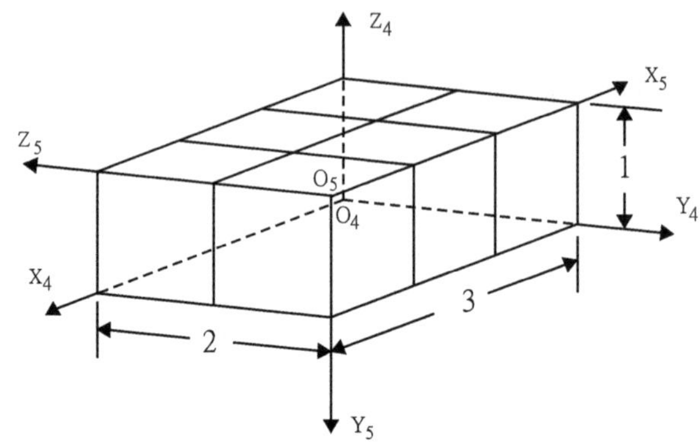


**Fig. 1.4** The illustration of the pose matrix $^4\bar{A}_5$

$$^4\bar{A}_5 = \begin{bmatrix} -1 & 0 & 0 & 3 \\ 0 & 0 & -1 & 2 \\ 0 & -1 & 0 & 1 \\ 0 & 0 & 0 & 1 \end{bmatrix}. \tag{1.11}$$

*Example 1.4* Given a vector $^5\overline{P}_1 = [-4 \quad 3 \quad 2 \quad 1]^T$ and the pose matrix $^4\bar{A}_5$ in Eq. (1.11), its transformation $^4\overline{P}_1$ is $^4\overline{P}_1 = {^4\bar{A}_5}{^5\overline{P}_1} = [7 \quad 0 \quad -2 \quad 1]^T$.

## 1.4 Translation and Rotation Matrices

The transformation matrices corresponding to the translations along vectors $t_x\bar{i}$, $t_y\bar{j}$, and $t_z\bar{k}$, with respect to a coordinate frame $(xyz)_h$, are respectively given by (see Figs. 1.5, 1.6, and 1.7)

$$^h\bar{A}_g = \text{tran}(t_x, 0, 0) = \begin{bmatrix} 1 & 0 & 0 & t_x \\ 0 & 1 & 0 & 0 \\ 0 & 0 & 1 & 0 \\ 0 & 0 & 0 & 1 \end{bmatrix}, \tag{1.12}$$

$$^h\bar{A}_g = \text{tran}(0, t_y, 0) = \begin{bmatrix} 1 & 0 & 0 & 0 \\ 0 & 1 & 0 & t_y \\ 0 & 0 & 1 & 0 \\ 0 & 0 & 0 & 1 \end{bmatrix}, \tag{1.13}$$

$$^h\bar{A}_g = \text{tran}(0, 0, t_z) = \begin{bmatrix} 1 & 0 & 0 & 0 \\ 0 & 1 & 0 & 0 \\ 0 & 0 & 1 & t_z \\ 0 & 0 & 0 & 1 \end{bmatrix}. \tag{1.14}$$

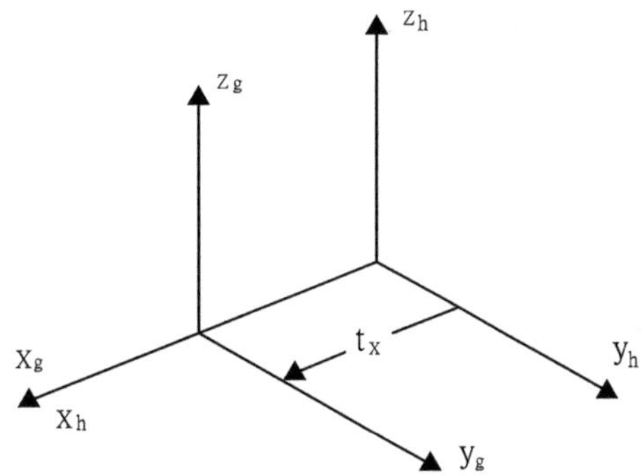


**Fig. 1.5** The translation along $x_h$ axis a distance $t_x$

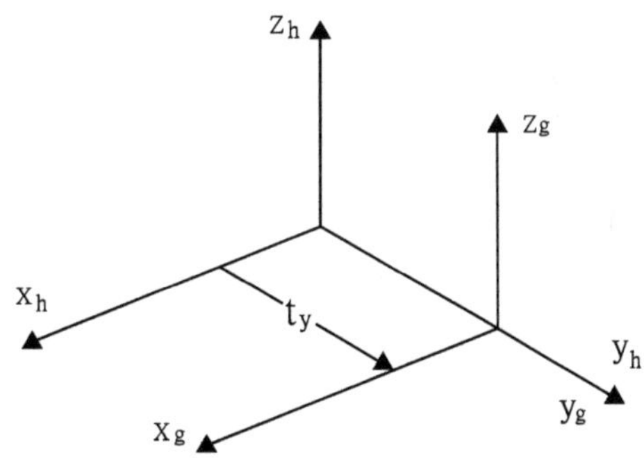


**Fig. 1.6** The translation along $y_h$ axis a distance $t_y$

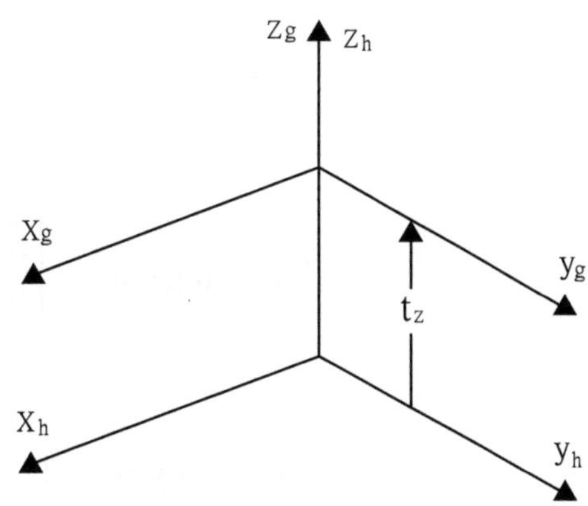


**Fig. 1.7** The translation along $z_h$ axis a distance $t_z$

Note that $\mathrm{tran}(t_x, 0, 0)$, $\mathrm{tran}(0, t_y, 0)$, and $\mathrm{tran}(0, 0, t_z)$ possess commutative property, i.e., the order of multiplication does not change the result.

The transformation matrices corresponding to rotations about the $x_h$, $y_h$, and $z_h$ axes of a coordinate frame $(xyz)_h$, respectively, by an angle $\theta$ are (see Figs. 1.8, 1.9, and 1.10)

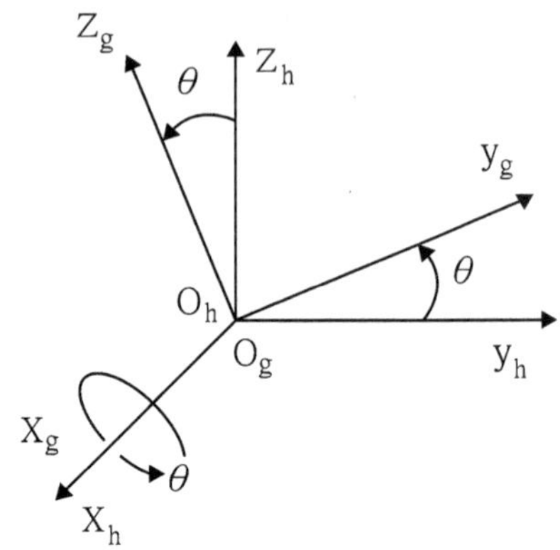


**Fig. 1.8** The illustration of rot($\bar{x}$, θ)

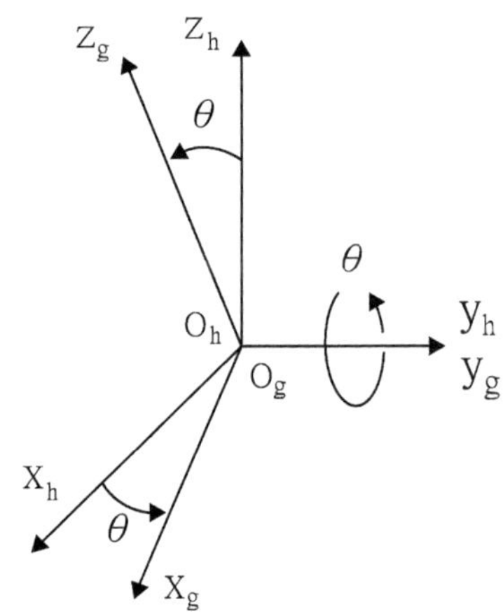


**Fig. 1.9** The illustration of rot($\bar{y}$, θ)

$$^{h}\bar{A}_{g} = \text{rot}(\bar{x}, \theta) = \begin{bmatrix} 1 & 0 & 0 & 0 \\ 0 & C\theta & -S\theta & 0 \\ 0 & S\theta & C\theta & 0 \\ 0 & 0 & 0 & 1 \end{bmatrix}, \tag{1.15}$$

$$^{h}\bar{A}_{g} = \text{rot}(\bar{y}, \theta) = \begin{bmatrix} C\theta & 0 & S\theta & 0 \\ 0 & 1 & 0 & 0 \\ -S\theta & 0 & C\theta & 0 \\ 0 & 0 & 0 & 1 \end{bmatrix}, \tag{1.16}$$

$$^{h}\bar{A}_{g} = \text{rot}(\bar{z}, \theta) = \begin{bmatrix} C\theta & -S\theta & 0 & 0 \\ S\theta & C\theta & 0 & 0 \\ 0 & 0 & 1 & 0 \\ 0 & 0 & 0 & 1 \end{bmatrix}, \tag{1.17}$$

where *S* and *C* denote sine and cosine, respectively. It is very important to remember that $\bar{x}$, $\bar{y}$, and $\bar{z}$ of Eqs. (1.15), (1.16), and (1.17) are the unit directional

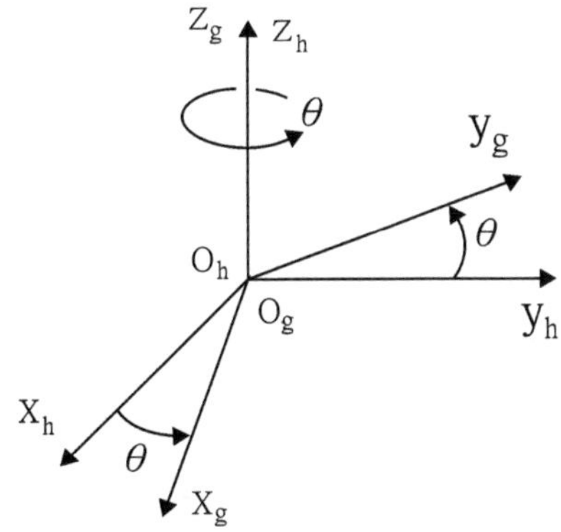


**Fig. 1.10** The illustration of $\text{rot}(\bar{z}, \theta)$

vectors of $x_h$, $y_h$, and $z_h$ axes, not of $x_g$, $y_g$, and $z_g$ axes. These transformation matrices have a simple geometric interpretation. For example, in the case of ${}^h\bar{A}_g = \text{rot}(\bar{z}, \theta)$, the third column representing z axis will remain constant, while the first and the second columns representing the x and y axes will vary as shown in Fig. 1.10.

The following properties may be useful for matrix manipulations:

$$\text{tran}(t_x, t_y, t_z)\text{tran}(p_x, p_y, p_z) = \text{tran}(t_x + p_x, t_y + p_y, t_z + p_z), \tag{1.18}$$

$$\text{rot}(\bar{x}, \theta)\text{rot}(\bar{x}, \Phi) = \text{rot}(\bar{x}, \theta+\Phi), \tag{1.19}$$

$$\text{rot}(\bar{y}, \theta)\text{rot}(\bar{y}, \Phi) = \text{rot}(\bar{y}, \theta+\Phi), \tag{1.20}$$

$$\text{rot}(\bar{z}, \theta)\text{rot}(\bar{z}, \Phi) = \text{rot}(\bar{z}, \theta+\Phi), \tag{1.21}$$

$$\text{rot}(\bar{x}, \theta)\text{tran}(t,0,0) = \text{tran}(t,0,0)\text{rot}(\bar{x}, \theta), \tag{1.22}$$

$$\text{rot}(\bar{y}, \theta)\text{tran}(0,t,0) = \text{tran}(0,t,0)\text{rot}(\bar{y}, \theta), \tag{1.23}$$

$$\text{rot}(\bar{z}, \theta)\text{tran}(0, 0, t) = \text{tran}(0, 0, t)\text{rot}(\bar{z}, \theta). \tag{1.24}$$

The transformation matrix representing a rotation around an arbitrary unit vector ${}^h\bar{\kappa} = \begin{bmatrix}\kappa_x & \kappa_y & \kappa_z & 0\end{bmatrix}^T$, where $\kappa_x^2 + \kappa_y^2 + \kappa_z^2 = 1$, located at the origin (see Fig. 1.11) is given by

$$\begin{aligned} {}^h\bar{A}_g &= \text{rot}(\bar{\kappa}, \theta) \\ &= \begin{bmatrix} \kappa_x^2(1 - C\theta) + C\theta & \kappa_x\kappa_y(1 - C\theta) - \kappa_z S\theta & \kappa_x\kappa_z(1 - C\theta) + \kappa_y S\theta & 0 \\ \kappa_x\kappa_y(1 - C\theta) + \kappa_z S\theta & \kappa_y^2(1 - C\theta) + C\theta & \kappa_y\kappa_z(1 - C\theta) - \kappa_x S\theta & 0 \\ \kappa_x\kappa_z(1 - C\theta) - \kappa_y S\theta & \kappa_y\kappa_z(1 - C\theta) + \kappa_x S\theta & \kappa_z^2(1 - C\theta) + C\theta & 0 \\ 0 & 0 & 0 & 1 \end{bmatrix}. \end{aligned} \tag{1.25}$$

**Fig. 1.11** $rot(\bar{\kappa}, \theta)$ is a rotation around an arbitrary unit vector ${}^{h}\bar{\kappa}$

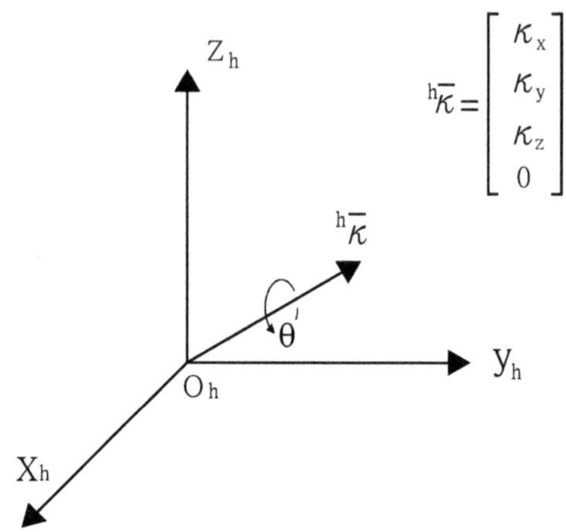


Note, again that the unit vector of Eq. (1.25) is ${}^{h}\bar{\kappa}$, not ${}^{g}\bar{\kappa}$. From this general rotation matrix we can obtain each of the elementary rotation matrices. For example, $rot(\bar{x}, \theta)$ is $rot(\bar{\kappa}, \theta)$ where ${}^{h}\bar{\kappa} = [1 \quad 0 \quad 0 \quad 0]^T$. Substituting these values of ${}^{h}\bar{\kappa}$ into Eq. (1.25) we obtain Eq. (1.15). Equation (1.25) provides an alternative method to specify the orientation of a pose matrix. Unfortunately, the axis ${}^{h}\bar{\kappa}$ of rotation to achieve some desired orientation is not intuitively obvious.

Another frequently used method to specify an orientation matrix is $RPY(\omega_z, \omega_y, \omega_x)$ by using three angles to describe any possible orientation:

$$
\begin{aligned}
{}^{h}\bar{A}_g &= RPY(\omega_z, \omega_y, \omega_x) = rot(\bar{z}, \omega_z) rot(\bar{y}, \omega_y) rot(\bar{x}, \omega_x) \\
&= \begin{bmatrix} C\omega_y C\omega_z & S\omega_x S\omega_y C\omega_z - C\omega_x S\omega_z & C\omega_x S\omega_y C\omega_z + S\omega_x S\omega_z & 0 \\ C\omega_y S\omega_z & C\omega_x C\omega_z + S\omega_x S\omega_y S\omega_z & -S\omega_x C\omega_z + C\omega_x S\omega_y S\omega_z & 0 \\ -S\omega_y & S\omega_x C\omega_y & C\omega_x C\omega_y & 0 \\ 0 & 0 & 0 & 1 \end{bmatrix}.
\end{aligned}
\tag{1.26}
$$

As shown in Fig. 1.12, $RPY(\omega_z, \omega_y, \omega_x)$describes the pose matrix ${}^{h}\bar{A}_g$ by a rotation of angle $\omega_z$ about $z_h$ to reach a intermediate coordinate frame $(xyz)_1$, followed by a rotation of angle $\omega_y$ about $y_1$ (not $y_h$) to reach another mediate coordinate frame $(xyz)_2$, and finally, a rotation $\omega_x$ about $x_2$ (neither $x_h$ nor $x_1$) to reach the target coordinate frame $(xyz)_g$. One also has to note that these two intermediate coordinate frames, $(xyz)_1$ and $(xyz)_2$, are only important in the interpretation, not in calculations, of these three angles.

The other frequently used method to specify orientation of two coordinate frames is called Euler transformation (Fig. 1.13).

**Fig. 1.12** Roll, pitch, and yaw angles

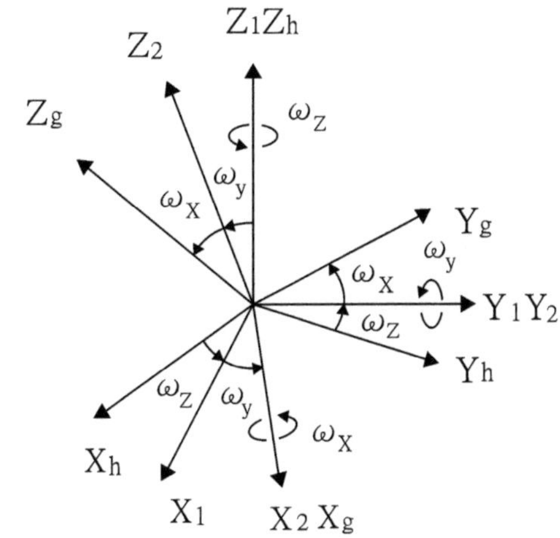


**Fig. 1.13** Euler angles

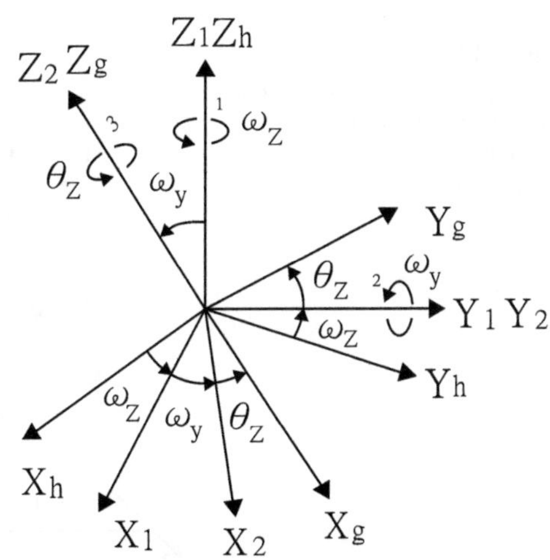


$$
\begin{aligned}
{}^{h}\bar{A}_{g} &= \text{Euler}(\omega_z, \omega_y, \theta_z) = \text{rot}(\bar{z}, \omega_z)\text{rot}(\bar{y}, \omega_y)\text{rot}(\bar{z}, \theta_z) \\
&= \begin{bmatrix} C\omega_z C\omega_y C\theta_z - S\omega_z S\theta_z & -C\omega_z C\omega_y S\theta_z - S\omega_z C\theta_z & C\omega_z S\omega_y & 0 \\ S\omega_z C\omega_y C\theta_z + C\omega_z S\theta_z & -S\omega_z C\omega_y S\theta_z + C\omega_z C\theta_z & S\omega_z S\omega_y & 0 \\ -S\omega_y C\theta_z & S\omega_y S\theta_z & C\omega_y & 0 \\ 0 & 0 & 0 & 1 \end{bmatrix}.
\end{aligned}
\tag{1.27}
$$

*Example 1.5* The pose matrix ${}^{4}\bar{A}_{5}$ shown in Fig. 1.14 is given by ${}^{4}\bar{A}_{5}$ = tran(1,0,0)tran(0,2,0)tran(0,0,3) = tran(1,2,3).

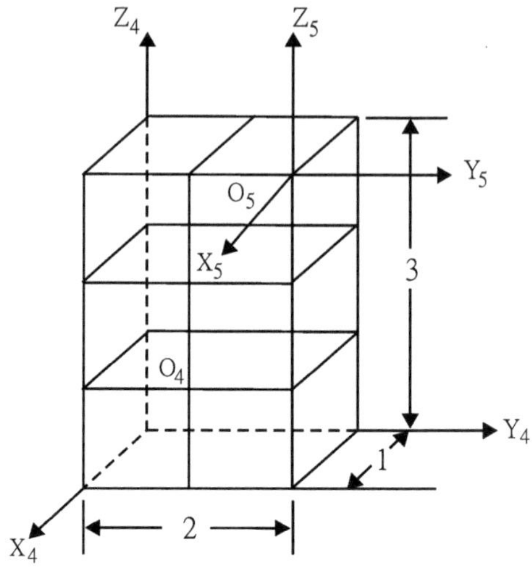


**Fig. 1.14** Two coordinate frames $(xyz)_4$ and $(xyz)_5$

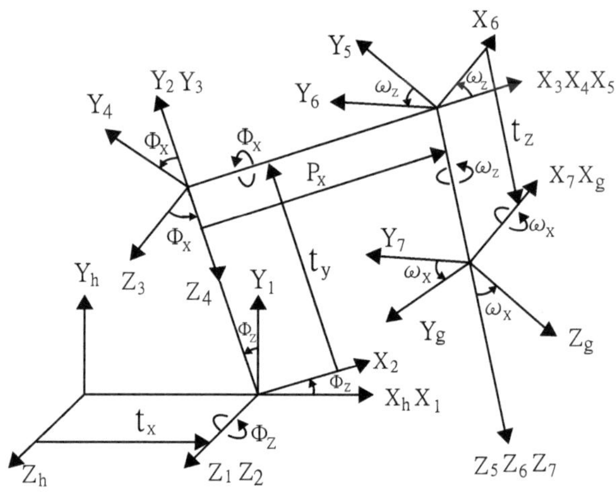


**Fig. 1.15** The pose matrix of $(xyz)_g$ with respect to $(xyz)_h$ can be specified by a sequence of rotations and translation

## 1.5 Specification of a Pose Matrix

There are different methods to specify ${}^{h}\bar{A}_{g}$ by using a sequence of rotations and translations. To describe the procedure for obtaining ${}^{h}\bar{A}_{g}$, one needs several intermediate coordinate frames, say $(xyz)_1$, $(xyz)_2$, $(xyz)_3\ldots$ (see Fig. 1.15). From Fig. 1.15 we have ${}^{h}\bar{A}_{1} = \mathrm{tran}(t_x, 0, 0)$, ${}^{1}\bar{A}_{2} = \mathrm{rot}(\bar{z}, \phi_z)$, ${}^{2}\bar{A}_{3} = \mathrm{tran}(0, t_y, 0)$, ${}^{3}\bar{A}_{4} = \mathrm{rot}(\bar{x}, \Phi_x)$, ${}^{4}\bar{A}_{5} = \mathrm{tran}(p_x, 0, 0)$, ${}^{5}\bar{A}_{6} = \mathrm{rot}(\bar{z}, \omega_z)$, ${}^{6}\bar{A}_{7} = \mathrm{tran}(0, 0, t_z)$, ${}^{7}\bar{A}_{g} = \mathrm{rot}(\bar{x}, \omega_x)$. Therefore, ${}^{h}\bar{A}_{g}$can be obtained by ${}^{h}\bar{A}_{g} = {}^{h}\bar{A}_{1}{}^{1}\bar{A}_{2}{}^{2}\bar{A}_{3}{}^{3}\bar{A}_{4}{}^{4}\bar{A}_{5}{}^{5}\bar{A}_{6}{}^{6}\bar{A}_{7}{}^{7}\bar{A}_{g}$ or just simply by:

$$\begin{aligned} {}^{h}\bar{A}_{g} = {} & \mathrm{tran}(t_x, 0, 0)\mathrm{rot}(\bar{z}, \Phi_z)\mathrm{tran}(0, t_y, 0)\mathrm{rot}(\bar{x}, \Phi_x)\mathrm{tran}(p_x, 0, 0)\mathrm{rot}(\bar{z}, \omega_z) \\ & \mathrm{tran}(0, 0, t_z)\mathrm{rot}(\bar{x}, \omega_x). \end{aligned} \tag{1.28}$$

Note that one may ignore the intermediate coordinate frames if one just uses a coordinate frame starting from the pose of $(xyz)_h$ (i.e. the coordinate frame defined by the leading super-script of ${}^h\bar{A}_g$) and then successively the translation and rotation operators to reach the final pose of $(xyz)_g$ (i.e. the coordinate frame defined by the post sub-script of ${}^h\bar{A}_g$). As in every case of a matrix multiplication, the order in which the rotations and translations are made is important. Different combinations of translation and rotation operators may be constructed for a given pose matrix ${}^h\bar{A}_g$. Example 1.6 may be helpful for the readers to understand how to form a pose matrix.

*Example 1.6* Referring to Fig. 1.4 again, the pose matrix ${}^4\bar{A}_5$ can be obtained by the following different procedures:

$$\begin{aligned}
{}^4\bar{A}_5 &= \text{tran}(3,0,0)\text{tran}(0,2,0)\text{tran}(0,0,1)\text{rot}(\bar{y},180^\circ)\text{rot}(\bar{x},90^\circ)\\
&= \text{tran}(3,0,0)\text{tran}(0,0,1)\text{tran}(0,2,0)\text{rot}(\bar{x},90^\circ)\text{rot}(\bar{z},180^\circ)\\
&= \text{tran}(3,0,0)\text{tran}(0,2,0)\text{tran}(0,0,1)\text{rot}(\bar{z},180^\circ)\text{rot}(\bar{x},-90^\circ)\\
&= \text{tran}(0,2,0)\text{tran}(0,0,1)\text{tran}(3,0,0)\text{rot}(\bar{z},180^\circ)\text{rot}(\bar{x},-90^\circ)\\
&= \text{rot}(\bar{y},180^\circ)\text{rot}(\bar{x},90^\circ)\text{tran}(-3,0,0)\text{tran}(0,-1,0)\text{tran}(0,0,-2)\\
&= \text{rot}(\bar{x},90^\circ)\text{rot}(\bar{z},180^\circ)\text{tran}(-3,0,0)\text{tran}(0,0,-2)\text{tran}(0,-1,0)\\
&= \text{rot}(\bar{z},180^\circ)\text{rot}(\bar{x},-90^\circ)\text{tran}(0,0,-2)\text{tran}(-3,0,0)\text{tran}(0,-1,0)\\
&= \text{rot}(\bar{x},90^\circ)\text{tran}(3,0,0)\text{rot}(\bar{z},180^\circ)\text{tran}(0,-1,0)\text{tran}(0,0,-2).
\end{aligned} \tag{1.29}$$

## 1.6 Inverse Transformation Matrix

We will develop the inverse matrix of ${}^h\bar{A}_g$, denoted as ${}^g\bar{A}_h$, as the transform which carries the transformed coordinate frame $(xyz)_g$ back to the original frame $(xyz)_h$. Given the pose matrix of Eq. (1.5), its inverse matrix has the simple form of:

$$\left({}^h\bar{A}_g\right)^{-1} = {}^g\bar{A}_h = \begin{bmatrix} I_x & I_y & I_z & -(t_xI_x + t_yI_y + t_zI_z) \\ J_x & J_y & J_z & -(t_xJ_x + t_yJ_y + t_zJ_z) \\ K_x & K_y & K_z & -(t_xK_x + t_yK_y + t_zK_z) \\ 0 & 0 & 0 & 1 \end{bmatrix}. \tag{1.30}$$

If the pose matrix ${}^h\bar{A}_g$ is given by a sequence of rotation and translation operators (e.g., ${}^h\bar{A}_g = \text{tran}(t_x,0,0)\text{tran}(0,t_y,0)\text{tran}(0,0,t_z)\text{rot}(\bar{z},\omega_z)\text{rot}(\bar{y},\omega_y)\text{rot}(\bar{x},\omega_x)$), then from the theory of matrix, one can have its inverse matrix as $\left({}^h\bar{A}_g\right)^{-1} = {}^g\bar{A}_h = \text{rot}(\bar{x},-\omega_x)\text{rot}(\bar{y},-\omega_y)\text{rot}(\bar{z},-\omega_z)\text{tran}(-t_x,-t_y,-t_z)$. This result is easily verified with post-multiplying ${}^h\bar{A}_g$ by ${}^g\bar{A}_h$.

*Example 1.7* Consider the pose matrix ${}^4\bar{A}_5 = \text{tran}(0,0,1)\text{tran}(0,2,0)\ \text{tran}(3,0,0)$ $\text{rot}(\bar{y},180°)\text{rot}(\bar{x},90°)$ of Example 1.3. Its inverse pose is

$$\left({}^4\bar{A}_5\right)^{-1} = {}^5\bar{A}_4 = \text{rot}(\bar{x},-90°)\text{rot}(\bar{y},-180°)\text{tran}(-3,0,0)\text{tran}(0,-2,0)\text{tran}(0,0,-1)$$
$$= \text{rot}(\bar{x},-90°)\text{rot}(\bar{y},-180°)\text{tran}(-3,-2,-1) = \begin{bmatrix} -1 & 0 & 0 & 3 \\ 0 & 0 & -1 & 1 \\ 0 & -1 & 0 & 2 \\ 0 & 0 & 0 & 1 \end{bmatrix}. \tag{1.31}$$

## 1.7 RPY Angle Solutions

In Sect. 1.5 we presented the different methods to specify the pose matrix of ${}^h\bar{A}_g$ by means of $\text{RPY}(\omega_z,\omega_y,\omega_x)$, $\text{Euler}(\omega_z,\omega_y,\theta_z)$, $\text{rot}(\bar{\kappa},\theta)$, etc. In the following, we will add another post-subscript "i" for all variables to indicate that these variables are defined for the ith boundary surface of an optical system. We will obtain solution of $t_{ix}$, $t_{iy}$, $t_{iz}$, $\omega_{ix}$, $\omega_{iy}$, and $\omega_{iz}$ for a given numeric values of ${}^h\bar{A}_g$ when ${}^h\bar{A}_g$ is specified by

$$\begin{aligned} {}^h\bar{A}_g &= \text{tran}(t_{ix},0,0)\text{tran}(0,t_{iy},0)\text{tran}(0,0,t_{iz})\text{rot}(\bar{z},\omega_{iz})\text{rot}(\bar{y},\omega_{iy})\text{rot}(\bar{x},\omega_{ix}) \\ &= \text{tran}(t_{ix},0,0)\text{tran}(0,t_{iy},0)\text{tran}(0,0,t_{iz})\text{RPY}(\omega_{iz},\omega_{iy},\omega_{ix}) \\ &= \text{tran}(t_{ix},t_{iy},t_{iz})\text{RPY}(\omega_{iz},\omega_{iy},\omega_{ix}), \end{aligned} \tag{1.32}$$

since it is the most frequently used method in this book. The reader can refer (p. 65 to 72 of [3]) if the orientation part of ${}^h\bar{A}_g$ is specified by $\text{Euler}(\omega_{iz},\omega_{iy},\theta_{iz})$ or $\text{rot}(\bar{\kappa},\theta)$.

If the numeric values of ${}^h\bar{A}_g$ are known, and

$$\begin{aligned} {}^h\bar{A}_g &= \text{tran}(t_{ix},0,0)\text{tran}(0,t_{iy},0)\text{tran}(0,0,t_{iz})\text{rot}(\bar{z},\omega_{iz})\text{rot}(\bar{y},\omega_{iy})\text{rot}(\bar{x},\omega_{ix}) \\ &= \begin{bmatrix} I_{ix} & J_{ix} & K_{ix} & t_{ix} \\ I_{iy} & J_{iy} & K_{iy} & t_{iy} \\ I_{iz} & J_{iz} & K_{iz} & t_{iz} \\ 0 & 0 & 0 & 1 \end{bmatrix} \\ &= \begin{bmatrix} C\omega_{iy}C\omega_{iz} & S\omega_{ix}S\omega_{iy}C\omega_{iz} - C\omega_{ix}S\omega_{iz} & C\omega_{ix}S\omega_{iy}C\omega_{iz} - S\omega_{ix}S\omega_{iz} & t_{ix} \\ C\omega_{iy}S\omega_{iz} & C\omega_{ix}C\omega_{iz} + S\omega_{ix}S\omega_{iy}S\omega_{iz} & -S\omega_{ix}C\omega_{iz} + C\omega_{ix}S\omega_{iy}S\omega_{iz} & t_{iy} \\ -S\omega_{iy} & S\omega_{ix}C\omega_{iy} & C\omega_{ix}C\omega_{iy} & t_{iz} \\ 0 & 0 & 0 & 1 \end{bmatrix}, \end{aligned} \tag{1.33}$$

from the following equation one can solve sequentially for the three angles, $\omega_{iz}$, $\omega_{iy}$, and $\omega_{ix}$, and the three position pose variables, $t_{ix}$, $t_{iy}$, and $t_{iz}$, by

$$\begin{cases} \omega_{iz} = \operatorname{atan2}(I_{iy}, I_{ix}) \text{ or } \omega_{iz} = \operatorname{atan2}(-I_{iy}, -I_{ix}) & \text{if } I_{ix}^2 + I_{iy}^2 \neq 0 \\ \omega_{iz} = 0 & \text{if } I_{ix}^2 + I_{iy}^2 = 0 \end{cases}, \tag{1.34}$$

$$\omega_{iy} = \operatorname{atan2}(-I_{iz}, I_{ix}C\omega_{iz} + I_{iy}S\omega_{iz}), \tag{1.35}$$

$$\omega_{ix} = \operatorname{atan2}(K_{ix}S\omega_{iz} - K_{iy}C\omega_{iz}, -J_{ix}S\omega_{iz} + J_{iy}C\omega_{iz}), \tag{1.36}$$

$$t_{ix} = t_{ix}, \tag{1.37}$$

$$t_{iy} = t_{iy}, \tag{1.38}$$

$$t_{iz} = t_{iz}. \tag{1.39}$$

Equation (1.34) shows that there are two solutions, each 180° apart, for $\omega_{iz}$ in the case of $I_{ix}^2 + I_{iy}^2 \neq 0$. The function atan2 returns the arctangent in the range $-\pi$ to $\pi$. It is usually necessary to add $2\pi$ to its solution in order to solve a variable in its required domain. Equations (1.37), (1.38), and (1.39) just show that the three position pose variables $t_{ix}$, $t_{iy}$, and $t_{iz}$ can be directly extracted from the (1,4)th, (2,4)th, and (3,4)th of ${}^{h}\bar{A}_{g}$.

*Example 1.8* atan2(1,1) = atan2(3,3) = 45°, atan2(−1, −1) = atan2(−3, −3) = −135°. atan2(−1, 1) = atan2(−3, 3) = −45° , atan2(1, −1) = atan2(3, −3) = 135°

*Example 1.9* Again, consider the pose matrix ${}^{4}\bar{A}_{5}$ of Eq. (1.11). The values of the six pose variables are $\omega_z = \operatorname{atan2}(0, -1) = 180°$, $\omega_y = \operatorname{atan2}(0, 1) = 0°$, $\omega_x = \operatorname{atan2}(-1, 0) = -90°$, $t_x = 3$, $t_y = 2$, and $t_z = 1$. Therefore, ${}^{4}\bar{A}_{5}$ can be obtained by ${}^{4}\bar{A}_{5} = \operatorname{tran}(3, 0, 0)\operatorname{tran}(0, 2, 0)\operatorname{tran}(0, 0, 1)\operatorname{rot}(\bar{z}, 180°)\operatorname{rot}(\bar{x}, -90°)$, which is shown in the third row of Eq. (1.29).

## 1.8 Jacobian and Hessian Matrices

In vector calculus, the Jacobian matrix (often simply called as the Jacobian) is the matrix of all first-order partial derivatives of a vector-valued function with respect to another vector. If $\overline{F} = \overline{F}(\overline{X}) = [f_1(\overline{X}) \quad f_2(\overline{X})]^T$ is a function of $\overline{X} = [x_1 \quad x_2 \quad x_3]^T$, then the Jacobian matrix of $\overline{F}$ with respect to $\overline{X}$ is written as

$$\frac{\partial \overline{F}}{\partial \overline{X}} = \frac{\partial(f_1, f_2)}{\partial(x_1, x_2, x_3)} = \begin{bmatrix} \frac{\partial f_1}{\partial x_1} & \frac{\partial f_1}{\partial x_2} & \frac{\partial f_1}{\partial x_3} \\ \frac{\partial f_2}{\partial x_1} & \frac{\partial f_2}{\partial x_2} & \frac{\partial f_2}{\partial x_3} \end{bmatrix} = \left[\frac{\partial f_u}{\partial x_v}\right] = [f_{u,v}]. \tag{1.40}$$

One has to remember that the notation $\partial x_v$ (or, v in $f_{u,v}$) indicates partial differentiation with respect to the vth variable of $\overline{X}$. It can be deduced from

Eq. (1.40) that $\partial f_1/\partial \overline{X}$ is a row matrix and $\partial \overline{F}/\partial x_1$ is a column matrix. Meanwhile, the Hessian matrix of a scalar function $f_u(\overline{X}) = f_u(x_1, x_2, x_3)$ is the second-order derivatives of $f_u(\overline{X})$ with respect to its variable vector, defined as

$$\frac{\partial^2 f_u}{\partial \overline{X}^2} = \frac{\partial^2 f_u}{\partial(x_1, x_2, x_3)^2} = \begin{bmatrix} \frac{\partial^2 f_u}{\partial x_1^2} & \frac{\partial^2 f_u}{\partial x_1 \partial x_2} & \frac{\partial^2 f_u}{\partial x_1 \partial x_3} \\ \frac{\partial^2 f_u}{\partial x_2 \partial x_1} & \frac{\partial^2 f_u}{\partial x_2^2} & \frac{\partial^2 f_u}{\partial x_2 \partial x_3} \\ \frac{\partial^2 f_u}{\partial x_3 \partial x_1} & \frac{\partial^2 f_u}{\partial x_3 \partial x_2} & \frac{\partial^2 f_u}{\partial x_3^2} \end{bmatrix} = \left[\frac{\partial^2 f_u}{\partial x_w \partial x_v}\right] = \left[f_{u,wv}\right], \quad (1.41)$$

The Hessian matrix of vector $\overline{F} = \left[f_1(\overline{X}) \quad f_2(\overline{X})\right]^T$ with respect to $\overline{X}$ is the second-order derivatives of $\overline{F}$, written as $\partial^2 \overline{F}/\partial \overline{X}^2 = \partial^2(f_1, f_2)/\partial(x_1, x_2, x_3)^2$. $\partial^2 \overline{F}/\partial \overline{X}^2$ is a $2 \times 3 \times 3$ tensor, which is essentially a 3D table of values. Hessian matrices are used in large-scale optimization problems within Newton-type methods because they contain the coefficients of the quadratic term of a local Taylor expansion of the function $f_u$. That is

$$f_u = f_u(\overline{X} + \Delta\overline{X}) \approx f_u(\overline{X}) + \frac{\partial f_u}{\partial \overline{X}} \Delta\overline{X} + \frac{1}{2} \Delta\overline{X}^T \frac{\partial^2 f_u}{\partial \overline{X}^2} \Delta\overline{X}. \quad (1.42)$$

In this book, $\|\ \|$ (two bars on each side of a vector) is used to denote the magnitude of vector, that is, the magnitude of ${}^g\overline{P}_i = [P_{ix} \quad P_{iy} \quad P_{iz} \quad 1]^T$ is denoted as

$$\|{}^g\overline{P}_i\| = \sqrt{P_{ix}^2 + P_{iy}^2 + P_{iz}^2}. \quad (1.43)$$

On the other hand, $|\ |$ (one bar on each side of the symbol) is reserved for denoting the absolute value of a scalar quantity. The determinant of a matrix ${}^h\bar{A}_g$ is denoted as det $({}^h\bar{A}_g)$ in this book.

## References

1. J.J. Uicker, On the dynamic analysis of spatial linkages using 4 × 4 Matrices, Ph.D. dissertation, Northwestern University, Evanston, ILL, 1965
2. J. Denavit, R.S. Hartenberg, A kinematic notation for lower pair mechanisms based on matrices. Trans. ASME J. Appl. Mech. **77**, 215–221 (1955)
3. R.P. Paul, *Robot Manipulators-Mathematics, Programming and Control* (MIT press, Cambridge, Mass, 1982)
4. F.L. Litvin Theory of Gearing (NASA Reference Publication, Chicago 1989)
5. J.D. Foley, A.V. Dam, S.K. Feiner, J.F. Hughes, *Computer Graphics, Principles and Practices*, 2nd edn. (Addision-Wesley Publishing Company, Massachusetts 1981)

# Chapter 2
# Skew-Ray Tracing at Boundary Surfaces

Geometrical optics, or ray optics, describes light propagation in terms of "rays". The "ray" is the tool that is used to predict the path of light [1]. Geometrical optics provides equations for propagating the rays through an optical system. Although due to its over simplification, geometrical optics could not account for optical effects such as diffraction and polarization, it is still a good approximation to describe ray propagation. Geometrical optics in this book can perform standard sequential raytracing and compute the first- and second-order derivative matrices of various optical quantities by formulating Snell's laws using homogeneous coordinate notation. Equations of geometrical optics involving homogeneous coordinate notation are often simpler and more symmetric than their vector counterparts, especially when it is used in practical applications.

## 2.1 Source Ray

In geometrical optics rays are assumed to move from left to right. It is necessary to re-emphasize that in this book the ith incidence point is designated as ${}^{g}\overline{P}_i = [P_{ix} \quad P_{iy} \quad P_{iz} \quad 1]^T$, where the pre-superscript "g" of the leading symbol ${}^{g}\overline{P}_i$ indicates that the components of this vector are referred with respect to the coordinate frame $(xyz)_g$. The reflected/or refracted ray at boundary surface i is denoted as ${}^{g}\overline{R}_i = [{}^{g}\overline{P}_i \quad {}^{g}\overline{\ell}_i]^T = [P_{ix} \quad P_{iy} \quad P_{iz} \quad \ell_{ix} \quad \ell_{iy} \quad \ell_{iz}]^T$, where ${}^{g}\overline{P}_i$ is the incident point of this ray on the surface and ${}^{g}\overline{\ell}_i$ is its unit directional vector. One should remember that if this vector ${}^{0}\overline{R}_i$ (or ${}^{0}\overline{P}_i$ or ${}^{0}\overline{\ell}_i$) is referred to the world coordinate frame $(xyz)_0$, then its pre-superscript "0" will be omitted for simplicity.

An optical lens or prism, which is referred as optical element hereafter, is a block of optical material with multiple boundary surfaces and possesses constant refractive index. To trace a ray in an optical system possessing n boundary surfaces, one has to mark the elements from j = 0 to j = k and label its boundary surfaces from i = 0 to i = n (e.g., Fig. 2.1 with values of variables in Table 2.1).

PD Lin, *New Computation Methods for Geometrical Optics*,
Springer Series in Optical Sciences 178, DOI: 10.1007/978-981-4451-79-6_2,

**Table 2.1** Values of variables of Petzval lens system in Fig. 2.1 (units:mm)

| j | Vj | $\xi_{ej}$ | $R_{2j-1}$ | $R_{2j}$ | $q_{ej}$ |
|---|---|---|---|---|---|
| 1 | 0.0000 | 1.65000 | 38.2219 | −56.0857 | 15.8496 |
| 2 | 0.0000 | 1.71736 | −56.0857 | −590.6820 | 5.9690 |
| 3 | 3.0226 | 1.00000 | | | 0.0000 |
| 4 | 14.0208 | 1.52583 | −41.7957 | 29.3446 | 2.5146 |
| 5 | 7.9248 | 1.65000 | 63.5635 | −56.8655 | 6.0960 |
| 6 | 49.6316 | | | | |

We always assign j = 0 and i = 0 to the source ray $\overline{R}_0 = \begin{bmatrix}\overline{P}_0 & \overline{\ell}_0\end{bmatrix}^T$, which originates at the source point

$$\overline{P}_0 = \begin{bmatrix}P_{0x} & P_{0y} & P_{0z} & 1\end{bmatrix}^T, \tag{2.1}$$

and travels along the unit directional vector

$$\overline{\ell}_0 = \begin{bmatrix}\ell_{0x} & \ell_{0y} & \ell_{0z} & 0\end{bmatrix}^T = \begin{bmatrix}C\beta_0 C(90^\circ + \alpha_0) & C\beta_0 S(90^\circ + \alpha_0) & S\beta_0 & 0\end{bmatrix}^T, \tag{2.2}$$

where $\beta_0$ is the angle between $\overline{\ell}_0$ and the projection of $\overline{\ell}_0$ onto the horizontal plane that pass through the origin of the unit sphere centered at the source point $\overline{P}_0$ (see Fig. 2.2). The angle $\alpha_0$ is the angle between the meridional plane and $\overline{\ell}_0$, measured along the zenith direction (which is parallel with the $z_0$ axis) of that unit sphere. Hereafter we will refer the cone shown in Fig. 2.2 as $\alpha_0(\beta_0)$ cone, which is generated by sweeping $\overline{\ell}_0$ with the constant value of $\beta_0$ around the zenith direction of that unit sphere. It is noted that $\overline{\ell}_0$ is parallel with $y_0$ axis when $\alpha_0 = 0^\circ$ and $\beta_0 = 0^\circ$. From Eqs. (2.1) and (2.2), we have the variable vector $\overline{X}_0$ of the source ray $\overline{R}_0$, given as

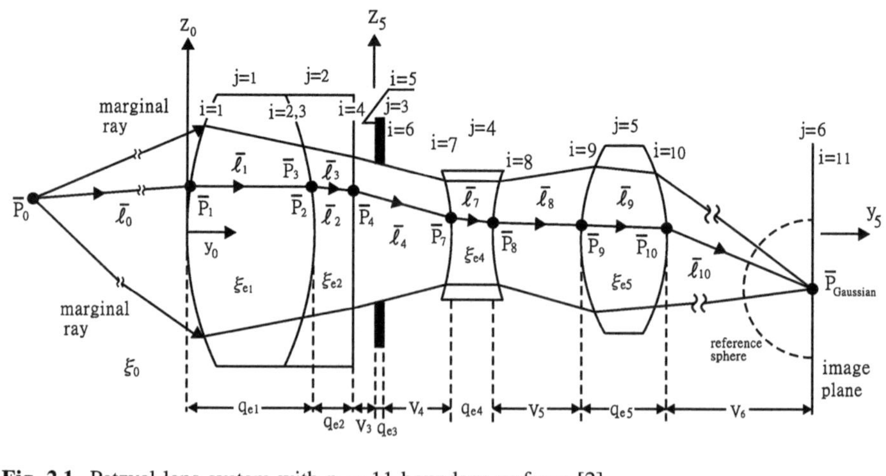


**Fig. 2.1** Petzval lens system with n = 11 boundary surfaces [2]

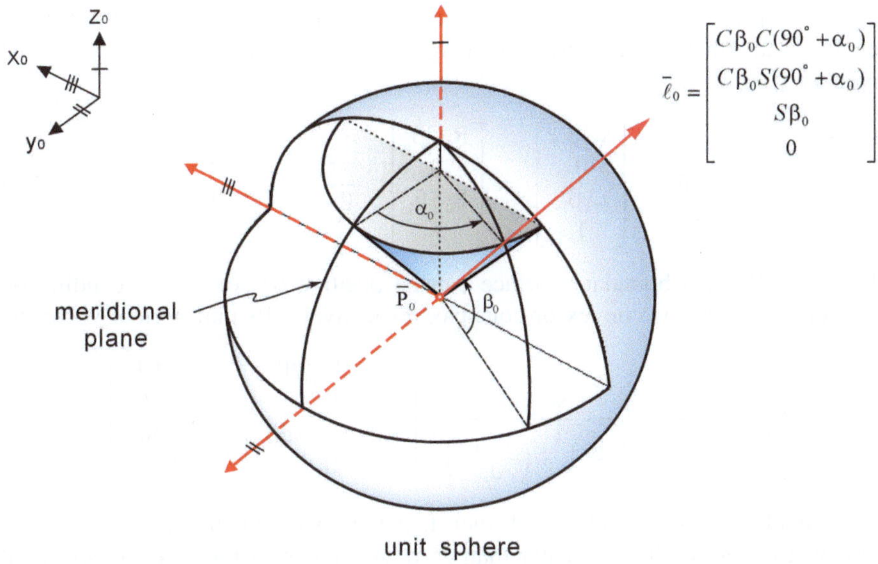


Fig. 2.2 Representation of the unit directional vector $\bar{\ell}_0$ originating from a source point $\bar{P}_0$

$$\bar{X}_0 = [P_{0x} \quad P_{0y} \quad P_{0z} \quad \alpha_0 \quad \beta_0]^T. \tag{2.3}$$

When a source point $\bar{P}_0$ is confined to the $y_0z_0$ plane (where $y_0$ points along the optical axis of the system) of an axis-symmetrical system, then a meridional ray (or tangential ray) is a ray lying on that plane. This $y_0z_0$ plane is called the meridional plane. One has to note that any source ray originating from source point $\bar{P}_0 = [0 \quad P_{0y} \quad h \quad 1]^T$ with $\alpha_0 = 0$ always travels on the meridional plane in an axis-symmetrical system. In other words, the meridional plane of a source point is a plane that is swept out by a source ray with $\alpha_0 = 0$ (see Fig. 2.2).

## 2.2 Skew-Ray Tracing at a Spherical Boundary Surface

The most common optical elements are spherical lenses. Their two boundary surfaces are parts of the surfaces of spheres, with the lens axis ideally perpendicular to both boundary surfaces. Each boundary surface can be convex (when the vertex lies to the left of the center), concave (if the vertex lies to the right of the center), or planar (flat). The line joining the centers of the spheres making up the lens surfaces is called the axis of the lens. The first step of geometrical optics is the establishment of boundary surfaces and then the study of skew-ray tracing technique among them.

As shown in Fig. 2.3, the generating curve ${}^{i}\bar{q}_i$ of a spherical boundary surface, denoted as the ith boundary surface in an optical system, is a curve in the $x_iz_i$ plane given as

$$ {}^{i}\bar{q}_i = \begin{bmatrix} x_i(\beta_i) \\ 0 \\ z_i(\beta_i) \\ 1 \end{bmatrix} = \begin{bmatrix} R_iC\beta_i \\ 0 \\ R_iS\beta_i \\ 1 \end{bmatrix}, \quad \frac{-\pi}{2} \le \beta_i \le \frac{\pi}{2}. \tag{2.4} $$

The radius $R_i$ of a boundary surface can be positive or negative depending on whether the surface is convex or concave, respectively. Its unit normal vector is:

$$ {}^{i}\bar{e}_i = \begin{bmatrix} e_{ix} \\ e_{iy} \\ e_{iz} \\ 0 \end{bmatrix} = \frac{s_i}{\sqrt{\left(\frac{dx_i(\beta_i)}{d\beta_i}\right)^2 + \left(\frac{dz_i(\beta_i)}{d\beta_i}\right)^2}} \begin{bmatrix} dz_i(\beta_i)/d\beta_i \\ 0 \\ -dx_i(\beta_i)/d\beta_i \\ 0 \end{bmatrix} = s_i \begin{bmatrix} C\beta_i \\ 0 \\ S\beta_i \\ 0 \end{bmatrix}. \tag{2.5} $$

In this book, $s_i$ is set to +1 or $-1$ such that the cosine of the incident angle $\theta_i$ computed by Eq. (2.20) is greater than zero. As shown in Fig. 2.4, the spherical boundary surface ${}^{i}\bar{r}_i$ and its unit normal vector ${}^{i}\bar{n}_i$ can be obtained respectively by rotating ${}^{i}\bar{q}_i$ and ${}^{i}\bar{e}_i$ about the $z_i$ axis by an angle $\alpha_i$ $(0 \le \alpha_i < 2\pi)$, i.e.

$$ {}^{i}\bar{r}_i = \begin{bmatrix} x_i \\ y_i \\ z_i \\ 1 \end{bmatrix} = \mathrm{rot}(\bar{z}, \alpha_i)\,{}^{i}\bar{q}_i = \begin{bmatrix} C\alpha_i & -S\alpha_i & 0 & 0 \\ S\alpha_i & C\alpha_i & 0 & 0 \\ 0 & 0 & 1 & 0 \\ 0 & 0 & 0 & 1 \end{bmatrix} \begin{bmatrix} R_iC\beta_i \\ 0 \\ R_iS\beta_i \\ 1 \end{bmatrix} = \begin{bmatrix} R_iC\beta_iC\alpha_i \\ R_iC\beta_iS\alpha_i \\ R_iS\beta_i \\ 1 \end{bmatrix}, \tag{2.6} $$

$$ {}^{i}\bar{n}_i = \mathrm{rot}(\bar{z}, \alpha_i)\,{}^{i}\bar{e}_i = s_i \begin{bmatrix} C\alpha_i & -S\alpha_i & 0 & 0 \\ S\alpha_i & C\alpha_i & 0 & 0 \\ 0 & 0 & 1 & 0 \\ 0 & 0 & 0 & 1 \end{bmatrix} \begin{bmatrix} C\beta_i \\ 0 \\ S\beta_i \\ 0 \end{bmatrix} = s_i \begin{bmatrix} C\beta_iC\alpha_i \\ C\beta_iS\alpha_i \\ S\beta_i \\ 0 \end{bmatrix}. \tag{2.7} $$

Equations (2.6) and (2.7) give the expressions of the boundary surface ${}^{i}\bar{r}_i$ and its unit normal vector ${}^{i}\bar{n}_i$ with parameters $\alpha_i$ and $\beta_i$. They are referred with the coordinate frame $(xyz)_i$. However, many derivations in this book are built relative to the world coordinate frame $(xyz)_o$. One thus needs the pose matrix of $(xyz)_i$ with respect to $(xyz)_0$, given as:

$$
\begin{aligned}
{}^0\bar{A}_i &= \mathrm{tran}(t_{ix},0,0)\mathrm{tran}(0,t_{iy},0)\mathrm{tran}(0,0,t_{iz})\mathrm{rot}(\bar{z},\omega_{iz})\mathrm{rot}(\bar{y},\omega_{iy})\mathrm{rot}(\bar{x},\omega_{ix}) \\
&= \mathrm{tran}(t_{ix},t_{iy},t_{iz})\mathrm{RPY}(\omega_{iz},\omega_{iy},\omega_{ix}) \\
&= \begin{bmatrix} C\omega_{iz}C\omega_{iy} & C\omega_{iz}S\omega_{iy}S\omega_{ix} - S\omega_{iz}C\omega_{ix} & C\omega_{iz}S\omega_{iy}C\omega_{ix} + S\omega_{iz}S\omega_{ix} & t_{ix} \\ S\omega_{iz}C\omega_{iy} & S\omega_{iz}S\omega_{iy}S\omega_{ix} + C\omega_{iz}C\omega_{ix} & S\omega_{iz}S\omega_{iy}C\omega_{ix} - C\omega_{iz}S\omega_{ix} & t_{iy} \\ -S\omega_{iy} & C\omega_{iy}S\omega_{ix} & C\omega_{iy}C\omega_{ix} & t_{iz} \\ 0 & 0 & 0 & 1 \end{bmatrix} \\
&= \begin{bmatrix} I_{ix} & J_{ix} & K_{ix} & t_{ix} \\ I_{iy} & J_{iy} & K_{iy} & t_{iy} \\ I_{iz} & J_{iz} & K_{iz} & t_{iz} \\ 0 & 0 & 0 & 1 \end{bmatrix},
\end{aligned}
\tag{2.8}
$$

where $t_{ix}, t_{iy}, t_{iz}, \omega_{iz}, \omega_{iy}$ and $\omega_{ix}$ are the boundary pose variables of this spherical boundary surface. Now the unit normal vector of this boundary surface expressed with respect to $(xyz)_o$ can be obtained by the following transformation:

$$
\begin{aligned}
\bar{n}_i &= \begin{bmatrix} n_{ix} & n_{iy} & n_{iz} & 0 \end{bmatrix}^T = {}^0\bar{A}_i\,{}^i\bar{n}_i \\
&= s_i \begin{bmatrix} I_{ix} & J_{ix} & K_{ix} & t_{ix} \\ I_{iy} & J_{iy} & K_{iy} & t_{iy} \\ I_{iz} & J_{iz} & K_{iz} & t_{iz} \\ 0 & 0 & 0 & 1 \end{bmatrix} \begin{bmatrix} C\beta_i C\alpha_i \\ C\beta_i S\alpha_i \\ S\beta_i \\ 0 \end{bmatrix} = s_i \begin{bmatrix} I_{ix}C\beta_i C\alpha_i + J_{ix}C\beta_i S\alpha_i + K_{ix}S\beta_i \\ I_{iy}C\beta_i C\alpha_i + J_{iy}C\beta_i S\alpha_i + K_{iy}S\beta_i \\ I_{iz}C\beta_i C\alpha_i + J_{iz}C\beta_i S\alpha_i + K_{iz}S\beta_i \\ 0 \end{bmatrix}.
\end{aligned}
\tag{2.9}
$$

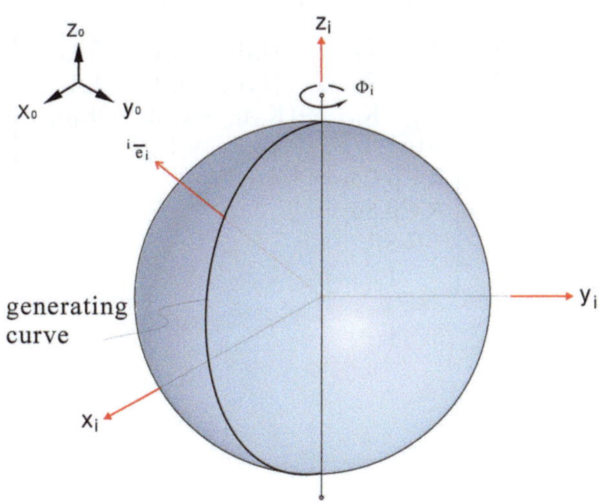


**Fig. 2.3** The generating curve and its unit normal vector of a spherical boundary surface

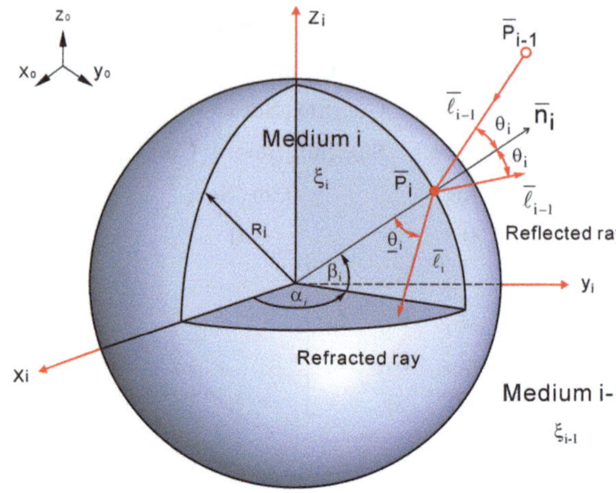


**Fig. 2.4** Ray trace at a spherical boundary surface

In Fig. 2.4 a ray originating from the incidence point $\bar{P}_{i-1} = [P_{i-1x} \quad P_{i-1y} \quad P_{i-1z} \quad 1]^T$ (at last boundary surface $\bar{r}_{i-1}$) directed along the unit directional vector $\bar{\ell}_{i-1} = [\ell_{i-1x} \quad \ell_{i-1y} \quad \ell_{i-1z} \quad 0]^T$ will be reflected/or refracted at current boundary surface $\bar{r}_i$. Any intermediate point $\bar{P}'_{i-1}$ along this ray as it travels from $\bar{P}_{i-1}$ is given by $\bar{P}'_{i-1} = \bar{P}_{i-1} + \lambda\bar{\ell}_{i-1}$. The parameter $\lambda = \lambda_i$, for which the incoming ray $\bar{R}_{i-1}$ hits the boundary surface $\bar{r}_i$ at the incidence point

$$\bar{P}_i = \begin{bmatrix} P_{ix} \\ P_{iy} \\ P_{iz} \\ 1 \end{bmatrix} = \begin{bmatrix} P_{i-1x} + \ell_{i-1x}\lambda_i \\ P_{i-1y} + \ell_{i-1y}\lambda_i \\ P_{i-1z} + \ell_{i-1z}\lambda_i \\ 1 \end{bmatrix}, \tag{2.10}$$

is obtained by equating Eq. (2.6) to ${}^i\bar{P}_i = {}^i\bar{A}_0\bar{P}_i = ({}^0\bar{A}_i)^{-1}\bar{P}_i$, yielding

$$\begin{aligned} {}^i\bar{P}_i &= \begin{bmatrix} I_{ix} & I_{iy} & I_{iz} & -(I_{ix}t_{ix} + I_{iy}t_{iy} + I_{iz}t_{iz}) \\ J_{ix} & J_{iy} & J_{iz} & -(J_{ix}t_{ix} + J_{iy}t_{iy} + J_{iz}t_{iz}) \\ K_{ix} & K_{iy} & K_{iz} & -(K_{ix}t_{ix} + K_{iy}t_{iy} + K_{iz}t_{iz}) \\ 0 & 0 & 0 & 1 \end{bmatrix} \begin{bmatrix} P_{i-1x} + \ell_{i-1x}\lambda_i \\ P_{i-1y} + \ell_{i-1y}\lambda_i \\ P_{i-1z} + \ell_{i-1z}\lambda_i \\ 1 \end{bmatrix} = \begin{bmatrix} \sigma_i \\ \rho_i \\ \tau_i \\ 1 \end{bmatrix} \\ &= {}^i\bar{r}_i = \begin{bmatrix} R_iC\beta_iC\alpha_i \\ R_iC\beta_iS\alpha_i \\ R_iS\beta_i \\ 1 \end{bmatrix}, \end{aligned} \tag{2.11}$$

where

$$\begin{aligned}\sigma_i &= I_{ix}(P_{i-1x} + \ell_{i-1x}\lambda_i) + I_{iy}(P_{i-1y} + \ell_{i-1y}\lambda_i) + I_{iz}(P_{i-1z} + \ell_{i-1z}\lambda_i) - (I_{ix}t_{ix} + I_{iy}t_{iy} + I_{iz}t_{iz}) \\ &= I_{ix}P_{ix} + I_{iy}P_{iy} + I_{iz}P_{iz} - (I_{ix}t_{ix} + I_{iy}t_{iy} + I_{iz}t_{iz}),\end{aligned} \tag{2.12}$$

$$\begin{aligned}\rho_i &= J_{ix}(P_{i-1x} + \ell_{i-1x}\lambda_i) + J_{iy}(P_{i-1y} + \ell_{i-1y}\lambda_i) + J_{iz}(P_{i-1z} + \ell_{i-1z}\lambda_i) - (J_{ix}t_{ix} + J_{iy}t_{iy} + J_{iz}t_{iz}) \\ &= J_{ix}P_{ix} + J_{iy}P_{iy} + J_{iz}P_{iz} - (J_{ix}t_{ix} + J_{iy}t_{iy} + J_{iz}t_{iz}),\end{aligned} \tag{2.13}$$

$$\begin{aligned}\tau_i &= K_{ix}(P_{i-1x} + \ell_{i-1x}\lambda_i) + K_{iy}(P_{i-1y} + \ell_{i-1y}\lambda_i) + K_{iz}(P_{i-1z} + \ell_{i-1z}\lambda_i) - (K_{ix}t_{ix} + K_{iy}t_{iy} + K_{iz}t_{iz}) \\ &= K_{ix}P_{ix} + K_{iy}P_{iy} + K_{iz}P_{iz} - (K_{ix}t_{ix} + K_{iy}t_{iy} + K_{iz}t_{iz}).\end{aligned} \tag{2.14}$$

From the sum of $\sigma_i^2$, $\rho_i^2$, and $\tau_i^2$, one has

$$\lambda_i = -D_i \pm \sqrt{D_i^2 - E_i}, \tag{2.15}$$

with

$$D_i = \ell_{i-1x}(P_{i-1x} - t_{ix}) + \ell_{i-1y}(P_{i-1y} - t_{iy}) + \ell_{i-1z}(P_{i-1z} - t_{iz}), \tag{2.16}$$

$$E_i = P_{i-1x}^2 + P_{i-1y}^2 + P_{i-1z}^2 - R_i^2 + t_{ix}^2 + t_{iy}^2 + t_{iz}^2 - 2(t_{ix}P_{i-1x} + t_{iy}P_{i-1y} + t_{iz}P_{i-1z}). \tag{2.17}$$

Note that the parameter $\lambda_i$ represents the geometrical length from point $\overline{P}_{i-1}$ to $\overline{P}_i$, and is positive. The $\pm$ sign in Eq. (2.15) indicates there may be two possible intersection points of this ray and a complete sphere. It is known that $\lambda_i = -D_i - \sqrt{D_i^2 - E_i}$ is for the nearer intersection point and $\lambda_i = -D_i + \sqrt{D_i^2 - E_i}$ is for the farther one. Clearly, only one of these points is useful, thus the appropriate sign must be chosen. When $\rho_i^2 + \sigma_i^2 \neq 0$, $\alpha_i$ and $\beta_i$ at the incidence point $\overline{P}_i$ can be solved from (see Eq. (2.11)) the following equations, since $C\beta_i > 0$ is always valid:

$$\alpha_i = \operatorname{atan2}(\rho_i, \sigma_i), \tag{2.18}$$

$$\beta_i = \operatorname{atan2}\left(\tau_i, \sqrt{\sigma_i^2 + \rho_i^2}\right). \tag{2.19}$$

It is noted from the 3rd component of Eq. (2.11) that an alternative approach of determination of $\beta_i$ is $\beta_i = \arcsin(\tau_i/|R_i|)$. However, one may not obtain the value of $\beta_i$ by using this equation when the incoming ray $\overline{R}_{i-1}$ do not intersect with the complete sphere of $\bar{r}_i$ when $|R_i| < \tau_i$. Equation (2.19) can avoid this difficulty, since it always gives solution when $\sigma_i^2 + \rho_i^2 + \tau_i^2 \neq 0$.

To trace the reflected/or refracted ray at the boundary surface, one needs the incidence angle $\theta_i$, which is defined as a non-negative acute angle (i.e., $0° \le \theta_i < 90°$) in geometrical optics, is determined by the dot product of $\bar{\ell}_i$ and the unit normal vector $\bar{n}_i$:

$$\begin{aligned} C\theta_i &= \left|\bar{\ell}_{i-1} \bullet \bar{n}_i\right| = \left|\bar{\ell}^T_{i-1}\bar{n}_i\right| = -(\ell_{i-1x}n_{ix} + \ell_{i-1y}n_{iy} + \ell_{i-1z}n_{iz}) \\ &= -s_i\left[\ell_{i-1x}(I_{ix}C\beta_iC\alpha_i + J_{ix}C\beta_iS\alpha_i + K_{ix}S\beta_i) + \ell_{i-1y}(I_{iy}C\beta_iC\alpha_i + J_{iy}C\beta_iS\alpha_i + K_{iy}S\beta_i)\right. \\ &\quad \left. + \ell_{i-1z}(I_{iz}C\beta_iC\alpha_i + J_{iz}C\beta_iS\alpha_i + K_{iz}S\beta_i)\right]. \end{aligned} \tag{2.20}$$

The refraction angle $\underline{\theta}_i$ between two optical media must satisfy Snell's law:

$$S\underline{\theta}_i = \frac{\xi_{i-1}}{\xi_i}S\theta_i = N_iS\theta_i, \tag{2.21}$$

where $\xi_i$ is the refractive index of medium i and $N_i = \xi_{i-1}/\xi_i$ is the refractive index of medium i − 1 relative to that of medium i. One also needs the following common unit normal vector $\bar{m}_i$ of $\bar{n}_i$ and $\bar{\ell}_{i-1}$:

$$\bar{m}_i = \begin{bmatrix} m_{ix} \\ m_{iy} \\ m_{iz} \\ 0 \end{bmatrix} = \frac{\bar{n}_i \times \bar{\ell}_{i-1}}{S\theta_i}. \tag{2.22}$$

It is useful to have the following equation, which is derived from Eq. (2.22), when we determine the unit directional vectors $\bar{\ell}_i$ of the reflected and refracted rays:

$$S\theta_i(\bar{m}_i \times \bar{n}_i) = (\bar{n}_i \times \bar{\ell}_{i-1}) \times \bar{n}_i = \bar{\ell}_{i-1} - (\bar{n}_i \bullet \bar{\ell}_{i-1})\bar{n}_i = \bar{\ell}_{i-1} + \bar{n}_i\, C\theta_i. \tag{2.23}$$

According to the reflection law of optics, the reflected unit directional vector $\bar{\ell}_i$ is obtained by rotating $\bar{n}_i$ about $\bar{m}_i$ by an angle $\theta_i$ (see Fig. 2.5 and use Eq. (1.25)). This leads to

$$\begin{aligned} \bar{\ell}_i &= [\ell_{ix} \quad \ell_{iy} \quad \ell_{iz} \quad 0]^T = \mathrm{rot}(\bar{m}_i, \theta_i)\bar{n}_i \\ &= \begin{bmatrix} m_{ix}^2(1 - C\theta_i) + C\theta_i & m_{iy}m_{ix}(1 - C\theta_i) - m_{iz}S\theta_i & m_{iz}m_{ix}(1 - C\theta_i) + m_{iy}S\theta_i & 0 \\ m_{ix}m_{iy}(1 - C\theta_i) + m_{iz}S\theta_i & m_{iy}^2(1 - C\theta_i) + C\theta_i & m_{iz}m_{iy}(1 - C\theta_i) - m_{ix}S\theta_i & 0 \\ m_{ix}m_{iz}(1 - C\theta_i) - m_{iy}S\theta_i & m_{iy}m_{iz}(1 - C\theta_i) + m_{ix}S\theta_i & m_{iz}^2(1 - C\theta_i) + C\theta_i & 0 \\ 0 & 0 & 0 & 1 \end{bmatrix} \begin{bmatrix} n_{ix} \\ n_{iy} \\ n_{iz} \\ 0 \end{bmatrix}. \end{aligned} \tag{2.24}$$

Further simplification of Eq. (2.24) is possible by utilizing Eq. (2.23), resulting in:

$$\bar{\ell}_i = \begin{bmatrix} \ell_{ix} \\ \ell_{iy} \\ \ell_{iz} \\ 0 \end{bmatrix} = \begin{bmatrix} n_{ix}C\theta_i + (n_{iz}m_{iy} - n_{iy}m_{iz})S\theta_i \\ n_{iy}C\theta_i + (n_{ix}m_{iz} - n_{iz}m_{ix})S\theta_i \\ n_{iz}C\theta_i + (n_{iy}m_{ix} - n_{ix}m_{iy})S\theta_i \\ 0 \end{bmatrix} = \begin{bmatrix} \ell_{i-1x} + 2n_{ix}C\theta_i \\ \ell_{i-1y} + 2n_{iy}C\theta_i \\ \ell_{i-1z} + 2n_{iz}C\theta_i \\ 0 \end{bmatrix}. \tag{2.25}$$

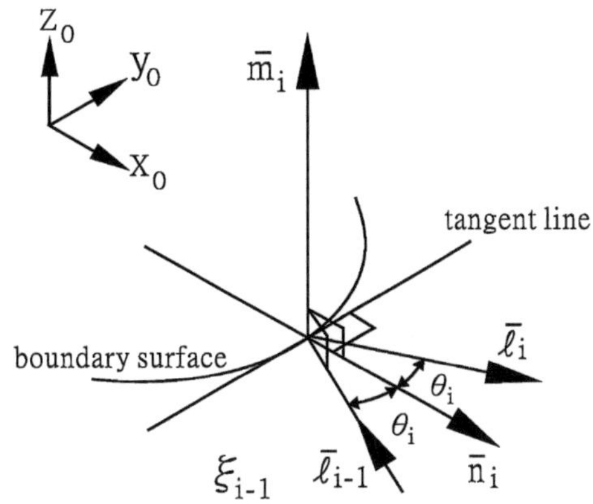


**Fig. 2.5** The reflected unit directional vector can be obtained by rotating $\bar{n}_i$ about $\bar{m}_i$ by an angle $\theta_i$

An alternative method of determination of $\bar{\ell}_i$ is by rotating $\bar{\ell}_{i-1}$ about $\bar{m}_i$ by an angle $\pi + 2\theta_i$, i.e., $\bar{\ell}_i = \text{rot}(\bar{m}_i, \pi + 2\theta_i)\bar{\ell}_{i-1}$.

According to the refraction law of optics, the refracted unit directional vector $\bar{\ell}_i$ can be obtained by rotating $\bar{n}_i$ about $\bar{m}_i$ by an angle $\pi - \underline{\theta}_i$ (see Fig. 2.6). This leads to

$$\bar{\ell}_i = [\ell_{ix} \quad \ell_{iy} \quad \ell_{iz} \quad 0]^T = \text{rot}(\bar{m}_i, \pi - \underline{\theta}_i)\bar{n}_i$$
$$= \begin{bmatrix} m_{ix}^2(1 + C\underline{\theta}_i) - C\underline{\theta}_i & m_{iy}m_{ix}(1 + C\underline{\theta}_i) - m_{iz}S\underline{\theta}_i & m_{iz}m_{ix}(1 + C\underline{\theta}_i) + m_{iy}S\underline{\theta}_i & 0 \\ m_{ix}m_{iy}(1 + C\underline{\theta}_i) + m_{iz}S\underline{\theta}_i & m_{iy}^2(1 + C\underline{\theta}_i) - C\underline{\theta}_i & m_{iz}m_{iy}(1 + C\underline{\theta}_i) - m_{ix}S\underline{\theta}_i & 0 \\ m_{ix}m_{iz}(1 + C\underline{\theta}_i) - m_{iy}S\underline{\theta}_i & m_{iy}m_{iz}(1 + C\underline{\theta}_i) + m_{ix}S\underline{\theta}_i & m_{iz}^2(1 + C\underline{\theta}_i) - C\underline{\theta}_i & 0 \\ 0 & 0 & 0 & 1 \end{bmatrix} \begin{bmatrix} n_{ix} \\ n_{iy} \\ n_{iz} \\ 0 \end{bmatrix}. \tag{2.26}$$

Again, further simplification of Eq. (2.26) is possible by utilizing Eq. (2.23) and Snell's law $S\underline{\theta}_i = N_i S\theta_i$. Thus one obtains the following refracted unit directional vector $\bar{\ell}_i$:

$$\bar{\ell}_i = \begin{bmatrix} \ell_{ix} \\ \ell_{iy} \\ \ell_{iz} \\ 0 \end{bmatrix} = \begin{bmatrix} -n_{ix}C\underline{\theta}_i + (n_{iz}m_{iy} - n_{iy}m_{iz})S\underline{\theta}_i \\ -n_{iy}C\underline{\theta}_i + (n_{ix}m_{iz} - n_{iz}m_{ix})S\underline{\theta}_i \\ -n_{iz}C\underline{\theta}_i + (n_{iy}m_{ix} - n_{ix}m_{iy})S\underline{\theta}_i \\ 0 \end{bmatrix} = \begin{bmatrix} -n_{ix}C\underline{\theta}_i + N_i(\ell_{i-1x} + n_{ix}C\theta_i) \\ -n_{iy}C\underline{\theta}_i + N_i(\ell_{i-1y} + n_{iy}C\theta_i) \\ -n_{iz}C\underline{\theta}_i + N_i(\ell_{i-1z} + n_{iz}C\theta_i) \\ 0 \end{bmatrix}$$
$$= \begin{bmatrix} -n_{ix}\sqrt{1 - N_i^2 + (N_iC\theta_i)^2} + N_i(\ell_{i-1x} + n_{ix}C\theta_i) \\ -n_{iy}\sqrt{1 - N_i^2 + (N_iC\theta_i)^2} + N_i(\ell_{i-1y} + n_{iy}C\theta_i) \\ -n_{iz}\sqrt{1 - N_i^2 + (N_iC\theta_i)^2} + N_i(\ell_{i-1z} + n_{iz}C\theta_i) \\ 0 \end{bmatrix}, \tag{2.27}$$

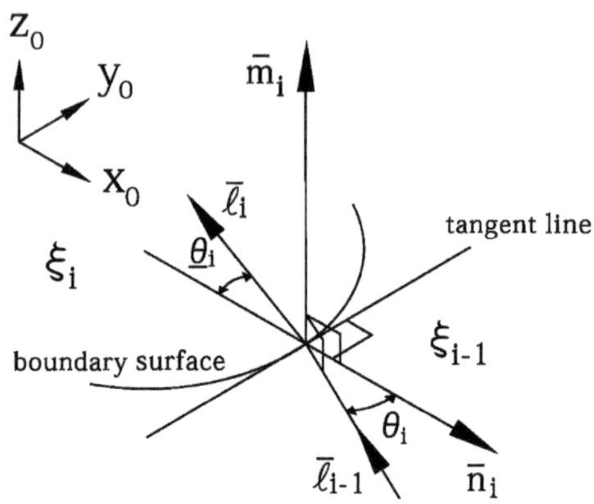


**Fig. 2.6** The refracted unit directional vector can be obtained by rotating $\bar{n}_i$ about $\bar{m}_i$ by an angle $\pi - \underline{\theta}_i$

where $\bar{n}_i = [n_{ix} \quad n_{iy} \quad n_{iz} \quad 0]^T$ and $C\theta_i$ are given by Eqs. (2.9 and 2.20), respectively.

One has to note that the points at $\beta_i = \pm\pi/2$ of a spherical boundary surface are two pseudosingular points (or irregular points), at which $\partial^i\bar{r}_i/\partial\beta_i$ and $\partial^i\bar{r}_i/\partial\alpha_i$ are not linearly independent, yielding the cross product $\partial^i\bar{r}_i/\partial\beta_i \times \partial^i\bar{r}_i/\partial\alpha_i$ cannot be performed. In order to avoid this problem, in this book we always let the $y_i$ axis of a spherical boundary surface coincides with the optical axis of an optical system.We designate (see Fig. 2.7)

$$\overline{X}_i = [t_{ix} \quad t_{iy} \quad t_{iz} \quad \omega_{ix} \quad \omega_{iy} \quad \omega_{iz} \quad \xi_{i-1} \quad \xi_i \quad R_i]^T, \tag{2.28}$$

as the boundary variable vector of a spherical boundary surface when the six pose variables of Eq. (2.8), the refractive indices $\xi_{i-1}, \xi_i$, and the radius $R_i$ of this boundary surface are included.

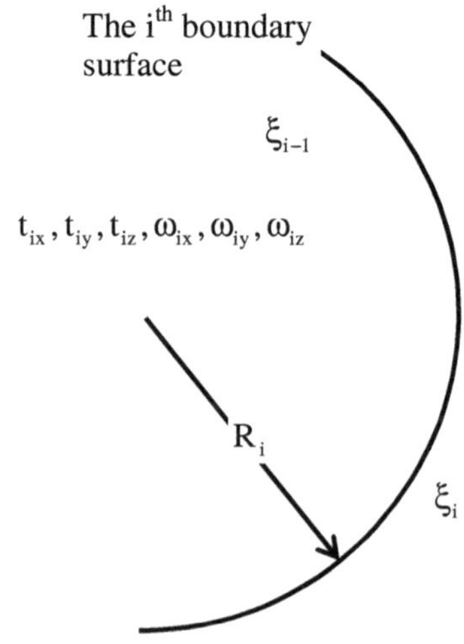


**Fig. 2.7** The variables of a spherical boundary surface

It is also noted from Eqs. (2.10, 2.25 and 2.27) that the ray $\overline{R}_i = [\overline{P}_i \quad \overline{\ell}_i]^T$ is a function of the incoming ray $\overline{R}_{i-1} = [\overline{P}_{i-1} \quad \overline{\ell}_{i-1}]^T$ with the given source ray $\overline{R}_0 = [\overline{P}_0 \quad \overline{\ell}_0]^T$. This shows that $\overline{R}_i$ is a recursive function, i.e., a function which operates in turn on other functions. It is like a Russian nesting doll. Each doll has a smaller and smaller doll inside it. To evaluate a recursive function, it is first necessary to evaluate the internal functions, and then to determine the outer function based on the results of the internal functions.

*Example* 2.1 Consider the biconvex element having convex-concave spherical boundary surfaces shown in Fig. 2.8, with $v_1 = 5$, thickness $q_{e1} = 10$, refractive index $\xi_{e1} = 1.5$, surface curvatures $\kappa_1 = 1/50$ and $\kappa_2 = -1/100$. (1) Assign boundary coordinate frames $(xyz)_1$ and $(xyz)_2$ to these two spherical boundary surfaces. (2) Find their unit normal vectors ${}^1\bar{n}_1$ and ${}^2\bar{n}_2$. (3) Determine the pose matrix ${}^0\overline{A}_1$ and boundary variable vector $\overline{X}_1$. (4) Give the pose matrix ${}^0\bar{A}_2$ and boundary variable vector $\overline{X}_2$. (5) Write out its source ray $\overline{R}_0$ when $\overline{P}_0 = [0 \quad -5 \quad 5 \quad 1]^T$ and $\alpha_0 = 0°$, $\beta_0 = 5°$. (6) Find $\lambda_1$, $\alpha_1$, $\beta_1$, unit normal vector $\bar{n}_1$, parameter $s_1$, incidence angle $\theta_1$, refractive index $N_1$, and the refracted ray $\overline{R}_1$ when the ray $\overline{R}_0$ is refracted by the 1st boundary surface $\bar{r}_1$. (7) Find $\lambda_2$, $\alpha_2$, $\beta_2$, unit normal vector $\bar{n}_2$, parameter $s_2$, incidence angle $\theta_1$, refractive index $N_2$, and the refracted ray $\overline{R}_2$ when ray $\overline{R}_1$ is refracted by the 2nd boundary surface $\bar{r}_2$.

(1) The assigned coordinate frames $(xyz)_1$ and $(xyz)_2$ are shown in Fig. 2.9. Note that the origins $o_1$ and $o_2$ are respectively located at the centers of $\bar{r}_1$ and $\bar{r}_2$, while $y_1$ and $y_2$ axes coincide with $y_0$ axis.

(2)

$$ {}^1\bar{n}_1 = s_1[C\beta_1 C\alpha_1 \quad C\beta_1 S\alpha_1 \quad S\beta_1 \quad 0]^T, \tag{2.29a} $$

$$ {}^2\bar{n}_2 = s_2[C\beta_2 C\alpha_2 \quad C\beta_2 S\alpha_2 \quad S\beta_2 \quad 0]^T. \tag{2.29b} $$

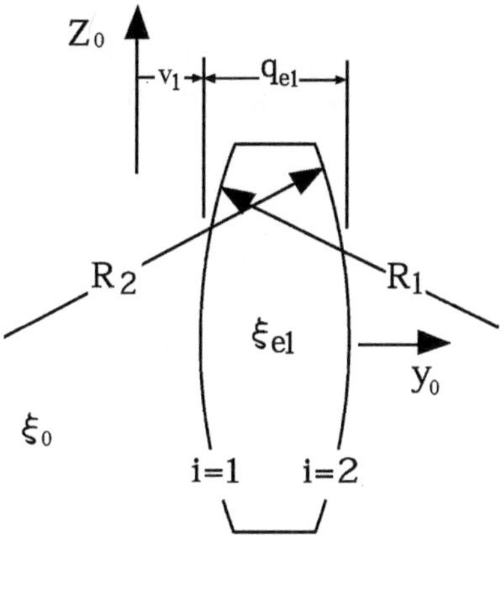


**Fig. 2.8** A biconvex element having convex–convex spherical boundary surfaces

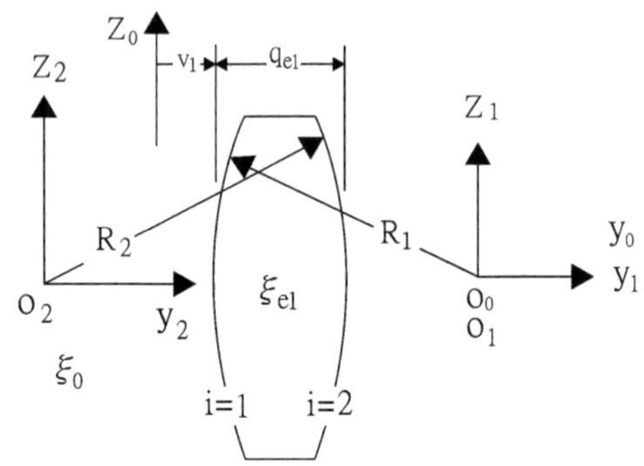


**Fig. 2.9** The assigned coordinate frames $(xyz)_1$ and $(xyz)_2$ for system of Fig. 2.8

(3)

$$^0\bar{A}_1 = \mathrm{tran}(0, v_1 + R_1, 0) = \mathrm{tran}(0, 5 + 50, 0) = \begin{bmatrix} 1 & 0 & 0 & 0 \\ 0 & 1 & 0 & 55 \\ 0 & 0 & 1 & 0 \\ 0 & 0 & 0 & 1 \end{bmatrix}, \tag{2.29c}$$

$$\begin{aligned} \overline{X}_1 &= \begin{bmatrix} t_{1x} & t_{1y} & t_{1z} & \omega_{1x} & \omega_{1y} & \omega_{1z} & \xi_0 & \xi_1 & R_1 \end{bmatrix}^T \\ &= \begin{bmatrix} t_{1x} & v_1 + R_1 & t_{1z} & \omega_{1x} & \omega_{1y} & \omega_{1z} & \xi_0 & \xi_{e1} & R_1 \end{bmatrix}^T \\ &= \begin{bmatrix} 0 & 55 & 0 & 0 & 0 & 0 & 1 & 1.5 & 50 \end{bmatrix}^T. \end{aligned} \tag{2.29d}$$

(4)

$$^0\bar{A}_2 = \mathrm{tran}(0, v_1 + q_{e1} + R_2, 0) = \mathrm{tran}(0, 5 + 10 - 100, 0) = \begin{bmatrix} 1 & 0 & 0 & 0 \\ 0 & 1 & 0 & -85 \\ 0 & 0 & 1 & 0 \\ 0 & 0 & 0 & 1 \end{bmatrix}. \tag{2.29e}$$

$$\begin{aligned} \overline{X}_2 &= \begin{bmatrix} t_{2x} & t_{2y} & t_{2z} & \omega_{2x} & \omega_{2y} & \omega_{2z} & \xi_1 & \xi_2 & R_2 \end{bmatrix}^T \\ &= \begin{bmatrix} t_{2x} & v_1 + q_{e1} + R_2 & t_{2z} & \omega_{2x} & \omega_{2y} & \omega_{2z} & \xi_{e1} & \xi_0 & R_2 \end{bmatrix}^T \\ &= \begin{bmatrix} 0 & -85 & 0 & 0 & 0 & 0 & 1.5 & 1 & -100 \end{bmatrix}^T. \end{aligned} \tag{2.29f}$$

(5)

$$\begin{aligned} \overline{R}_0 &= \begin{bmatrix} \overline{P}_0 & \overline{\ell}_0 \end{bmatrix}^T \\ &= \begin{bmatrix} 0 & -5 & 5 & C5^\circ C90^\circ & C5^\circ S90^\circ & S5^\circ \end{bmatrix}^T \\ &= \begin{bmatrix} 0 & -5 & 5 & 0 & 0.99619 & 0.08716 \end{bmatrix}^T. \end{aligned} \tag{2.29g}$$

(6)

$$\lambda_1 = 10.38951, \tag{2.29h}$$

$$\alpha_1 = -90^\circ, \tag{2.29i}$$

$$\beta_1 = 6.78304^\circ, \tag{2.29j}$$

$$\begin{aligned}\bar{n}_1 = {}^0\bar{A}_1{}^1\bar{n}_1 = {}^1\bar{n}_1 = s_1[C\beta_1C\alpha_1 \quad C\beta_1S\alpha_1 \quad S\beta_1 \quad 0]^T \\ = s_1[0 \quad -0.99300 \quad 0.11811 \quad 0]^T,\end{aligned} \tag{2.29k}$$

$$s_1 = 1, \tag{2.29l}$$

$$\theta_1 = 11.78304^\circ, \tag{2.29m}$$

$$N_1 = \xi_0/\xi_1 = 1/1.5, \tag{2.29n}$$

$$\overline{R}_1 = [0 \quad 5.34997 \quad 5.90551 \quad 0 \quad 0.99983 \quad 0.01817]^T. \tag{2.29o}$$

(7)

$$\lambda_2 = 9.46673, \tag{2.29p}$$

$$\alpha_2 = 90^\circ, \tag{2.29q}$$

$$\beta_2 = 3.48433^\circ. \tag{2.29r}$$

$$\begin{aligned}\bar{n}_2 = {}^0\bar{A}_2{}^2\bar{n}_2 = {}^2\bar{n}_2 = s_2[C\beta_2C\alpha_2 \quad C\beta_2S\alpha_2 \quad S\beta_2 \quad 0]^T \\ = s_2[0 \quad -0.99815 \quad -0.06708 \quad 0]^T,\end{aligned} \tag{2.29s}$$

$$s_2 = -1, \tag{2.29t}$$

$$\theta_2 = 2.44297^\circ, \tag{2.29u}$$

$$N_2 = \xi_1/\xi_2 = 1.5, \tag{2.29v}$$

$$\overline{R}_2 = [0 \quad 14.81515 \quad 6.07756 \quad 0 \quad 0.99999 \quad -0.00317]^T. \tag{2.29w}$$

## 2.3 Skew-Ray Tracing at a Flat Boundary Surface

Many optical tasks require optical elements having flat boundary surfaces. Examples of such elements are plano-convex lenses, plano-concave lenses, optical flats, beam-splitters, and flat first-surface mirrors. Raytracing in a flat boundary surface is also every important to be discussed. Most literature deals a flat boundary surface as a spherical one with its curvature to be zero (e.g., p. 312 of [3]). Unfortunately, we found that some important useful equations of prisms (which will be given in Chap. 5) can not be obtained, if we use this approach. Therefore, it is necessary to have a different methodology to treat flat boundary surfaces.

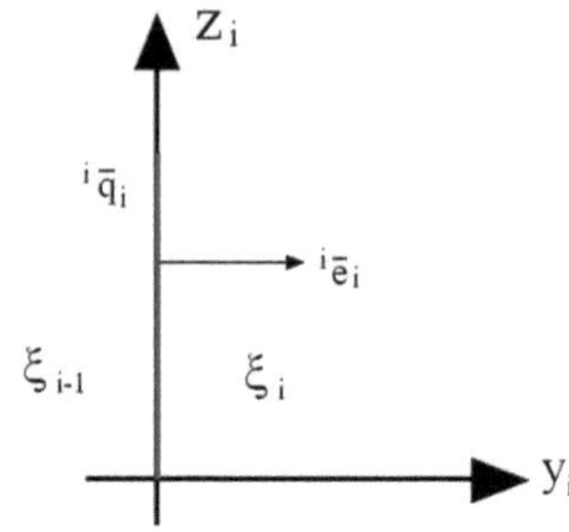


**Fig. 2.10** The generating line and its unit normal vector of a flat boundary surface

Recall that in Sect. 2.2 we always let the $y_i$ axis of a boundary coordinate frame $(xyz)_i$ coincides with the optical axis of an axis-symmetrical optical system. For the purpose of convenient programming, we again let the $y_i$ axis of the boundary coordinate frame $(xyz)_i$ of a flat boundary surface coincides with the optical axis. This leads that a flat boundary surface ${}^i\bar{r}_i$ has to be established by rotating its generating line (see Fig. 2.10)

$$ {}^i\bar{q}_i = \begin{bmatrix} 0 \\ 0 \\ z_i(\beta_i) \\ 1 \end{bmatrix} = \begin{bmatrix} 0 \\ 0 \\ \beta_i \\ 1 \end{bmatrix} \quad (\beta_i \geq 0), \tag{2.30} $$

about the optical $y_i$ axis by an angle $\alpha_i$ $(0 \leq \alpha_i < 2\pi)$, i.e.

$$ {}^i\bar{r}_i = \begin{bmatrix} x_i \\ y_i \\ z_i \\ 1 \end{bmatrix} = \text{rot}(\bar{y}, \alpha_i)\,{}^i\bar{q}_i = \begin{bmatrix} C\alpha_i & 0 & S\alpha_i & 0 \\ 0 & 1 & 0 & 0 \\ -S\alpha_i & 0 & C\alpha_i & 0 \\ 0 & 0 & 0 & 1 \end{bmatrix} \begin{bmatrix} 0 \\ 0 \\ \beta_i \\ 1 \end{bmatrix} = \begin{bmatrix} \beta_i S\alpha_i \\ 0 \\ \beta_i C\alpha_i \\ 1 \end{bmatrix}. \tag{2.31} $$

The unit normal vector ${}^i\bar{e}_i$ of the generating curve ${}^i\bar{q}_i$ is:

$$ {}^i\bar{e}_i = \begin{bmatrix} e_{ix} \\ e_{iy} \\ e_{iz} \\ 0 \end{bmatrix} = s_i \begin{bmatrix} 0 \\ -1 \\ 0 \\ 0 \end{bmatrix}. \tag{2.32} $$

Again, $s_i$ is set to +1 or −1 such that the cosine of the incident angle $\theta_i$ computed by Eq. (2.43) will be greater than zero. The unit normal vector ${}^i\bar{n}_i$ of ${}^i\bar{r}_i$ is determined by rotating ${}^i\bar{e}_i$ about the $y_i$ axis by an angle $\alpha_i$, i.e.:

$$ {}^i\bar{n}_i = \text{rot}(\bar{y}, \alpha_i)\,{}^i\bar{e}_i = s_i \begin{bmatrix} C\alpha_i & 0 & S\alpha_i & 0 \\ 0 & 1 & 0 & 0 \\ -S\alpha_i & 0 & C\alpha_i & 0 \\ 0 & 0 & 0 & 1 \end{bmatrix} \begin{bmatrix} 0 \\ -1 \\ 0 \\ 1 \end{bmatrix} = s_i \begin{bmatrix} 0 \\ -1 \\ 0 \\ 0 \end{bmatrix}. \tag{2.33} $$

Equations (2.31) and (2.33) give the expressions of the boundary surface ${}^{i}\bar{r}_i$ and its unit normal vector ${}^{i}\bar{n}_i$ with respect to the boundary coordinate frame $(xyz)_i$. However, many derivations in this book are built relative to the world coordinate frame $(xyz)_o$. One thus needs the pose matrix of $(xyz)_i$ with respect to $(xyz)_o$, given as:

$$
\begin{aligned}
{}^{0}\bar{A}_i &= \text{tran}(t_{ix},0,0)\text{tran}(0,t_{iy},0)\text{tran}(0,0,t_{iz})\text{rot}(\bar{z},\omega_{iz})\text{rot}(\bar{y},\omega_{iy})\text{rot}(\bar{x},\omega_{ix}) \\
&= \text{tran}(t_{ix},t_{iy},t_{iz})\text{RPY}(\omega_{iz},\omega_{iy},\omega_{ix}) \\
&= \begin{bmatrix} C\omega_{iz}C\omega_{iy} & C\omega_{iz}S\omega_{iy}S\omega_{ix} - S\omega_{iz}C\omega_{ix} & C\omega_{iz}S\omega_{iy}C\omega_{ix} + S\omega_{iz}S\omega_{ix} & t_{ix} \\ S\omega_{iz}C\omega_{iy} & S\omega_{iz}S\omega_{iy}S\omega_{ix} + C\omega_{iz}C\omega_{ix} & S\omega_{iz}S\omega_{iy}C\omega_{ix} - C\omega_{iz}S\omega_{ix} & t_{iy} \\ -S\omega_{iy} & C\omega_{iy}S\omega_{ix} & C\omega_{iy}C\omega_{ix} & t_{iz} \\ 0 & 0 & 0 & 1 \end{bmatrix} \\
&= \begin{bmatrix} I_{ix} & J_{ix} & K_{ix} & t_{ix} \\ I_{iy} & J_{iy} & K_{iy} & t_{iy} \\ I_{iz} & J_{iz} & K_{iz} & t_{iz} \\ 0 & 0 & 0 & 1 \end{bmatrix},
\end{aligned}
\tag{2.34}
$$

where $t_{ix}, t_{iy}, t_{iz}, \omega_{iz}, \omega_{iy}$ and $\omega_{ix}$ are the pose variables of this flat boundary surface. Now the unit normal vector $\bar{n}_i$ expressed with respect to $(xyz)_0$ can be obtained as:

$$
\bar{n}_i = \begin{bmatrix} n_{ix} \\ n_{iy} \\ n_{iz} \\ 0 \end{bmatrix} = {}^{0}\bar{A}_i{}^{i}\bar{n}_i = s_i \begin{bmatrix} I_{ix} & J_{ix} & K_{ix} & t_{ix} \\ I_{iy} & J_{iy} & K_{iy} & t_{iy} \\ I_{iz} & J_{iz} & K_{iz} & t_{iz} \\ 0 & 0 & 0 & 1 \end{bmatrix} \begin{bmatrix} 0 \\ -1 \\ 0 \\ 0 \end{bmatrix} = s_i \begin{bmatrix} -J_{ix} \\ -J_{iy} \\ -J_{iz} \\ 0 \end{bmatrix}. \tag{2.35}
$$

In Fig. 2.11 a ray originating at the incidence point $\bar{P}_{i-1} = [P_{i-1x} \quad P_{i-1y} \quad P_{i-1z} \quad 1]^T$ (at last boundary surface $\bar{r}_{i-1}$) directed along the unit directional vector $\bar{\ell}_{i-1} = [\ell_{i-1x} \quad \ell_{i-1y} \quad \ell_{i-1z} \quad 0]^T$ will be reflected/or refracted at current flat boundary surface $\bar{r}_i$. Any intermediate point $\bar{P}'_{i-1}$ along this ray as it travels from $\bar{P}_{i-1}$ is given by $\bar{P}'_{i-1} = \bar{P}_{i-1} + \lambda\bar{\ell}_{i-1}$. The incidence point $\bar{P}_i$, at which the ray hits this flat boundary surface, is

$$
\bar{P}_i = \begin{bmatrix} P_{ix} \\ P_{iy} \\ P_{iz} \\ 1 \end{bmatrix} = \begin{bmatrix} P_{i-1x} + \ell_{i-1x}\lambda_i \\ P_{i-1y} + \ell_{i-1y}\lambda_i \\ P_{i-1z} + \ell_{i-1z}\lambda_i \\ 1 \end{bmatrix}, \tag{2.36}
$$

where the parameter $\lambda_i$ is given by equating (2.31) to ${}^{i}\bar{P}_i = {}^{i}\bar{A}_0\bar{P}_i = ({}^{0}\bar{A}_i)^{-1}\bar{P}_i$, yielding

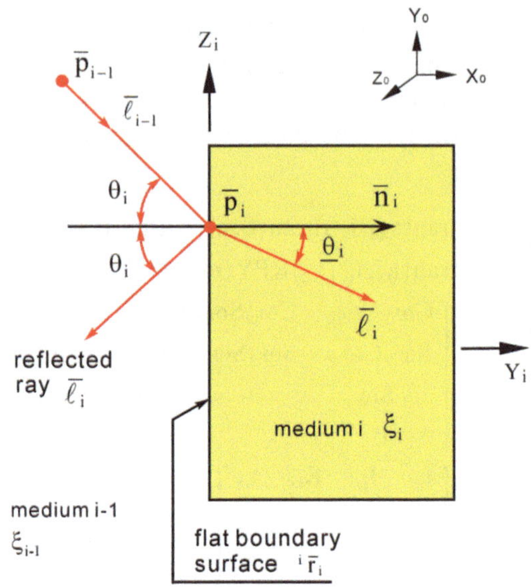


**Fig. 2.11** Raytracing at a flat boundary surface

$$
{}^{i}\bar{P}_i = \begin{bmatrix} I_{ix} & I_{iy} & I_{iz} & -(I_{ix}t_{ix} + I_{iy}t_{iy} + I_{iz}t_{iz}) \\ J_{ix} & J_{iy} & J_{iz} & -(J_{ix}t_{ix} + J_{iy}t_{iy} + J_{iz}t_{iz}) \\ K_{ix} & K_{iy} & K_{iz} & -(K_{ix}t_{ix} + K_{iy}t_{iy} + K_{iz}t_{iz}) \\ 0 & 0 & 0 & 1 \end{bmatrix} \begin{bmatrix} P_{i-1x} + \ell_{i-1x}\lambda_i \\ P_{i-1y} + \ell_{i-1y}\lambda_i \\ P_{i-1z} + \ell_{i-1z}\lambda_i \\ 1 \end{bmatrix} = \begin{bmatrix} \sigma_i \\ \rho_i \\ \tau_i \\ 1 \end{bmatrix}
$$
$$
= {}^{i}\bar{r}_i = \begin{bmatrix} \beta_i S\alpha_i \\ 0 \\ \beta_i C\alpha_i \\ 1 \end{bmatrix}, \tag{2.37}
$$

where $\sigma_i$, $\rho_i$, and $\tau_i$ are given in Eqs. (2.12), (2.13) and (2.14), respectively. One can solve for $\lambda_i$ from the second component of Eq. (2.37) (i.e., $\rho_i = 0$) to give

$$
\lambda_i = \frac{-[J_{ix}P_{i-1x} + J_{iy}P_{i-1y} + J_{iz}P_{i-1z} - (J_{ix}t_{ix} + J_{iy}t_{iy} + J_{iz}t_{iz})]}{J_{ix}\ell_{i-1x} + J_{iy}\ell_{i-1y} + J_{iz}\ell_{i-1z}} = \frac{-D_i}{E_i}, \tag{2.38}
$$

where

$$
D_i = J_{ix}P_{i-1x} + J_{iy}P_{i-1y} + J_{iz}P_{i-1z} - (J_{ix}t_{ix} + J_{iy}t_{iy} + J_{iz}t_{iz}), \tag{2.39}
$$

$$
E_i = J_{ix}\ell_{i-1x} + J_{iy}\ell_{i-1y} + J_{iz}\ell_{i-1z}. \tag{2.40}
$$

The parameters $\alpha_i$ $(0 \le \alpha_i < 2\pi)$ and $\beta_i$ $(0 \le \beta_i)$, where the incidence point $\bar{P}_i$ hits the $x_iz_i$ plane, are determined respectively by

$$
\beta_i = \sqrt{\sigma_i^2 + \tau_i^2}, \tag{2.41}
$$

$$\alpha_i \begin{cases} = \operatorname{atan2}(\sigma_i, \tau_i) & \text{when } \beta_i \neq 0 \\ = \text{any value} & \text{when } \beta_i = 0 \end{cases}. \tag{2.42}$$

One has to note that although the point at $\beta_i = 0$ of a flat boundary surface is pseudosingular point, it does not cause any difficulty in computations.

The incidence angle $\theta_i$, a non-negative acute angle in geometrical optics, can be computed by

$$C\theta_i = \left|\bar{\ell}_{i-1} \bullet \bar{n}_i\right| = \left|\bar{\ell}_{i-1}^T \bar{n}_i\right| = s_i\left(J_{ix}\ell_{i-1x} + J_{iy}\ell_{i-1y} + J_{iz}\ell_{i-1z}\right) = s_i E_i. \tag{2.43}$$

The refraction angle $\underline{\theta}_i$ between two optical media must satisfy Snell's law: $S\underline{\theta}_i = (\xi_{i-1}/\xi_i)S\theta_i = N_i S\theta_i$. Similar with the case of a spherical boundary surface, one can determine the reflected unit directional vector $\bar{\ell}_i$ by rotating the unit normal vector $\bar{n}_i$ about the unit common normal vector $\bar{m}_i$ by an angle $\theta_i$, yielding

$$\bar{\ell}_i = \begin{bmatrix} \ell_{ix} \\ \ell_{iy} \\ \ell_{iz} \\ 0 \end{bmatrix} = \begin{bmatrix} \ell_{i-1x} + 2n_{ix}C\theta_i \\ \ell_{i-1y} + 2n_{iy}C\theta_i \\ \ell_{i-1z} + 2n_{iz}C\theta_i \\ 0 \end{bmatrix} = \begin{bmatrix} \ell_{i-1x} - 2J_{ix}E_i \\ \ell_{i-1y} - 2J_{iy}E_i \\ \ell_{i-1z} - 2J_{iz}E_i \\ 0 \end{bmatrix}. \tag{2.44}$$

The refracted unit directional vector $\bar{\ell}_i$ can be obtained by rotating the unit normal vector $\bar{n}_i$ about the unit common vector $\bar{m}_i$ by an angle $\pi - \underline{\theta}_i$, giving

$$\bar{\ell}_i = \begin{bmatrix} \ell_{ix} \\ \ell_{iy} \\ \ell_{iz} \\ 0 \end{bmatrix} = \begin{bmatrix} -n_{ix}\sqrt{1 - N_i^2 + (N_i C\theta_i)^2} + N_i(\ell_{i-1x} + n_{ix}C\theta_i) \\ -n_{iy}\sqrt{1 - N_i^2 + (N_i C\theta_i)^2} + N_i\left(\ell_{i-1y} + n_{iy}C\theta_i\right) \\ -n_{iz}\sqrt{1 - N_i^2 + (N_i C\theta_i)^2} + N_i(\ell_{i-1z} + n_{iz}C\theta_i) \\ 0 \end{bmatrix}$$

$$= \begin{bmatrix} s_i J_{ix}\sqrt{1 - N_i^2 + (N_i E_i)^2} + N_i(\ell_{i-1x} - J_{ix}E_i) \\ s_i J_{iy}\sqrt{1 - N_i^2 + (N_i E_i)^2} + N_i\left(\ell_{i-1y} - J_{iy}E_i\right) \\ s_i J_{iz}\sqrt{1 - N_i^2 + (N_i E_i)^2} + N_i(\ell_{i-1z} - J_{iz}E_i) \\ 0 \end{bmatrix}. \tag{2.45}$$

It is noted that only eight variables (i.e., $t_{ix}$, $t_{iy}$, $t_{iz}$, $\omega_{ix}$, $\omega_{iy}$, $\omega_{iz}$, $\xi_{i-1}$, $\xi_i$) are needed to completely describe the effects of a flat boundary surface in geometrical optics. However, for convenience, we also use Eq. (2.28) to designate the variable vector of a flat boundary surface by ignoring $R_i$.

This chapter illustrates how to find the path of a reflected/or refracted ray $\bar{R}_i$ relative to a single boundary surface. One can apply this approach successively to trace rays with respect to an optical system possessing n boundary surfaces. To trace a ray reflected/or refracted at the gth boundary surface, one should label the boundary surfaces of an optical system sequentially from 1 to n, and then use the preceding algorithm with $i = 1$, $i = 2$,…, until $i = g$ to obtain the ray $\bar{R}_g$ refracted/or reflected by the gth boundary surface (see Fig. 2.12), where:

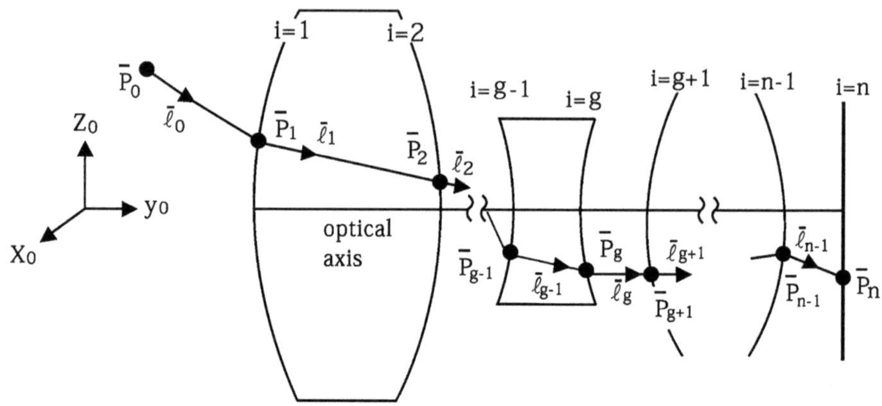


**Fig. 2.12** The ray $\bar{R}_g$ can be traced by using the proposed approach

$$\bar{P}_g = \begin{bmatrix} P_{gx} \\ P_{gy} \\ P_{gz} \\ 1 \end{bmatrix} = \begin{bmatrix} P_{g-1x} + \ell_{g-1x}\lambda_g \\ P_{g-1y} + \ell_{g-1y}\lambda_g \\ P_{g-1z} + \ell_{g-1z}\lambda_g \\ 1 \end{bmatrix} = \begin{bmatrix} P_{g-2x} + \ell_{g-2x}\lambda_{g-1} + \ell_{g-1x}\lambda_g \\ P_{g-2y} + \ell_{g-2y}\lambda_{g-1} + \ell_{g-1y}\lambda_g \\ P_{g-2z} + \ell_{g-2z}\lambda_{g-1} + \ell_{g-1z}\lambda_g \\ 1 \end{bmatrix} = \ldots$$
$$= \begin{bmatrix} P_{0x} + \ell_{0x}\lambda_1 + \ldots + \ell_{g-1x}\lambda_g \\ P_{0y} + \ell_{0y}\lambda_1 + \ldots + \ell_{g-1y}\lambda_g \\ P_{0z} + \ell_{0z}\lambda_1 + \ldots + \ell_{g-1z}\lambda_g \\ 1 \end{bmatrix}. \tag{2.46}$$

*Example* 2.2 Consider the rectangular optical flat of Fig. 2.13 with $v_1 = 5$, thickness $q_{e1} = 10$, and refractive index $\xi_{e1} = 1.5$. (1) Assign boundary coordinate frames $(xyz)_1$ and $(xyz)_2$ to these two flat boundary surfaces. (2) Find their unit normal vectors ${}^1\bar{n}_1$ and ${}^2\bar{n}_2$. (3) Determine the pose matrix ${}^0\bar{A}_1$ and boundary variable vector $\bar{X}_1$. (4) Give the pose matrix ${}^0\bar{A}_2$ and boundary variable vector $\bar{X}_2$. (5) Write out its source ray $\bar{R}_0$ when $\bar{P}_0 = [0 \quad -5 \quad 5 \quad 1]^T$ and $\alpha_0 = 0°$, $\beta_0 = 5°$. (6) Find $\lambda_1$, $\alpha_1$, $\beta_1$, unit normal vector $\bar{n}_1$, parameter $s_1$, incidence angle $\theta_1$, refractive index $N_1$, and the refracted ray $\bar{R}_1$ when ray $\bar{R}_0$ is refracted by the 1st boundary surface $\bar{r}_1$. (7) Find $\lambda_2$, $\alpha_2$, $\beta_2$, unit normal vector $\bar{n}_2$, parameter $s_2$, incidence angle $\theta_1$, refractive index $N_2$, and the refracted ray $\bar{R}_2$ when ray $\bar{R}_1$ is refracted by the 2nd boundary surface $\bar{r}_2$.

(1) The assigned coordinate frames $(xyz)_1$ and $(xyz)_2$ are shown in Fig. 2.14 with their origins, $o_1$ and $o_2$, are respectively located at any convenient in-plane points of boundary surfaces $\bar{r}_1$ and $\bar{r}_2$. $y_1$ and $y_2$ axes are in the same direction of $y_0$ axis of the world coordinate frame $(xyz)_0$.

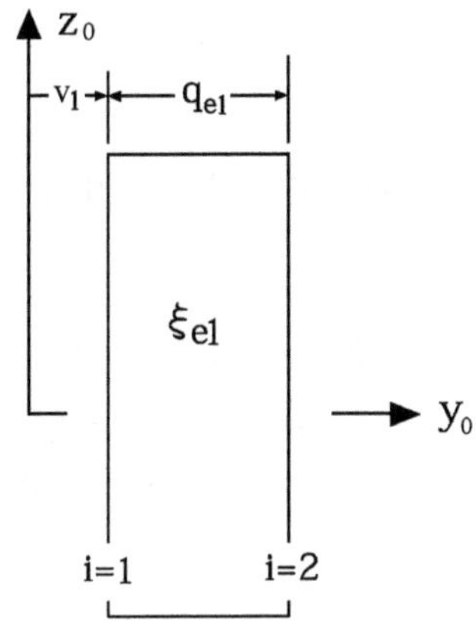


**Fig. 2.13** A rectangular optical flat with thickness $q_{e1} = 10$ and refractive index $\xi_{e1} = 1.5$

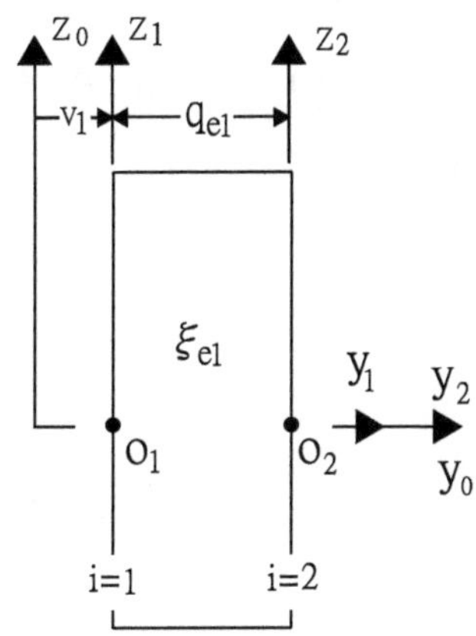


**Fig. 2.14** The assigned coordinate frames $(xyz)_1$ and $(xyz)_2$ for the system in Fig. 2.13

(2)

$$^1\bar{n}_1 = s_1[0 \quad -1 \quad 0 \quad 0]^T, \tag{2.47a}$$

$$^2\bar{n}_2 = s_2[0 \quad -1 \quad 0 \quad 0]^T. \tag{2.47b}$$

(3) If $o_1$ lies on the $y_0$ axis, then

$$^0\bar{A}_1 = \text{tran}(0, v_1, 0) = \text{tran}(0, 5, 0) = \begin{bmatrix} 1 & 0 & 0 & 0 \\ 0 & 1 & 0 & 5 \\ 0 & 0 & 1 & 0 \\ 0 & 0 & 0 & 1 \end{bmatrix}, \tag{2.47c}$$

$$\begin{aligned} \bar{X}_1 &= [t_{1x} \quad t_{1y} \quad t_{1z} \quad \omega_{1x} \quad \omega_{1y} \quad \omega_{1z} \quad \xi_0 \quad \xi_1 \quad R_1]^T \\ &= [t_{1x} \quad v_1 \quad t_{1z} \quad \omega_{1x} \quad \omega_{1y} \quad \omega_{1z} \quad \xi_0 \quad \xi_{e1} \quad R_1]^T. \\ &= [0 \quad 5 \quad 0 \quad 0 \quad 0 \quad 0 \quad 1 \quad 1.5 \quad \infty]^T \end{aligned} \tag{2.47d}$$

(4) If $o_2$ lies on the $y_0$ axis, then

$$^{0}\bar{A}_2 = \text{tran}(0, v_1 + q_{e1}, 0) = \text{tran}(0, 5 + 10, 0) = \begin{bmatrix} 1 & 0 & 0 & 0 \\ 0 & 1 & 0 & 15 \\ 0 & 0 & 1 & 0 \\ 0 & 0 & 0 & 1 \end{bmatrix}, \tag{2.47e}$$

$$\begin{aligned} \overline{X}_2 &= \begin{bmatrix} t_{2x} & t_{2y} & t_{2z} & \omega_{2x} & \omega_{2y} & \omega_{2z} & \xi_1 & \xi_2 & R_2 \end{bmatrix}^T \\ &= \begin{bmatrix} t_{2x} & v_1 + q_{e1} & t_{2z} & \omega_{2x} & \omega_{2y} & \omega_{2z} & \xi_{e1} & \xi_0 & R_2 \end{bmatrix}^T \\ &= \begin{bmatrix} 0 & 15 & 0 & 0 & 0 & 0 & 1.5 & 1 & \infty \end{bmatrix}^T. \end{aligned} \tag{2.47f}$$

(5)

$$\begin{aligned} \overline{R}_0 &= \begin{bmatrix} \overline{P}_0 & \bar{\ell}_0 \end{bmatrix}^T = \begin{bmatrix} C\beta_0 C(90^\circ + \alpha_0) & C\beta_0 S(90^\circ + \alpha_0) & S\beta_0 & 0 \end{bmatrix}^T \\ &= \begin{bmatrix} 0 & -5 & 5 & C5^\circ C90^\circ & C5^\circ S90^\circ & S5^\circ \end{bmatrix}^T \\ &= \begin{bmatrix} 0 & -5 & 5 & 0 & 0.99619 & 0.08716 \end{bmatrix}^T \end{aligned} \tag{2.47g}$$

(6)

$$\lambda_1 = 10.03820, \tag{2.47h}$$

$$\alpha_1 = -90^\circ, \tag{2.47i}$$

$$\beta_1 = 5.87489, \tag{2.47j}$$

$$\bar{n}_1 = {}^{0}\bar{A}_1 {}^{1}\bar{n}_1 = {}^{1}\bar{n}_1 = s_1 \begin{bmatrix} 0 & -1 & 0 & 0 \end{bmatrix}^T, \tag{2.47k}$$

$$s_1 = 1, \tag{2.47l}$$

$$\theta_1 = 5^\circ, \tag{2.47m}$$

$$N_1 = \xi_0 / \xi_1 = 1/1.5, \tag{2.47n}$$

$$\overline{R}_1 = \begin{bmatrix} 0 & 5 & 5.87489 & 0 & 0.99831 & 0.05810 \end{bmatrix}^T. \tag{2.47o}$$

(7)

$$\lambda_2 = 10.01692, \tag{2.47p}$$

$$\alpha_2 = 90^\circ, \tag{2.46q}$$

$$\beta_2 = 6.45691, \tag{2.47r}$$

$$\bar{n}_2 = {}^{0}\bar{A}_2 {}^{2}\bar{n}_2 = {}^{2}\bar{n}_2 = s_2 \begin{bmatrix} 0 & -1 & 0 & 0 \end{bmatrix}^T, \tag{2.47s}$$

$$s_2 = 1, \tag{2.47t}$$

$$\theta_2 = 3.33098^\circ, \tag{2.47u}$$

$$N_2 = \xi_1/\xi_2 = 1.5, \tag{2.47v}$$

$$\bar{R}_2 = [0 \quad 15 \quad 6.45691 \quad 0 \quad 0.99619 \quad 0.08716]^T. \tag{2.47w}$$

## 2.4 Skew-Ray Tracing at General Aspherical Surfaces

Most optical systems comprise flat or spherical boundary surfaces since such surfaces are easily manufactured with low cost. In this section, the modeling method is extended to the aspherical boundary surfaces ([4–6], p. 312 of [3]). In practice, the ability to trace rays at aspherical boundary surfaces is highly important since even though such surfaces are difficult and expensive to manufacture, there are cases where elements with aspherical boundary surfaces have significant advantages over those with spherical boundary surfaces. While in principle aspherical surfaces can take a wide variety of forms, an aspherical boundary surface can be designed with the generating curve ${}^i\bar{q}_i$ of the form (Fig. 2.15, in which $(xyz)_i$ is the boundary coordinate frame)

$${}^i\bar{q}_i = \begin{bmatrix} 0 \\ y_i(\beta_i) \\ z_i(\beta_i) \\ 1 \end{bmatrix} \quad (0 \le z_i(\beta_i)), \tag{2.48}$$

where the optical axis is presumed to lie in the $y_i$ axis, and $y_i(\beta_i)$ is the sag (the $y_i$ component of the displacement from the vertex at distance $z_i(\beta_i)$ from the optical axis). Meanwhile, the unit normal vector of the generating curve has the form

$${}^i\bar{e}_i = \begin{bmatrix} e_{ix} \\ e_{iy} \\ e_{iz} \\ 0 \end{bmatrix} = \frac{s_i}{\sqrt{\left(\frac{y_i(\beta_i)}{d\beta_i}\right)^2 + \left(\frac{z_i(\beta_i)}{d\beta_i}\right)^2}} \begin{bmatrix} 0 \\ -z_i(\beta_i)/d\beta_i \\ y_i(\beta_i)/d\beta_i \\ 0 \end{bmatrix} = \frac{s_i}{\sqrt{y_i'^2 + z_i'^2}} \begin{bmatrix} 0 \\ -z_i' \\ y_i' \\ 0 \end{bmatrix}, \tag{2.49}$$

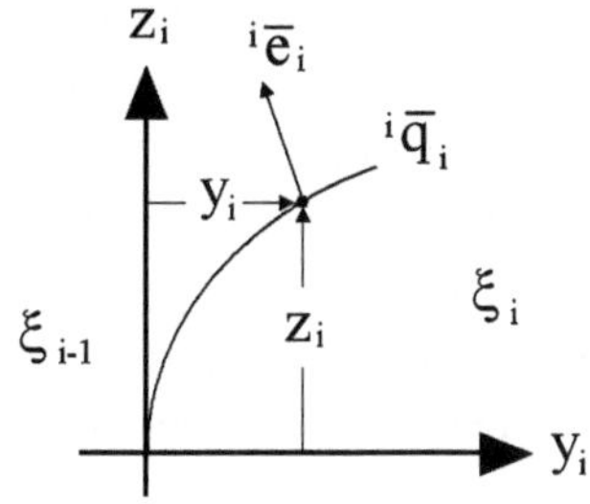


**Fig. 2.15** The generating curve at an aspherical boundary surface

where $s_i$ is set to +1 or −1 such that the cosine of the incidence angle $\theta_i$ (computed similar with Eq. (2.20)) is greater than zero. The aspherical boundary surface ${}^i\bar{r}_i$ and its unit normal vector ${}^i\bar{n}_i$ can be obtained by rotating ${}^i\bar{q}_i$ and ${}^i\bar{e}_i$ about the $y_i$ axis through an angle $\alpha_i$ ($0 \le \alpha_i < 2\pi$), i.e.,

$$
{}^i\bar{r}_i = \begin{bmatrix} x_i \\ y_i \\ z_i \\ 1 \end{bmatrix} = \mathrm{rot}(\bar{y}, \alpha_i)\, {}^i\bar{q}_i = \begin{bmatrix} C\alpha_i & 0 & S\alpha_i & 0 \\ 0 & 1 & 0 & 0 \\ -S\alpha_i & 0 & C\alpha_i & 0 \\ 0 & 0 & 0 & 1 \end{bmatrix} \begin{bmatrix} 0 \\ y_i(\beta_i) \\ z_i(\beta_i) \\ 1 \end{bmatrix} = \begin{bmatrix} z_i(\beta_i)\, S\alpha_i \\ y_i(\beta_i) \\ z_i(\beta_i)\, C\alpha_i \\ 1 \end{bmatrix}, \tag{2.50}
$$

$$
{}^i\bar{n}_i = \mathrm{rot}(\bar{y}, \alpha_i)\, {}^i\bar{e}_i = \begin{bmatrix} C\alpha_i & 0 & S\alpha_i & 0 \\ 0 & 1 & 0 & 0 \\ -S\alpha_i & 0 & C\alpha_i & 0 \\ 0 & 0 & 0 & 1 \end{bmatrix} \begin{bmatrix} e_{ix} \\ e_{iy} \\ e_{iz} \\ 0 \end{bmatrix} = \frac{s_i}{\sqrt{y_i'^2 + z_i'^2}} \begin{bmatrix} y'_i\, S\alpha_i \\ -z'_i \\ y'_i\, C\alpha_i \\ 0 \end{bmatrix}. \tag{2.51}
$$

Equations (2.50) and (2.51) give ${}^i\bar{r}_i$ and ${}^i\bar{n}_i$ with respect to the coordinate frame $(xyz)_i$. However, many derivations in this book are built relative to the world coordinate frame $(xyz)_o$. One thus needs the pose matrix ${}^0\bar{A}_i$ of $(xyz)_i$ with respect to $(xyz)_o$, given as in Eq. (2.8). Now the unit normal vector of this aspherical surface relative to the world coordinate frame $(xyz)_0$ can be obtained from the following transformation:

$$
\begin{aligned}
\bar{n}_i = \begin{bmatrix} n_{ix} \\ n_{iy} \\ n_{iz} \\ 0 \end{bmatrix} = {}^0\bar{A}_i {}^i\bar{n}_i &= \frac{s_i}{\sqrt{y_i'^2 + z_i'^2}} \begin{bmatrix} I_{ix} & J_{ix} & K_{ix} & t_{ix} \\ I_{iy} & J_{iy} & K_{iy} & t_{iy} \\ I_{iz} & J_{iz} & K_{iz} & t_{iz} \\ 0 & 0 & 0 & 1 \end{bmatrix} \begin{bmatrix} y_i'\, S\alpha_i \\ -z_i' \\ y_i'\, C\alpha_i \\ 0 \end{bmatrix} \\
&= \frac{s_i}{\sqrt{y_i'^2 + z_i'^2}} \begin{bmatrix} I_{ix}y_i'\, S\alpha_i - J_{ix}z_i' + K_{ix}y_i'\, C\alpha_i \\ I_{iy}y_i'\, S\alpha_i - J_{iy}z_i' + K_{iy}y_i'\, C\alpha_i \\ I_{iz}y_i'\, S\alpha_i - J_{iz}z_i' + K_{iz}y_i'\, C\alpha_i \\ 0 \end{bmatrix}.
\end{aligned} \tag{2.52}
$$

Figure 2.16 shows a typical ray path at an aspherical boundary surface. As shown, the ray originates at point $\bar{P}_{i-1} = [P_{i-1x} \quad P_{i-1y} \quad P_{i-1z} \quad 1]^T$ and travels along the unit directional vector $\bar{\ell}_{i-1} = [\ell_{i-1x} \quad \ell_{i-1y} \quad \ell_{i-1z} \quad 0]^T$ until it is reflected/or refracted at the boundary surface. Any intermediate point $\bar{P}'_{i-1}$ along this ray as it travels from $\bar{P}_{i-1}$ is given by $\bar{P}'_{i-1} = \bar{P}_{i-1} + \lambda\bar{\ell}_{i-1}$. Let $\lambda_i$ represent the geometrical distance from point $\bar{P}_{i-1}$ at last boundary surface to the incidence point $\bar{P}_i$ at the current boundary surface $\bar{r}_i$. The incidence point $\bar{P}_i$ on the boundary surface is $\bar{P}_i = \bar{P}_{i-1} + \lambda_i\bar{\ell}_{i-1}$ given in Eq. (2.10). To solve for $\lambda_i$, $\alpha_i$ and $\beta_i$, again, it is necessary to transform $\bar{P}_i$ (Eq. (2.10)) to coordinate frame $(xyz)_i$ and then equate the result with the boundary surface ${}^i\bar{r}_i$ (Eq. (2.50)), i.e.,

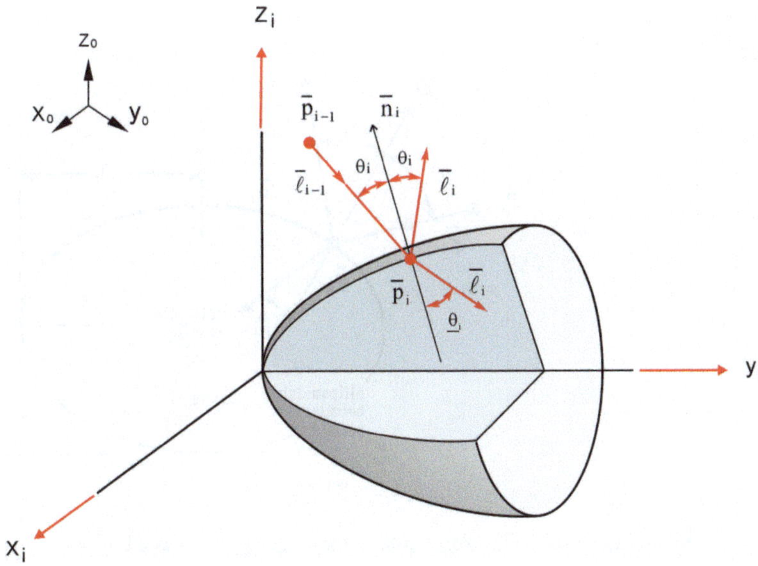


**Fig. 2.16** Raytracing at an aspherical boundary surface

$$
{}^{i}\bar{P}_{i} = {}^{i}\bar{A}_{0}\bar{P}_{i} = \left({}^{0}\bar{A}_{i}\right)^{-1}\bar{P}_{i} = \begin{bmatrix} \sigma_i \\ \rho_i \\ \tau_i \\ 1 \end{bmatrix} = {}^{i}\bar{r}_{i} = \begin{bmatrix} z_i(\beta_i)\, S\alpha_i \\ y_i(\beta_i) \\ z_i(\beta_i)\, C\alpha_i \\ 1 \end{bmatrix}, \tag{2.53}
$$

where $\sigma_i$, $\rho_i$, and $\tau_i$ are defined in Eqs. (2.12), 2.13) and (2.14), respectively. The parameter $\alpha_i$ at the incidence point $\bar{P}_i$ can now be obtained from the first and third components of Eq. (2.53) by using the following function

$$
\begin{cases} \alpha_i = \operatorname{atan2}(\sigma_i, \tau_i) & \text{when } z_i(\beta_i) > 0 \\ \alpha_i = \text{any value} & \text{when } z_i(\beta_i) = 0 \end{cases}. \tag{2.54}
$$

The parameters $\lambda_i$ and $\beta_i$ at incidence point $\bar{P}_i$ can be determined from the following two independent equations:

$$
z_i(\beta_i)^2 = \sigma_i^2 + \tau_i^2, \tag{2.55}
$$

$$
y_i(\beta_i) = \rho_i. \tag{2.56}
$$

The difficulty in tracing a skew-ray at a general aspherical surface lies in determining $\lambda_i$ from Eq. (2.55) and/or Eq. (2.56) since the solution cannot usually be determined directly. Some form of numerical method is required (e.g., p. 314 of [3]). By following similar procedures of Sect. 2.2, one can obtain the expressions for reflected and refracted unit directional vectors $\bar{\ell}_i$ respectively given in Eqs. (2.25) and (2.27).The boundary variable vector $\bar{X}_i$ of the aspherical boundary surface is

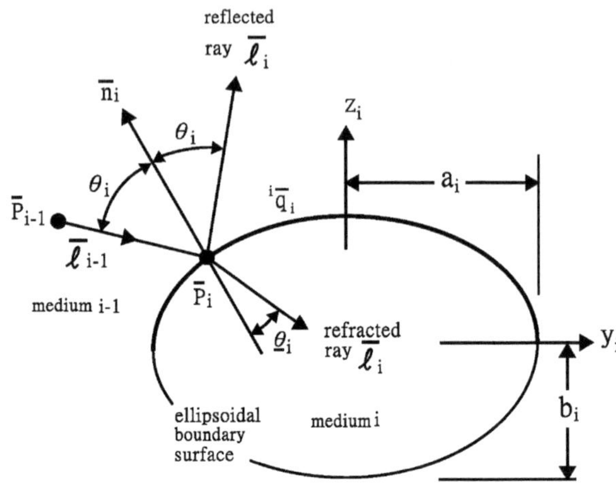


**Fig. 2.17** Raytracing at an ellipsoidal boundary surface

$$\overline{X}_i = \begin{bmatrix} t_{ix} & t_{iy} & t_{iz} & \omega_{ix} & \omega_{iy} & \omega_{iz} & \xi_{i-1} & \xi_i & \overline{coef_i} \end{bmatrix}^T, \tag{2.57}$$

in which $t_{ix}$, $t_{iy}$, $t_{iz}$, $\omega_{iz}$, $\omega_{iy}$ and $\omega_{ix}$ are the six pose variables of the aspherical boundary surface; $\xi_{i-1}$ and $\xi_i$ are the refractive indices of medium i-1 and medium i, respectively; and $\overline{coef_i}$ contains independent coefficients of $y_i(\beta_i)$and $z_i(\beta_i)$.

*Example* 2.3 The geometrical length $\lambda_i$, $\beta_i$, and boundary variable vector $\overline{X}_i$ of the ellipsoidal boundary surface defined by the generating curve (see Fig. 2.17)

$$\begin{gathered} {}^i\overline{q}_i = [0 \quad y_i(\beta_i) \quad z_i(\beta_i) \quad 1]^T = [0 \quad a_i S\beta_i \quad b_i C\beta_i \quad 1]^T \\ \left(0 < a_i,\ 0 < b_i,\ -\frac{\pi}{2} \le \beta_i \le \frac{\pi}{2}\right) \end{gathered} \tag{2.58a}$$

are given by

$$\lambda_i = \frac{-D_i \pm \sqrt{D_i^2 - H_i\, E_i}}{H_i}, \tag{2.58b}$$

$$\beta_i = \operatorname{atan2}\left(\frac{\rho_i}{a_i}, \frac{\sqrt{\sigma_i^2 + \tau_i^2}}{b_i}\right), \tag{2.58c}$$

$$\overline{X}_i = \begin{bmatrix} t_{ix} & t_{iy} & t_{iz} & \omega_{ix} & \omega_{iy} & \omega_{iz} & \xi_{i-1} & \xi_i & a_i\ b_i \end{bmatrix}^T, \tag{2.58d}$$

where

$$\begin{aligned} H_i = {} & 2\left(\frac{1}{a_i^2} - \frac{1}{b_i^2}\right)\left(J_{ix}J_{iy}\ell_{i-1x}\ell_{i-1y} + J_{ix}J_{iz}\ell_{i-1x}\ell_{i-1z} + J_{iy}J_{iz}\ell_{i-1y}\ell_{i-1z}\right) \\ & + \frac{1}{b_i^2} + \left(\frac{1}{a_i^2} - \frac{1}{b_i^2}\right)\left(J_{ix}^2\ell_{i-1x}^2 + J_{iy}^2\ell_{i-1y}^2 + J_{iz}^2\ell_{i-1z}^2\right), \end{aligned} \tag{2.58e}$$

$$D_i = \left(\frac{1}{b_i^2} - \frac{J_{ix}^2}{b_i^2} + \frac{J_{ix}^2}{a_i^2}\right) P_{i-1x}\ell_{i-1x} + \left(\frac{1}{b_i^2} - \frac{J_{iy}^2}{b_i^2} + \frac{J_{iy}^2}{a_i^2}\right) P_{i-1y}\ell_{i-1y} + \left(\frac{1}{b_i^2} - \frac{J_{iz}^2}{b_i^2} + \frac{J_{iz}^2}{a_i^2}\right) P_{i-1z}\ell_{i-1z}$$
$$+ \left(\frac{1}{a_i^2} - \frac{1}{b_i^2}\right)\left[J_{ix}J_{iy}\left(P_{i-1x}\ell_{i-1y} + \ell_{i-1x}P_{i-1y}\right) + J_{ix}J_{iz}\left(P_{i-1x}\ell_{i-1z} + \ell_{i-1x}P_{i-1z}\right) + J_{iy}J_{iz}\left(P_{i-1y}\ell_{i-1z} + \ell_{i-1y}P_{i-1z}\right)\right]$$
$$- \frac{\left(t_{ix}\ell_{i-1x} + t_{iy}\ell_{i-1y} + t_{iz}\ell_{i-1z}\right)}{b_i^2} + \left(\frac{1}{b_i^2} - \frac{1}{a_i^2}\right)\left(J_{ix}t_{ix} + J_{iy}t_{iy} + J_{iz}t_{iz}\right)\left(J_{ix}\ell_{i-1x} + J_{iy}\ell_{i-1y} + J_{iz}\ell_{i-1z}\right), \tag{2.58f}$$

$$E_i = \left(\frac{1}{b_i^2} - \frac{J_{ix}^2}{b_i^2} + \frac{J_{ix}^2}{a_i^2}\right) P_{i-1x}^2 + \left(\frac{1}{b_i^2} - \frac{J_{iy}^2}{b_i^2} + \frac{J_{iy}^2}{a_i^2}\right) P_{i-1y}^2 + \left(\frac{1}{b_i^2} - \frac{J_{iz}^2}{b_i^2} + \frac{J_{iz}^2}{a_i^2}\right) P_{i-1z}^2$$
$$+ \frac{t_{ix}^2}{b_i^2} + \frac{t_{iy}^2}{a_i^2} + \frac{t_{iz}^2}{b_i^2} - \frac{2}{b_i^2}\left(t_{ix}P_{i-1x} + t_{iy}P_{i-1y} + t_{iz}P_{i-1z}\right) - 1$$
$$+ 2\left(\frac{1}{a_i^2} - \frac{1}{b_i^2}\right)\left(J_{ix}J_{iy}P_{i-1x}P_{i-1y} + J_{ix}J_{iz}P_{i-1x}P_{i-1z} + J_{iy}J_{iz}P_{i-1y}P_{i-1z}\right)$$
$$- 2\left(\frac{1}{a_i^2} - \frac{1}{b_i^2}\right)\left(t_{ix}J_{ix} + t_{iy}J_{iy} + t_{iz}J_{iz}\right)\left(J_{ix}P_{i-1x} + J_{iy}P_{i-1y} + J_{iz}P_{i-1z}\right). \tag{2.58g}$$

$\beta_i = \pm\pi/2$ are the two pseudo-singular points of this ellipsoidal boundary surface. The cross product $\partial^i\bar{r}_i/\partial\beta_i \times \partial^i\bar{r}_i/\partial\alpha_i$ cannot be obtained at these two points, and thus some difficulties occur in computing the point spread function and modulation transfer function.

*Example* 2.4 The geometrical length $\lambda_i$, and boundary variable vector $\overline{X}_i$ of the paraboloidal boundary surface defined by generating curve (see Fig. 2.18)

$$^i\bar{q}_i = \begin{bmatrix}0 & y_i(\beta_i) & z_i(\beta_i) & 1\end{bmatrix}^T = \begin{bmatrix}0 & a_i\beta_i^2 & \beta_i & 1\end{bmatrix}^T \quad (0 \le \beta_i,\ 0 < a_i) \tag{2.59a}$$

are given by:

(a) If $J_{ix}\ell_{i-1x} + J_{iy}\ell_{i-1y} + J_{iz}\ell_{i-1z} \neq \pm 1$, then

$$\lambda_i = \frac{-D_i \pm \sqrt{D_i^2 - H_iE_i}}{H_i}, \tag{2.59b}$$

where

$$H_i = 1 - \left(J_{ix}\ell_{i-1x} + J_{iy}\ell_{i-1y} + J_{iz}\ell_{i-1z}\right)^2 \tag{2.59c}$$

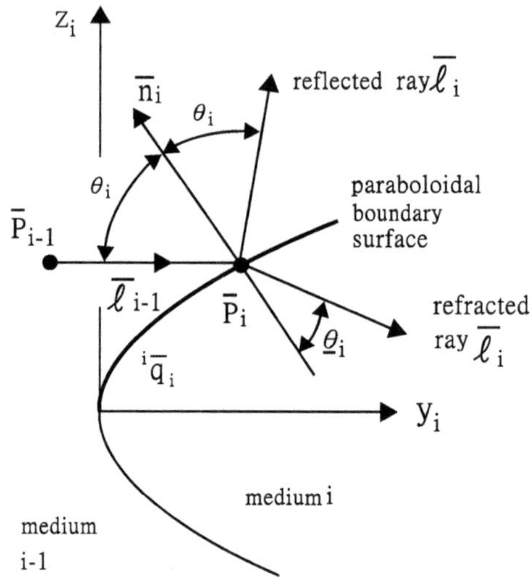


**Fig. 2.18** Raytracing at a paraboloidal boundary surface

$$\begin{aligned} D_i = {} & (1 - J_{ix}^2)P_{i-1x}\ell_{i-1x} + \left(1 - J_{iy}^2\right)P_{i-1y}\ell_{i-1y} + (1 - J_{iz}^2)P_{i-1z}\ell_{i-1z} - \left(t_{ix}\ell_{i-1x} + t_{iy}\ell_{i-1y} + t_{iz}\ell_{i-1z}\right) \\ & - J_{ix}J_{iy}\left(P_{i-1x}\ell_{i-1y} + P_{i-1y}\ell_{i-1x}\right) - J_{ix}J_{iz}(P_{i-1x}\ell_{i-1z} + P_{i-1z}\ell_{i-1x}) - J_{iy}J_{iz}\left(P_{i-1y}\ell_{i-1z} + P_{i-1z}\ell_{i-1y}\right) \\ & + \left(J_{ix}t_{ix} + J_{iy}t_{iy} + J_{iz}t_{iz}\right)\left(J_{ix}\ell_{i-1x} + J_{iy}\ell_{i-1y} + J_{iz}\ell_{i-1z}\right) - \frac{1}{2a_i}\left(J_{ix}\ell_{i-1x} + J_{iy}\ell_{i-1y} + J_{iz}\ell_{i-1z}\right), \end{aligned} \tag{2.59d}$$

$$\begin{aligned} E_i = {} & \left(P_{i-1x}^2 + P_{i-1y}^2 + P_{i-1z}^2\right) + t_{ix}^2 + t_{iy}^2 + t_{iz}^2 - 2\left(t_{ix}P_{i-1x} + t_{iy}P_{i-1y} + t_{iz}P_{i-1z}\right) \\ & + \left[J_{ix}(t_{ix} + P_{i-1x}) + J_{iy}\left(t_{iy} + P_{i-1y}\right) + J_{iz}(t_{iz} + P_{i-1z})\right]^2 \\ & + \frac{1}{a_i}\left[J_{ix}(t_{ix} - P_{i-1x}) + J_{iy}\left(t_{iy} - P_{i-1y}\right) + J_{iz}(t_{iz} - P_{i-1z})\right]. \end{aligned} \tag{2.59f}$$

(b) If $J_{ix}\ell_{i-1x} + J_{iy}\ell_{i-1y} + J_{iz}\ell_{i-1z} = \pm 1$, then

$$\lambda_i = \frac{D_i}{H_i}, \tag{2.59g}$$

where

$$\begin{aligned} D_i = {} & J_{ix}(t_{ix} - P_{i-1x}) + J_{iy}\left(t_{iy} - P_{i-1y}\right) + J_{iz}(t_{iz} - P_{i-1z}) \\ & + a_i\left[(t_{ix} - P_{i-1x})^2 + \left(t_{iy} - P_{i-1y}\right)^2 + (t_{iz} - P_{i-1z})^2\right] \\ & - a_i\left[J_{ix}(t_{ix} - P_{i-1x}) + J_{iy}\left(t_{iy} - P_{i-1y}\right) + J_{iz}(t_{iz} - P_{i-1z})\right]^2 \end{aligned} \tag{2.59h}$$

$$H_i = J_{ix}\ell_{i-1x} + J_{iy}\ell_{i-1y} + J_{iz}\ell_{i-1z}, \tag{2.59i}$$

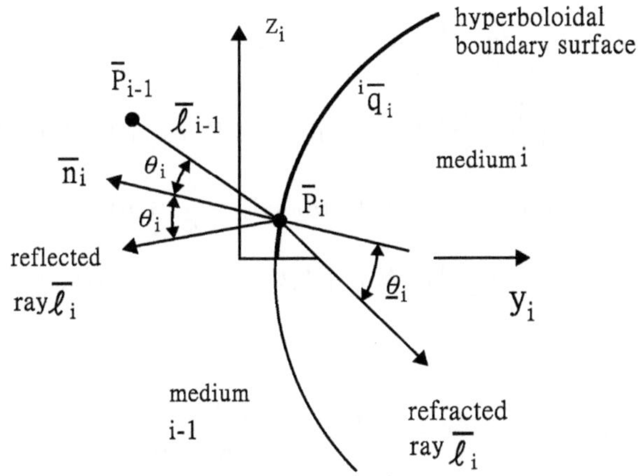


**Fig. 2.19** Raytracing at a hyperboloidal boundary surface

$$\bar{X}_i = [t_{ix} \quad t_{iy} \quad t_{iz} \quad \omega_{ix} \quad \omega_{iy} \quad \omega_{iz} \quad \xi_{i-1} \quad \xi_i \quad a_i]^T. \tag{2.59j}$$

$\beta_i = 0$ is the only pseudo-singular point on the paraboloidal boundary surface.

*Example* 2.5 The geometrical length $\lambda_i$, and boundary variable vector $\bar{X}_i$ of the hyperboloidal boundary surface defined by the generating curve (see Fig. 2.19)

$$\begin{aligned} {}^i\bar{q}_i &= [0 \quad y_i(\beta_i) \quad z_i(\beta_i) \quad 1]^T \\ &= \left[0 \quad \frac{a_i}{C\beta_i} \quad \frac{b_i S\beta_i}{C\beta_i} \quad 1\right]^T \ (0 \le \beta_i < \pi/2,\ 0 < a_i,\ 0 < b_i) \end{aligned} \tag{2.60a}$$

are given by

$$\lambda_i = \frac{-D_i \pm \sqrt{D_i^2 - H_i E_i}}{H_i}, \tag{2.60b}$$

where

$$\begin{aligned} H_i = {} & \left(\frac{1}{b_i^2} - \frac{J_{ix}^2}{b_i^2} - \frac{J_{ix}^2}{a_i^2}\right)\ell_{i-1x}^2 + \left(\frac{1}{b_i^2} - \frac{J_{iy}^2}{b_i^2} - \frac{J_{iy}^2}{a_i^2}\right)\ell_{i-1y}^2 + \left(\frac{1}{b_i^2} - \frac{J_{iz}^2}{b_i^2} - \frac{J_{iz}^2}{a_i^2}\right)\ell_{i-1z}^2 \\ & - 2\left(\frac{1}{a_i^2} + \frac{1}{b_i^2}\right)\left(J_{ix}J_{iy}\ell_{i-1x}\ell_{i-1y} + J_{ix}J_{iz}\ell_{i-1x}\ell_{i-1z} + J_{iy}J_{iz}\ell_{i-1y}\ell_{i-1z}\right), \end{aligned} \tag{2.60c}$$

$$\begin{aligned}D_i = &\left(\frac{1}{b_i^2}-\frac{J_{ix}^2}{b_i^2}-\frac{J_{ix}^2}{a_i^2}\right)P_{i-1x}\ell_{i-1x}+\left(\frac{1}{b_i^2}-\frac{J_{iy}^2}{b_i^2}-\frac{J_{iy}^2}{a_i^2}\right)P_{i-1y}\ell_{i-1y}+\left(\frac{1}{b_i^2}-\frac{J_{iz}^2}{b_i^2}-\frac{J_{iz}^2}{a_i^2}\right)P_{i-1z}\ell_{i-1z}\\&-\left(\frac{1}{a_i^2}+\frac{1}{b_i^2}\right)\left[J_{ix}J_{iy}\left(P_{i-1x}\ell_{i-1y}+P_{i-1y}\ell_{i-1x}\right)+J_{ix}J_{iz}(P_{i-1x}\ell_{i-1z}+P_{i-1z}\ell_{i-1x})\right.\\&\left.+J_{iy}J_{iz}\left(P_{i-1y}\ell_{i-1z}+P_{i-1z}\ell_{i-1y}\right)\right]-\frac{1}{b_i^2}\left(\ell_{i-1x}t_{ix}+\ell_{i-1y}t_{iy}+\ell_{i-1z}t_{iz}\right)\\&+\left(\frac{1}{a_i^2}+\frac{1}{b_i^2}\right)\left(J_{ix}\ell_{i-1x}+J_{iy}\ell_{i-1y}+J_{iz}\ell_{i-1z}\right)\left(J_{ix}t_{ix}+J_{iy}t_{iy}+J_{iz}t_{iz}\right),\end{aligned} \tag{2.60d}$$

$$\begin{aligned}E_i = &\left(\frac{1}{b_i^2}-\frac{J_{ix}^2}{b_i^2}-\frac{J_{ix}^2}{a_i^2}\right)P_{i-1x}^2+\left(\frac{1}{b_i^2}-\frac{J_{iy}^2}{b_i^2}-\frac{J_{iy}^2}{a_i^2}\right)P_{i-1y}^2+\left(\frac{1}{b_i^2}-\frac{J_{iz}^2}{b_i^2}-\frac{J_{iz}^2}{a_i^2}\right)P_{i-1z}^2\\&-2\left(\frac{1}{a_i^2}+\frac{1}{b_i^2}\right)\left(J_{ix}J_{iy}P_{i-1x}P_{i-1y}+J_{ix}J_{iz}P_{i-1x}P_{i-1z}+J_{iy}J_{iz}P_{i-1y}P_{i-1z}\right)\\&+2\left(\frac{1}{a_i^2}+\frac{1}{b_i^2}\right)\left(J_{ix}P_{i-1x}+J_{iy}P_{i-1y}+J_{iz}P_{i-1z}\right)\left(J_{ix}t_{ix}+J_{iy}t_{iy}+J_{iz}t_{iz}\right)\\&-\frac{2}{b_i^2}\left(P_{i-1x}t_{ix}+P_{i-1y}t_{iy}+P_{i-1z}t_{iz}\right)+\frac{t_{ix}^2+t_{iy}^2+t_{iz}^2}{a_i^2}+1,\end{aligned} \tag{2.60e}$$

$$\overline{X}_i = \begin{bmatrix} t_{ix} & t_{iy} & t_{iz} & \omega_{ix} & \omega_{iy} & \omega_{iz} & \xi_{i-1} & \xi_i & a_i\ b_i \end{bmatrix}^T. \tag{2.60f}$$

$\beta_i = 0$ is the only pseudo-singular point on the hyperboloidal boundary surface.

*Example* 2.6 The geometrical length $\lambda_i$, and boundary variable vector $\overline{X}_i$ of the cylindrical boundary surface defined by the generating curve (see Fig. 2.20)

$${}^i\bar{q}_i = \begin{bmatrix}0 & y_i(\beta_i) & z_i(\beta_i) & 1\end{bmatrix}^T = \begin{bmatrix}0 & \beta_i & R_i & 1\end{bmatrix}^T \ (0 \leq \beta_i,\ 0 < R_i) \tag{2.61a}$$

are given by

$$\lambda_i = \frac{-D_i \pm \sqrt{D_i^2 - H_i\ E_i}}{H_i}, \tag{2.61b}$$

where

$$H_i = 1 - \left(J_{ix}\ell_{i-1x} + J_{iy}\ell_{i-1y} + J_{iz}\ell_{i-1z}\right)^2, \tag{2.61c}$$

$$\begin{aligned}D_i = &(P_{i-1x}-t_{ix})\ell_{i-1x}+\left(P_{i-1y}-t_{iy}\right)\ell_{i-1y}+(P_{i-1z}-t_{iz})\ell_{i-1z}\\&+\left[J_{ix}(t_{ix}-P_{i-1x})+J_{iy}\left(t_{iy}-P_{i-1y}\right)+J_{iz}(t_{iz}-P_{i-1z})\right]\left(J_{ix}\ell_{i-1x}+J_{iy}\ell_{i-1y}+J_{iz}\ell_{i-1z}\right),\end{aligned} \tag{2.61d}$$

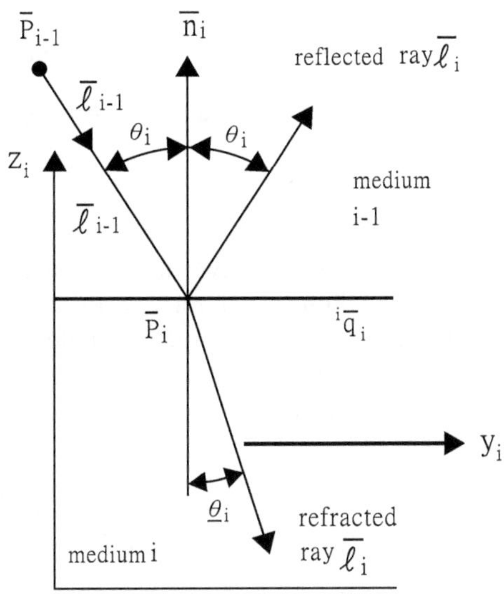


**Fig. 2.20** Raytracing at a cylindrical boundary surface

$$E_i = P_{i-1x}^2 + P_{i-1y}^2 + P_{i-1z}^2 + t_{ix}^2 + t_{iy}^2 + t_{iz}^2 - R_i^2 - 2\left(t_{ix}P_{i-1x} + t_{iy}P_{i-1y} + t_{iz}P_{i-1z}\right) - \left[J_{ix}(t_{ix} - P_{i-1x}) + J_{iy}\left(t_{iy} - P_{i-1y}\right) + J_{iz}(t_{iz} - P_{i-1z})\right]^2, \tag{2.61e}$$

$$\bar{X}_i = \left[t_{ix} \quad t_{iy} \quad t_{iz} \quad \omega_{ix} \quad \omega_{iy} \quad \omega_{iz} \quad \xi_{i-1} \quad \xi_i \quad R_i\right]^T. \tag{2.61f}$$

There are no pseudo-singular points on the cylindrical boundary surface.

## 2.5 Unit Normal Vector at Incidence Point on a Refractive/or Reflective Boundary Surfaces

Snell's law describes the relationship between the incidence angle and reflection (or refraction) angle of a light ray impinging on the interface between two different isotropic media. In this section, Snell's law is used to derive the unit normal vectors of a boundary surface given the unit directional vectors of the incoming and outgoing rays. The proposed method has important applications in the design and fabrication of aspherical surfaces, since the surface normal vectors determine not only the optical performance of the surface, but also the cutting tool angles required to machine the surface.

As shown in Fig. 2.6, the design of a refractive surface requires a knowledge of the incidence angle $\theta_i$ and refraction angle $\underline{\theta}_i$ corresponding to the unit directional vectors of the incoming and outgoing rays, i.e., $\bar{\ell}_{i-1}$ and $\bar{\ell}_i$, respectively. In accordance with Snell's law, the refraction phenomenon at the interface between two different isotropic media can be modeled as follows:

$$\bar{\ell}_{i-1} \bullet \bar{\ell}_i = C(\theta_i - \underline{\theta}_i) = C\theta_i C\underline{\theta}_i + S\theta_i S\underline{\theta}_i, \tag{2.62}$$

$$S\underline{\theta}_i = \frac{\xi_{i-1}}{\xi_i} S\theta_i = N_i S\theta_i, \tag{2.21}$$

where $\xi_i$ is the refractive index of medium i and $N_i = \xi_{i-1}/\xi_i$ is the refractive index of medium i − 1 relative to that of medium i. In geometrical optics, the incidence angle $\theta_i$ and refraction angle $\underline{\theta}_i$ are confined to the range of $0° \le \theta_i < 90°$ and $0° \le \underline{\theta}_i < 90°$. Therefore one can have the following four equations from Eqs. (2.62) and (2.21), since the trigonometric functions of $\theta_i$ and $\underline{\theta}_i$ are positive:

$$S\theta_i = \frac{\sqrt{1 - (\bar{\ell}_{i-1} \bullet \bar{\ell}_i)^2}}{\sqrt{N_i^2 + 1 - 2N_i(\bar{\ell}_{i-1} \bullet \bar{\ell}_i)}}, \tag{2.63a}$$

$$C\theta_i = \frac{|N_i - (\bar{\ell}_{i-1} \bullet \bar{\ell}_i)|}{\sqrt{N_i^2 + 1 - 2N_i(\bar{\ell}_{i-1} \bullet \bar{\ell}_i)}}, \tag{2.63b}$$

$$S\underline{\theta}_i = \frac{N_i\sqrt{1 - (\bar{\ell}_{i-1} \bullet \bar{\ell}_i)^2}}{\sqrt{N_i^2 + 1 - 2N_i(\bar{\ell}_{i-1} \bullet \bar{\ell}_i)}}, \tag{2.63c}$$

$$C\underline{\theta}_i = \frac{|1 - N_i(\bar{\ell}_{i-1} \bullet \bar{\ell}_i)|}{\sqrt{N_i^2 + 1 - 2N_i(\bar{\ell}_{i-1} \bullet \bar{\ell}_i)}}. \tag{2.63d}$$

Referring to Fig. 2.6, to determine the unit normal vector $\bar{n}_i$ at any incidence point on a refractive surface, it is first necessary to compute the common unit normal $\bar{m}_i$ of the unit directional vectors $\bar{\ell}_{i-1}$ and $\bar{\ell}_i$ of the incoming and outgoing rays, respectively. Given the assumption $\theta_i \ne \underline{\theta}_i$ when $N_i \ne 1$, $\bar{m}_i$, can be obtained as

$$\bar{m}_i = [m_{ix} \quad m_{iy} \quad m_{iz} \quad 0]^T = \frac{\bar{\ell}_{i-1} \times \bar{\ell}_i}{S(\theta_i - \underline{\theta}_i)}. \tag{2.64}$$

To simplify the expression of the unit normal vector $\bar{n}_i$, it is useful to have the following equation obtained by taking the post cross product of Eq. (2.64) by $S(\theta_i - \underline{\theta}_i)\bar{\ell}_{i-1}$, i.e.,

$$\begin{aligned} S(\theta_i - \underline{\theta}_i)(\bar{m}_i \times \bar{\ell}_{i-1}) &= (\bar{\ell}_{i-1} \times \bar{\ell}_i) \times \bar{\ell}_{i-1} = \bar{\ell}_i(\bar{\ell}_{i-1} \bullet \bar{\ell}_{i-1}) - \bar{\ell}_{i-1}(\bar{\ell}_i \bullet \bar{\ell}_{i-1}) \\ &= \bar{\ell}_i - \bar{\ell}_{i-1}\, C(\theta_i - \underline{\theta}_i). \end{aligned} \tag{2.65}$$

As shown in Fig. 2.6, the unit normal vector $\bar{n}_i$ can be obtained by rotating $-\bar{\ell}_{i-1}$ about $\bar{m}_i$ by an angle $\theta_i$ Eq. (1.25) This leads to

$$\bar{n}_i = \begin{bmatrix} n_{ix} \\ n_{iy} \\ n_{iz} \\ 0 \end{bmatrix} = \begin{bmatrix} m_{ix}^2(1-C\theta_i)+C\theta_i & m_{iy}m_{ix}(1-C\theta_i)-m_{iz}S\theta_i & m_{iz}m_{ix}(1-C\theta_i)+m_{iy}S\theta_i & 0 \\ m_{ix}m_{iy}(1-C\theta_i)+m_{iz}S\theta_i & m_{iy}^2(1-C\theta_i)+C\theta_i & m_{iz}m_{iy}(1-C\theta_i)-m_{ix}S\theta_i & 0 \\ m_{ix}m_{iz}(1-C\theta_i)-m_{iy}S\theta_i & m_{iy}m_{iz}(1-C\theta_i)+m_{ix}S\theta_i & m_{iz}^2(1-C\theta_i)+C\theta_i & 0 \\ 0 & 0 & 0 & 1 \end{bmatrix} \begin{bmatrix} -\ell_{i-1x} \\ -\ell_{i-1y} \\ -\ell_{i-1z} \\ 0 \end{bmatrix}. \tag{2.66}$$

Utilizing Eq. (2.65), Eq. (2.66) can be simplified as follows:

$$\begin{aligned} \bar{n}_i = \begin{bmatrix} n_{ix} \\ n_{iy} \\ n_{iz} \\ 0 \end{bmatrix} &= \left[\frac{C(\theta_i-\underline{\theta}_i)S\theta_i}{S(\theta_i-\underline{\theta}_i)} - C\theta_i\right] \begin{bmatrix} \ell_{i-1x} \\ \ell_{i-1y} \\ \ell_{i-1z} \\ 0 \end{bmatrix} - \left[\frac{S\theta_i}{S(\theta_i-\underline{\theta}_i)}\right] \begin{bmatrix} \ell_{ix} \\ \ell_{iy} \\ \ell_{iz} \\ 0 \end{bmatrix} \\ &= \left[\frac{(C\theta_i C\underline{\theta}_i + S\theta_i S\underline{\theta}_i)S\theta_i}{S\theta_i C\underline{\theta}_i - C\theta_i S\underline{\theta}_i} - C\theta_i\right] \begin{bmatrix} \ell_{i-1x} \\ \ell_{i-1y} \\ \ell_{i-1z} \\ 0 \end{bmatrix} - \left[\frac{S\theta_i}{S\theta_i C\underline{\theta}_i - C\theta_i S\underline{\theta}_i}\right] \begin{bmatrix} \ell_{ix} \\ \ell_{iy} \\ \ell_{iz} \\ 0 \end{bmatrix}, \end{aligned} \tag{2.67}$$

where $S\theta_i$, $C\theta_i$, $S\underline{\theta}_i$, and $C\underline{\theta}_i$ are given in Eqs. (2.63a), (2.63b), (2.63c) and (2.63d), respectively.

Referring to Fig. 2.5, to determine the unit normal vector $\bar{n}_i$ at any incidence point on a reflective surface, it is first necessary to calculate the common normal vector $\bar{m}_i$ of the unit directional vectors of the incoming and outgoing rays, i.e., $\bar{\ell}_{i-1}$ and $\bar{\ell}_i$, respectively. Given the assumption $\theta_i \neq 0$, $\bar{m}_i$, can be obtained as

$$\bar{m}_i = \begin{bmatrix} m_{ix} & m_{iy} & m_{iz} & 0 \end{bmatrix}^T = \frac{\bar{\ell}_i \times \bar{\ell}_{i-1}}{S(2\theta_i)}, \tag{2.68}$$

where the incidence angle $\theta_i$ is determined by

$$C(2\theta_i) = \left|\bar{\ell}_{i-1} \bullet \bar{\ell}_i\right|. \tag{2.69}$$

Note that Eq. (2.69) is also applicable when the incidence angle is equal to zero.

To simplify the expression of the unit normal vector $\bar{n}_i$, it is useful to have the following equation obtained by taking the post cross product of Eq. (2.68) by $S(2\theta_i)\bar{\ell}_{i-1}$, i.e.,

$$\begin{aligned} S(2\theta_i)\left(\bar{m}_i \times \bar{\ell}_{i-1}\right) &= \left(\bar{\ell}_i \times \bar{\ell}_{i-1}\right) \times \bar{\ell}_{i-1} \\ &= \bar{\ell}_{i-1}\left(\bar{\ell}_i \bullet \bar{\ell}_{i-1}\right) - \bar{\ell}_i\left(\bar{\ell}_{i-1} \bullet \bar{\ell}_{i-1}\right) = -C(2\theta_i)\bar{\ell}_{i-1} - \bar{\ell}_i. \end{aligned} \tag{2.70}$$

As shown in Fig. 2.5, the unit normal vector $\bar{n}_i$ can be determined by rotating $-\bar{\ell}_{i-1}$ about $\bar{m}_i$ by an angle $\theta_i$ (i.e., Eq. (2.66)). Simplifying Eq. (2.66) using Eq. (2.70), the following expression is obtained for the unit normal vector $\bar{n}_i$ at any incidence point on the reflective surface when $\bar{\ell}_{i-1}$ and $\bar{\ell}_i$ are given:

$$\bar{n}_i = \begin{bmatrix} n_{ix} \\ n_{iy} \\ n_{iz} \\ 0 \end{bmatrix} = \frac{1}{2C\theta_i}\left(\begin{bmatrix} \ell_{ix} \\ \ell_{iy} \\ \ell_{iz} \\ 0 \end{bmatrix} - \begin{bmatrix} \ell_{i-1x} \\ \ell_{i-1y} \\ \ell_{i-1z} \\ 0 \end{bmatrix}\right) = \frac{1}{\sqrt{2\left(1 - \bar{\ell}_{i-1} \bullet \bar{\ell}_i\right)}}\left(\begin{bmatrix} \ell_{ix} \\ \ell_{iy} \\ \ell_{iz} \\ 0 \end{bmatrix} - \begin{bmatrix} \ell_{i-1x} \\ \ell_{i-1y} \\ \ell_{i-1z} \\ 0 \end{bmatrix}\right). \tag{2.71}$$

The fabrication of aspherical surfaces requires the use of high-precision manufacturing techniques in order to achieve the necessary surface accuracy and smoothness. Large aspherical lenses are typically produced using grinding and polishing techniques. Single-point diamond turning [4] provides an alternative for fabricating large aspherical surfaces, and it typically results in a better metallurgical structure than that produced by polishing and lapping. However, in using such a method, precise tool angle settings must be determined in advance in order to obtain the desired surface profile [7, 8]. The setting angles are determined by both the tool geometry and the normal vectors of the aspherical surface. Thus, Eqs. (2.67) and (2.71) are important not only in predicting the optical performance of the surface, but also in formulating the numerical codes required to machine the surface during the fabrication process.

## References

1. S. Cornbleet, Geometrical optics reviewed: a new light on an old subject, in *Proceedings of the IEEE*, vol. 71 (April 1983), pp. 471–502
2. W.J. Smith, *Modern Optical Engineering*, 3rd edn. (Edmund Industrial Optics, Barrington, 2001)
3. J. Haisma, E. Hugues, C. Babolat, Realization of a bi-aspherical objective lens for the Philips video play system. Opt. Lett. **4**, 70–72 (1979)
4. P.D. Lin, C.Y. Tsai, Determination of first-order derivatives of skew-ray at aspherical surface. J. Opt. Soc. Am. A **29**, 1141–1153 (2012)
5. W.A. Allen, J.R. Snyder, Ray tracing through uncentered and aspherical surfaces. J. Opt. Soc. Am. A **42**, 243–249 (1952)
6. G. Boothroyd, *Fundamentals of Metal Machining and Machine Tools* (McGraw-Hill, New York, 1975), Chapter 7
7. A.Q. Biddut, M. Rahman, K.S. Neo, K.M. Rezaur, M. Sawa, Y. Maeda, Performance of single crystal diamond tools with different rake angles during micro-grooving on electroless nickel plated die materials. Int. J. Adv. Manuf. Technol. **33,** 891–899 (2007)
8. M. Laikin, *Lens Design* (Marcel Dekker, Inc., New York, 1995), pp. 71–72

# Chapter 3
# Modeling an Optical System

In geometrical optics, how to efficiently and systematically model an optical system is highly challenging. This chapter addresses recent developments in mathematical modeling in geometrical optics by using homogeneous coordinate notation. Particular emphasis is on the determination of the boundary variable vectors and system variable vectors. The stops, aperture, and some special rays in an axis-symmetrical optical system will be reviewed in Sect. 3.1. In Sect. 3.2 the modeling technique of an axis-symmetrical optical system will be addressed. In Sect. 3.3 that modeling technique will be extended to optical systems containing prisms. The determinations of spot size, point spread function, modulation transfer function by using raytracing data are addressed in the rest sections. Although some of the subject matter is classical, the approaches presented are new and represent the latest developments in geometrical optics by the author.

## 3.1 Stops, Aperture, and Various Rays in an Axis-Symmetrical System

An optical element (e.g., dove prism) is a block of optical material possessing constant refractive index. It usually possess multiple boundary surfaces; each having an own unique poses. If an optical element is labeled as the jth element in an optical system, and has $L_j$ boundary surfaces, its boundary surfaces will be marked by $i = m_j - L_j + 1$ to $i = m_j$ (see Fig. 3.1 and Table 3.1). One has to note that in this book the feature of parameters used for an element is that we always use "e" to lead the element number "j". For examples, $\xi_{e3}$ and $q_{e3}$ represent respectively the refractive index and thickness of the 3rd element in an optical system. A number of optical elements built together with fixed relative pose matrices form an optical assembly. Several optical assemblies functioning together for particular purpose can be an optical system. To model an optical system possessing multiple optical elements and optical assemblies, one has to label its

PD Lin, *New Computation Methods for Geometrical Optics*,
Springer Series in Optical Sciences 178, DOI: 10.1007/978-981-4451-79-6_3,

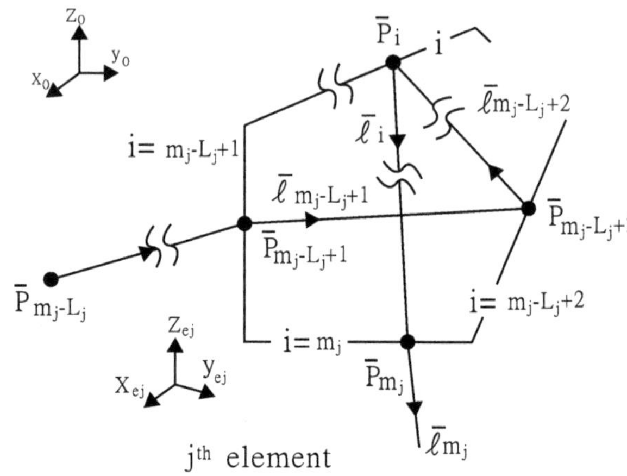


**Fig. 3.1** An optical element is a block of optical material having constant refractive index $\xi_{ej}$ of medium. Each element j contains $L_j$ number of boundary surfaces labeled from $i = m_j - L_j + 1$ to $i = m_j$

optical elements sequentially from $j = 1$ to $j = k$ and boundary surfaces from $i = 1$ to $i = n$ (e.g., Fig. 2.1 and Table 2.1).

An optical system typically has many openings, or structures that limit the ray bundles. These structures may be the edge of a lens or mirror, or a ring or other fixture that holds an optical element in place to limit the light admitted by the system. In general, these structures are called stops, and the aperture is the stop that determines the ray cone angle, or equivalently the brightness, at an image point. More specifically, the aperture of an optical system is the opening that determines the cone angle of a bundle of rays that come to a focus in the image plane. The aperture also determines how many of the incoming rays are actually admitted and thus how much light reaches the image plane. When an aperture is a hole or an opening (such as a diaphragm, e.g., element $j = 3$ of Fig. 2.1) through which light travels, it is convenient to treat such aperture as a ghost element, i.e., an element with constant refractive index $\xi_{ej} = \xi_{air} = 1$ and thickness $q_{ej} = 0$ (e.g., element $j = 3$ of system in Fig. 2.1).

There are several special rays that are frequently used in analyzing axis-symmetrical systems. They are defined and described below.

a. When a source point $\bar{P}_0$ is confined to the $y_0z_0$ plane (where $y_0$ points along the optical axis of the system) of an axis-symmetrical system, then a meridional ray (or tangential ray) is a ray lying on that plane. This $y_0z_0$ plane is called the meridional plane (see Fig. 2.1) of this system. One has to note that, when $\bar{P}_0$ is on $y_0z_0$ plane, any source ray with $\alpha_0 = 0$ (see Eq. (2.2)) always travels on meridional plane in an axis-symmetrical system.
b. The marginal ray in an axis-symmetrical system is a ray that starts at a source point $\bar{P}_0$ and will touch the edge of the aperture of the system when it propagates in the system.
c. The chief ray $\bar{R}_{i/chief} = \begin{bmatrix} \bar{P}_{i/chief} & \bar{\ell}_{i/chief} \end{bmatrix}^T$ at the ith boundary surface in an axis-symmetrical system is the meridional ray that starts from a source point $\bar{P}_0$

**Table 3.1** Each element j (j = 1 to j = 5) in the system shown in Fig. 2.1 contains two ($L_j = 2$) boundary surfaces labeled by i = 2j − 1 and i = 2j

| Element number | j = 1 | j = 2 | j = 3 | j = 4 | j = 5 | j = k = 6 |
|---|---|---|---|---|---|---|
| Lj (number of boundary surfaces in element j) | 2 | 2 | 2 | 2 | 2 | 1 |
| The first boundary surface of element j is $i = m_j - L_j + 1$ | 1 | 3 | 5 | 7 | 9 | 11 |
| The last boundary surface of element j is $i = m_j$ | 2 | 4 | 6 | 8 | 10 | 11 |

and will passed (or passed) through the center of the aperture (e.g., $x_5 = z_5 = 0$ of Fig. 2.1). Usually the spherical coordinates $\alpha_{0/\text{chief}}$ and $\beta_{0/\text{chief}}$ of $\bar{\ell}_{0/\text{chief}} = \begin{bmatrix} \ell_{0x/\text{chief}} & \ell_{0y/\text{chief}} & \ell_{0z/\text{chief}} & 0 \end{bmatrix}^T = \begin{bmatrix} C\beta_{0/\text{chief}}C(90^\circ + \alpha_{0/\text{chief}}) & C\beta_{0/\text{chief}}S(90^\circ + \alpha_{0/\text{chief}})S\beta_{0/\text{chief}} & 0 \end{bmatrix}^T$ from the source point $\bar{P}_0$ has to be determined by numerical methods.

d. A skew-ray is a ray that does not propagate in the plane which contains both the source point and the optical axis. Such rays neither cross nor are parallel to the optical axis anywhere.
e. In conventional textbooks, a sagittal ray (or transverse ray) from an off-axis source point $\bar{P}_0$ is a ray that propagates in the plane that is perpendicular to the meridional plane and contains the chief ray. As shown in Fig. 2.1, if the optical axis is the $y_0$ axis and the meridional plane is the $y_0z_0$ plane, sagittal rays intersect the aperture at $z_5 = 0$. In the future, instead of sagittal plane, the sagittal cone $\alpha_0(\beta_{0/\text{chief}})$ will be used with the meridional plane to discuss the aberrations of an optical system. The sagittal cone $\alpha_0(\beta_{0/\text{chief}})$ of an off-axis source point $\bar{P}_0$ in an axis-symmetrical optical system is generated by sweeping the unit directional vector $\bar{\ell}_0$ with $\beta_0 = \beta_{0/\text{chief}}$ by fixing $\bar{P}_0$ (i.e., the $\alpha_0$ cone with $\beta_0 = \beta_{0/\text{chief}}$ shown in Fig. 2.2). Accordingly, the sagittal cone is actually different from the sagittal plane, yielding a ray originating from sagittal cone may be not a sagittal ray. The chief ray is both sagittal and meridional ray. All other sagittal rays originating from sagittal cone are skew-rays.
f. A ray that travels along the optical axis of an axis-symmetrical system is called axial ray.
g. A paraxial ray is a ray that makes a small angle to the optical axis $y_0$ of axis-symmetrical system, and lies close to the axis throughout the system. Such rays can be modeled reasonably well by using the paraxial approximation. When discussing raytracing this definition is often reversed: a "paraxial ray" is then a ray that is modeled using the paraxial approximation and that does not necessarily remain close to the axis.
h. A finite ray or real ray is a ray that is traced without the paraxial approximation.

Traditionally, people estimate the unit directional vector $\bar{\ell}_{0/\text{chief}}$ of the chief ray of a source point $\bar{P}_0$ by connecting $\bar{P}_0$ to the center of the entrance pupil. Similarly, this estimation also used to construct the marginal rays or a ray passing through a particular point of the aperture. We can always accurately determine any particular

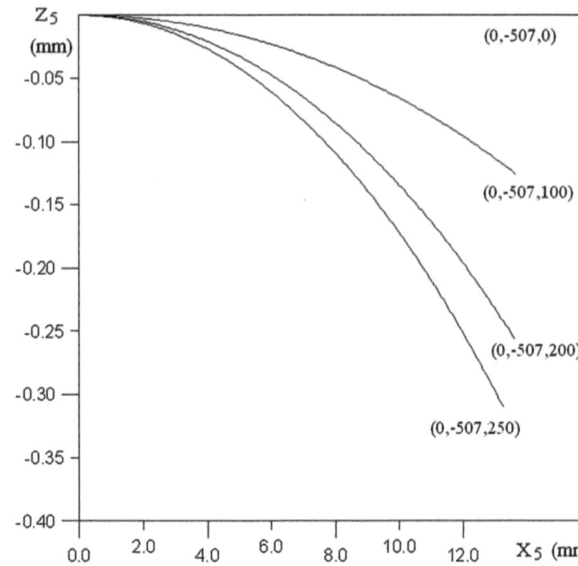


**Fig. 3.2** The rays emitted from the sagittal cone of any source point $\bar{P}_0$ of Fig. 2.1 always hits the neighborhood of $z_5 = 0$, indicating one may use polar coordinates $\alpha_0$ and $\beta_0$ to discuss the aberrations of an optical system

ray by using Newton-Ranson method, if we have the Jacobian matrix of that particular ray with respect to $\alpha_0$ and $\beta_0$.

*Example 3.1* As shown in Fig. 2.1, if the optical axis is the $y_0$ axis and the meridional plane is the $y_0z_0$ plane, any sagittal ray will intersect the aperture at $z_5 = 0$. Figure 3.2 shows the loci of incidence points on the aperture plane (i.e., $x_5z_5$ plane) for different source points $\bar{P}_0$ when $\bar{\ell}_0$ is emitted from its sagittal cone of system in Fig. 2.1. It is noted from this figure that any ray emitted from the sagittal cone of any source point $\bar{P}_0$ always hits the neighborhood of $z_5 = 0$. This indicates that, instead of using coordinates of the aperture (e.g., $x_5$ and $z_5$ of Fig. 2.1), one may use polar coordinates $\alpha_0$ and $\beta_0$ to discuss the aberrations.

## 3.2 Axis-Symmetrical Optical Systems

The discussion of this section is limited to axis-symmetrical systems which have a straight-line optical axis. An optical axis is a line along which there is some degree of rotational symmetry in an optical system. For a system composed of simple lenses, the optical axis passes through the center of curvature of each surface and coincides with the axis of rotational symmetry. Most elements used in an axis-symmetrical system are spherical elements; their two boundary surfaces are parts of the spheres. The curvature of a boundary surface can be positive, zero, or negative depending on whether the surface is convex, flat, or concave, respectively. There are nine possible types of such elements if they are classified by the geometries of their two boundary surfaces and orientation (see Figs. 3.3–3.11).

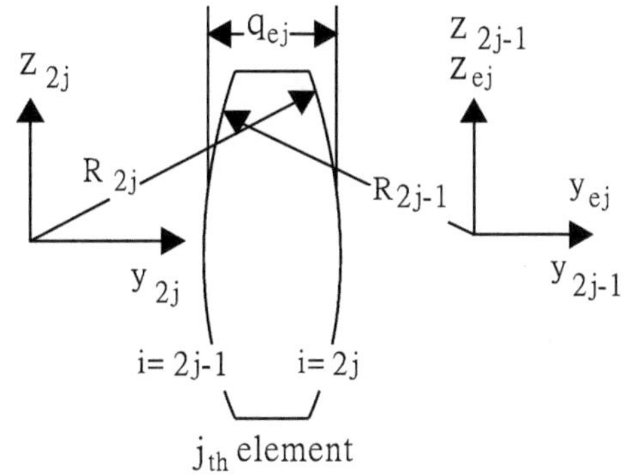


**Fig. 3.3** A biconvex element having convex—convex spherical boundary surfaces

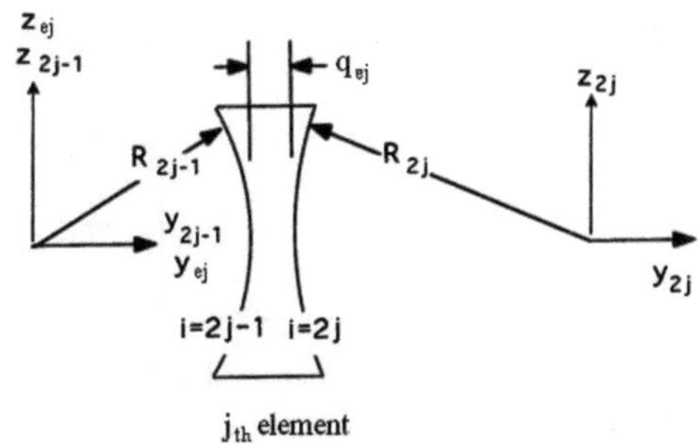


**Fig. 3.4** A biconcave element possessing concave—concave spherical boundary surfaces

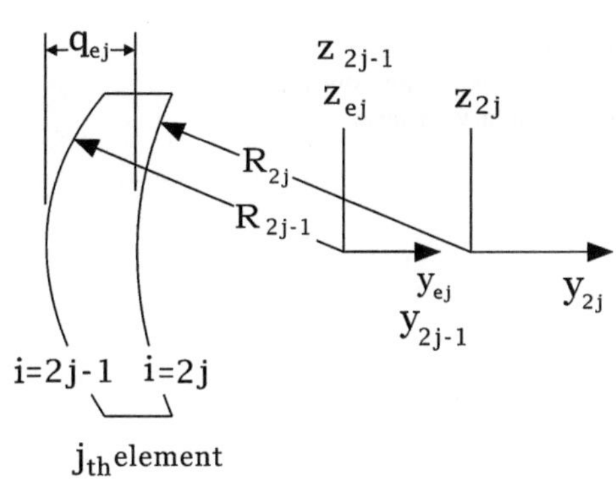


**Fig. 3.5** A meniscus element having convex—concave spherical boundary surfaces

One feature of such elements is that each element (denoted as the $j^{th}$ element or element j) is composed of two boundary surfaces $\bar{r}_{2j-1}$ and $\bar{r}_{2j}$, whose boundary coordinate frames are respectively $(xyz)_{2j-1}$ and $(xyz)_{2j}$. In 3-D we need an element coordinate frame $(xyz)_{ej}$ to define the pose of this jth optical element. For

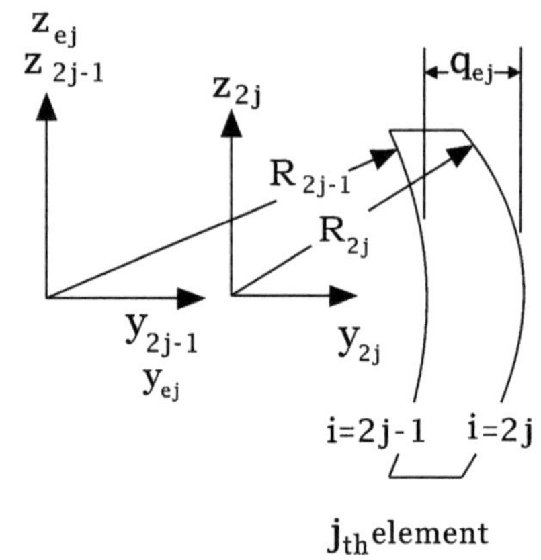


**Fig. 3.6** A meniscus element having concave—convex spherical boundary surfaces

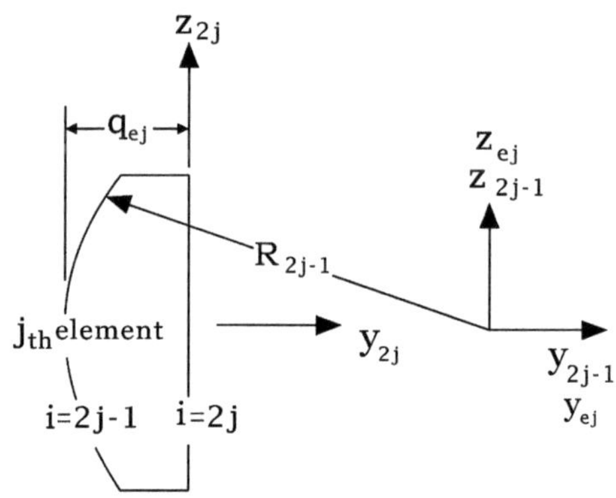


**Fig. 3.7** A plano-convex element having convex-flat boundary surfaces

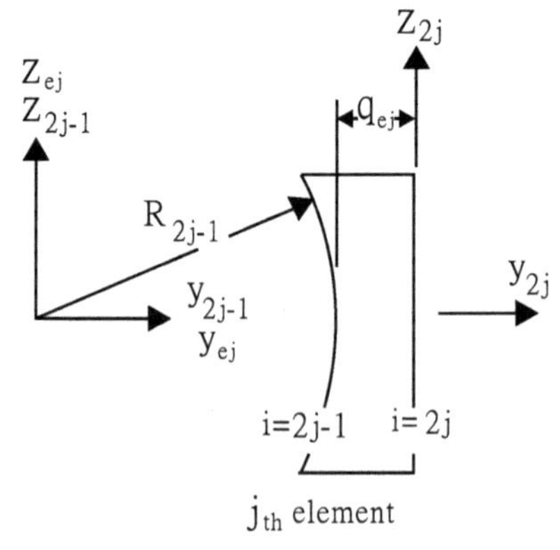


**Fig. 3.8** A plano-concave element possessing concave-flat boundary surfaces

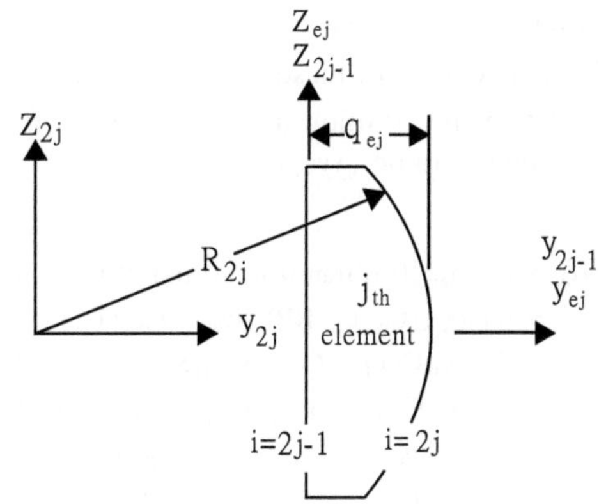


**Fig. 3.9** A plano-convex element with flat-convex boundary surfaces

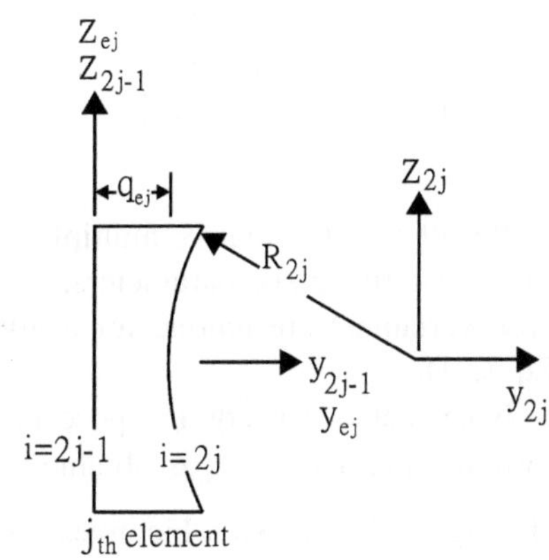


**Fig. 3.10** A plano-concave element having flat-concave boundary surfaces

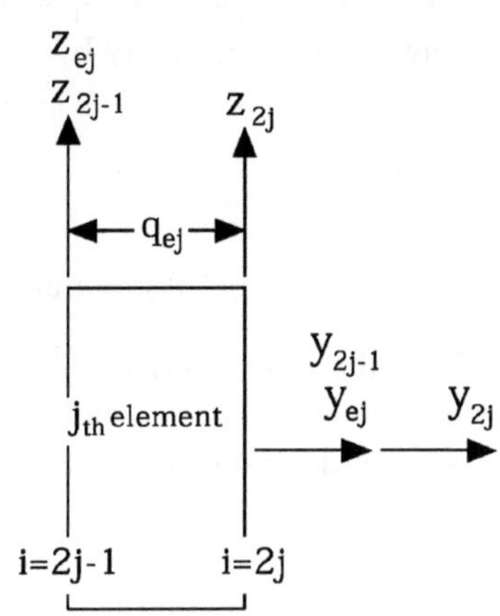


**Fig. 3.11** An optical element having two flat boundary surfaces

simplicity, we always let $(xyz)_{ej}$ coincides with the first boundary coordinate frame $(xyz)_{2j-1}$ of these nine elements. We can use the following pose matrix ${}^{0}\bar{A}_{ej}$ to define the position and orientation of this element with respect to the world coordinate frame $(xyz)_0$:

$$
\begin{aligned}
{}^{0}\bar{A}_{ej} &= \text{tran}(t_{ejx},0,0)\text{tran}(0,t_{ejy},0)\text{tran}(0,0,t_{ejz})\text{rot}(\bar{z},\omega_{ejz})\text{rot}(\bar{y},\omega_{ejy})\text{rot}(\bar{x},\omega_{ejx}) \\
&= \text{tran}(t_{ejx},t_{ejy},t_{ejz})\text{RPY}(\omega_{ejz},\omega_{ejy},\omega_{ejx}) \\
&= \begin{bmatrix} C\omega_{ejz}C\omega_{ejy} & C\omega_{ejz}S\omega_{ejy}S\omega_{ejx}-S\omega_{ejz}C\omega_{ejx} & C\omega_{ejz}S\omega_{ejy}C\omega_{ejx}+S\omega_{ejz}S\omega_{ejx} & t_{ejx} \\ S\omega_{ejz}C\omega_{ejy} & S\omega_{ejz}S\omega_{ejy}S\omega_{ejx}+C\omega_{ejz}C\omega_{ejx} & S\omega_{ejz}S\omega_{ejy}C\omega_{ejx}-C\omega_{ejz}S\omega_{ejx} & t_{ejy} \\ -S\omega_{ejy} & C\omega_{ejy}S\omega_{ejx} & C\omega_{ejy}C\omega_{ejx} & t_{ejz} \\ 0 & 0 & 0 & 1 \end{bmatrix} \\
&= \begin{bmatrix} I_{ejx} & J_{ejx} & K_{ejx} & t_{ejx} \\ I_{ejy} & J_{ejy} & K_{ejy} & t_{ejy} \\ I_{ejz} & J_{ejz} & K_{ejz} & t_{ejz} \\ 0 & 0 & 0 & 1 \end{bmatrix}.
\end{aligned}
\tag{3.1}
$$

Recall that the matrix multiplication is not commutative. This property indicates that these six parameters, $t_{ejx}, t_{ejy}, t_{ejz}, \omega_{ejx}, \omega_{ejy}, \omega_{ejz}$ (referred as element pose variables) are unique for a given element if the pose matrix is defined by Eq. (3.1).

Note that one needs the pose matrix ${}^{0}\bar{A}_{i}$ (see Eqs. (2.8) and (2.34)) for performing raytracing. ${}^{0}\bar{A}_{i}$ is obtained by the following methodology:

(1) Figure 3.3 shows a biconvex element with thickness $q_{ej}$. If $\kappa_{2j-1}$ and $\kappa_{2j}$ are the curvatures of its first and second boundary surfaces, then the pose matrices of $(xyz)_{2j-1}$ and $(xyz)_{2j}$ with respect to its element coordinate frame $(xyz)_{ej}$ are given respectively by

$$ {}^{ej}\bar{A}_{2j-1} = \bar{I}_{4\times 4}, \tag{3.2}$$

$$ {}^{ej}\bar{A}_{2j} = \text{tran}(0, -1/\kappa_{2j-1} + q_{ej} + 1/\kappa_{2j}, 0). \tag{3.3}$$

Now the pose matrices ${}^{0}\bar{A}_{i}$ ($i = 2j-1$ and $i = 2j$) of its two boundary coordinate frames, $(xyz)_{2j-1}$ and $(xyz)_{2j}$, with respect to the world coordinate frame $(xyz)_0$ are determined respectively by

$$
\begin{aligned}
{}^{0}\bar{A}_{2j-1} &= {}^{0}\bar{A}_{ej}\,{}^{ej}\bar{A}_{2j-1} = {}^{0}\bar{A}_{ej} \\
&= \text{tran}(t_{ejx}, t_{ejy}, t_{ejz})\,\text{rot}(\bar{z}, \omega_{ejz})\,\text{rot}(\bar{y}, \omega_{ejy})\,\text{rot}(\bar{x}, \omega_{ejx}),
\end{aligned}
\tag{3.4}
$$

$$
\begin{aligned}
{}^{0}\overline{A}_{2j} &= {}^{0}\overline{A}_{ej}\,{}^{ej}\overline{A}_{2j} \\
&= \text{tran}(t_{ejx}, t_{ejy}, t_{ejz})\,\text{rot}(\overline{z}, \omega_{ejz})\,\text{rot}(\overline{y}, \omega_{ejy})\,\text{rot}(\overline{x}, \omega_{ejx})\,\text{tran}(0, -1/\kappa_{2j-1} \\
&\quad + q_{ej} + 1/\kappa_{2j}, 0).
\end{aligned} \tag{3.5}
$$

Equations (3.4) and (3.5) provide the required pose matrices, ${}^{0}\bar{A}_{i} = {}^{0}\bar{A}_{2j-1}$ and ${}^{0}\bar{A}_{i} = {}^{0}\bar{A}_{2j}$, in performing the skew-ray tracing described in Chap. 2. From Eqs. (3.4) and (3.5) we know the variable vector $\overline{X}_{ej}$ of this biconvex element should include the six pose variables, $t_{ejx}$, $t_{ejy}, t_{ejz}, \omega_{ejx}, \omega_{ejy}, \omega_{ejz}$, the curvatures $\kappa_{2j-1}$ and $\kappa_{2j}$ of the two spherical boundary surfaces, the element thickness $q_{ej}$, and the element refractive index $\xi_{ej}$. When the refractive index of air $\xi_{air}$ is included, the variable vector of this biconvex element j is

$$\overline{X}_{ej} = \begin{bmatrix} t_{ejx} & t_{ejy} & t_{ejz} & \omega_{ejx} & \omega_{ejy} & \omega_{ejz} & \xi_{air} & \xi_{ej} & \kappa_{2j-1} & q_{ej} & \kappa_{2j} \end{bmatrix}^{T}. \tag{3.6}$$

Figures 3.4, 3.5, and 3.6 illustrate respectively a biconcave element and two meniscus elements. Equations (3.2–3.6) are still valid for these elements, since the meniscus, biconcave, and biconvex elements are also composed by two spherical boundary surfaces. The appropriate signs of their curvatures have to be used.

(2) Figure 3.7 illustrates a plano-convex element with $i = 2j - 1$ and $i = 2j$ representing respectively convex and flat boundary surfaces. The pose matrices, ${}^{ej}\bar{A}_{2j-1}$ and ${}^{ej}\bar{A}_{2j}$, of the first and the second boundary coordinate frames, $(xyz)_{2j-1}$ and $(xyz)_{2j}$, with respect to the element coordinate frame $(\mathbf{xyz})_{ej}$ are given respectively by

$${}^{ej}\bar{A}_{2j-1} = \bar{I}_{4\times4}, \tag{3.7}$$

$${}^{ej}\overline{A}_{2j} = \text{tran}(0, -1/\kappa_{2j-1} + q_{ej}, 0). \tag{3.8}$$

Here $q_{ej}$ and $\kappa_{2j-1}$ are respectively its thickness and the curvature of the first boundary surface. The pose matrices, ${}^{0}\bar{A}_{2j-1}$ and ${}^{0}\bar{A}_{2j}$, of the first and the second boundary coordinate frames, $(xyz)_{2j-1}$ and $(xyz)_{2j}$, with respect to the world coordinate frame $(xyz)_{0}$ can be determined respectively by

$$
\begin{aligned}
{}^{0}\overline{A}_{2j-1} &= {}^{0}\overline{A}_{ej}\,{}^{ej}\overline{A}_{2j-1} = {}^{0}\overline{A}_{ej} \\
&= \text{tran}(t_{ejx}, t_{ejy}, t_{ejz})\text{rot}(\bar{z}, \omega_{ejz})\text{rot}(\bar{y}, \omega_{ejy})\text{rot}(\bar{x}, \omega_{ejx}),
\end{aligned} \tag{3.9}
$$

$$
\begin{aligned}
{}^{0}\bar{A}_{2j} &= {}^{0}\bar{A}_{ej}\,{}^{ej}\bar{A}_{2j} \\
&= \text{tran}(t_{ejx}, t_{ejy}, t_{ejz})\text{rot}(\bar{z}, \omega_{ejz})\text{rot}(\bar{y}, \omega_{ejy})\text{rot}(\bar{x}, \omega_{ejx})\text{tran}(0, -1/\kappa_{2j-1} + q_{ej}, 0).
\end{aligned} \tag{3.10}
$$

One has the following element variable vector by collecting the variables in Eqs. (3.9) and (3.10) and including the refractive indices of air and the element:

$$\bar{X}_{ej} = \begin{bmatrix} t_{ejx} & t_{ejy} & t_{ejz} & \omega_{ejx} & \omega_{ejy} & \omega_{ejz} & \xi_{air} & \xi_{ej} & \kappa_{2j-1} & q_{ej} \end{bmatrix}^T. \tag{3.11}$$

Compared with biconvex or biconcave element, there are only 10 components in its variable vector $\bar{X}_{ej}$, since its second boundary surface is a flat one:

Figure 3.8 gives a plano-concave element with $i = 2j-1$ and $i = 2j$ representing concave and flat boundary surfaces respectively. Equations (3.7–3.11) are still valid for such element, since the first boundary surfaces of Figs. 3.7 and 3.8 are spherical.

(3) Figure 3.9 shows another arrangement of a plano-convex element with $i = 2j-1$ and $i = 2j$ representing respectively flat and convex spherical boundary surfaces. The pose matrices, ${}^{ej}\bar{A}_{2j-1}$ and ${}^{ej}\bar{A}_{2j}$, of the first and the second boundary coordinate frames, $(xyz)_{2j-1}$ and $(xyz)_{2j}$, with respect to the element coordinate frame $(xyz)_{ej}$ are given respectively by

$${}^{ej}\bar{A}_{2j-1} = \bar{I}_{4\times4}, \tag{3.12}$$

$${}^{ej}\bar{A}_{2j} = \text{tran}(0, q_{ej} + 1/\kappa_{2j}, 0). \tag{3.13}$$

Here $q_{ej}$ and $\kappa_{2j}$ are respectively its thickness and the curvature of the second boundary surface of this element. The pose matrices, ${}^{0}\bar{A}_{2j-1}$ and ${}^{0}\bar{A}_{2j}$, of $(xyz)_{2j-1}$ and $(xyz)_{2j}$ with respect to the world coordinate frame $(xyz)_0$ are given respectively by

$$\begin{aligned} {}^{0}\bar{A}_{2j-1} &= {}^{0}\bar{A}_{ej}\,{}^{ej}\bar{A}_{2j-1} = {}^{0}\bar{A}_{ej} \\ &= \text{tran}(t_{ejx}, t_{ejy}, t_{ejz})\text{rot}(\bar{z}, \omega_{ejz})\text{rot}(\bar{y}, \omega_{ejy})\text{rot}(\bar{x}, \omega_{ejx}), \end{aligned} \tag{3.14}$$

$$\begin{aligned} {}^{0}\bar{A}_{2j} &= {}^{0}\bar{A}_{ej}\,{}^{ej}\bar{A}_{2j} \\ &= \text{tran}(t_{ejx}, t_{ejy}, t_{ejz})\,\text{rot}(\bar{z}, \omega_{ejz})\,\text{rot}(\bar{y}, \omega_{ejy})\,\text{rot}(\bar{x}, \omega_{ejx})\text{tran}(0, q_{ej} + 1/\kappa_{2j}, 0). \end{aligned} \tag{3.15}$$

Although this is also an plano-convex element, its variable vector is different from that described in case (2), and is given by

$$\bar{X}_{ej} = \begin{bmatrix} t_{ejx} & t_{ejy} & t_{ejz} & \omega_{ejx} & \omega_{ejy} & \omega_{ejz} & \xi_{air} & \xi_{ej} & q_{ej} & \kappa_{2j} \end{bmatrix}^T. \tag{3.16}$$

Figure 3.10 shows another arrangement of a plano-concave element with $i = 2j-1$ and $i = 2j$ representing flat and concave spherical boundary surfaces respectively. Equations (3.12–3.16) are still valid for such element, since the first and the second boundary surfaces of this case and Fig. 3.9 are flat and spherical, respectively. Again, the appropriate sign of $\kappa_{2j}$ has to be used.

(4) Figure 3.11 illustrates an optical element possessing two flat boundary surfaces with thickness $q_{ej}$. The pose matrices, ${}^{ej}\bar{A}_{2j-1}$ and ${}^{ej}\bar{A}_{2j}$, of the first and the second boundary coordinate frames, $(xyz)_{2j-1}$ and $(xyz)_{2j}$, with respect to the element coordinate frame $(xyz)_{ej}$ are given respectively by

$$ {}^{ej}\bar{A}_{2j-1} = \bar{I}_{4\times4}, \tag{3.17}$$

$$ {}^{ej}\bar{A}_{2j} = \text{tran}(0, q_{ej}, 0). \tag{3.18}$$

The pose matrices, ${}^{0}\bar{A}_{2j-1}$ and ${}^{0}\bar{A}_{2j}$, of the first and the second boundary coordinate frames, $(xyz)_{2j-1}$ and $(xyz)_{2j}$, with respect to the world coordinate frame $(xyz)_0$ can be given respectively by

$$\begin{aligned} {}^{0}\overline{A}_{2j-1} &= {}^{0}\overline{A}_{ej}\,{}^{ej}\overline{A}_{2j-1} = {}^{0}\overline{A}_{ej} \\ &= \text{tran}(t_{ejx}, t_{ejy}, t_{ejz})\,\text{rot}(\bar{z}, \omega_{ejz})\,\text{rot}(\bar{y}, \omega_{ejy})\,\text{rot}(\bar{x}, \omega_{ejx}), \end{aligned} \tag{3.19}$$

$$\begin{aligned} {}^{0}\bar{A}_{2j} &= {}^{0}\bar{A}_{ej}\,{}^{ej}\bar{A}_{2j} \\ &= \text{tran}(t_{ejx}, t_{ejy}, t_{ejz})\text{rot}(\bar{z}, \omega_{ejz})\text{rot}(\bar{y}, \omega_{ejy})\text{rot}(\bar{x}, \omega_{ejx})\text{tran}(0, q_{ej}, 0). \end{aligned} \tag{3.20}$$

Since we do not need $\kappa_{2j-1}$ and $\kappa_{2j}$ to designate the curvatures of its two boundary surfaces, there are only 9 components of the variable vector:

$$\overline{X}_{ej} = \begin{bmatrix} t_{ejx} & t_{ejy} & t_{ejz} & \omega_{ejx} & \omega_{ejy} & \omega_{ejz} & \xi_{air} & \xi_{ej} & q_{ej} \end{bmatrix}^T. \tag{3.21}$$

It is noted that Figs. 3.4–3.11 can be deduced from Fig. 3.3, provided the appropriate signs of curvatures are used. Consequently, in design stage of optical systems we can use an element of Fig. 3.3 to represent any one of these nine possible elements, and let its two curvatures changing from lower bound (which may be negative value) to upper bound for optimization purpose.

*Example 3.2* The use of multiple elements in a system allows more optical aberrations to be corrected than is possible with single element. Referring the system shown in Fig. 2.1 and Table 2.1, one has the following pose matrices ${}^{0}\bar{A}_{ej}$ ($j = 1$ to $j = 6$) for its six elements:

$$\begin{aligned} {}^{0}\overline{A}_{e1} &= \text{tran}(t_{e1x}, v_1 + 1/\kappa_1, t_{e1z})\,\text{rot}(\bar{z}, \omega_{e1z})\,\text{rot}(\bar{y}, \omega_{e1y})\,\text{rot}(\bar{x}, \omega_{e1x}) \\ &= \text{tran}(0, 38.2219, 0), \end{aligned} \tag{3.22a}$$

$$\begin{aligned} {}^{0}\bar{A}_{e2} &= \text{tran}(t_{e2x}, v_1 + q_{e1} + v_2 + 1/\kappa_3, t_{e2z})\,\text{rot}(\bar{z}, \omega_{e2z})\,\text{rot}(\bar{y}, \omega_{e2y})\,\text{rot}(\bar{x}, \omega_{e2x}) \\ &= \text{tran}(0, -40.23610, 0), \end{aligned} \tag{3.22b}$$

$$
\begin{aligned}
{}^{0}\bar{A}_{e3} &= \mathrm{tran}(t_{e3x}, v_1 + q_{e1} + v_2 + q_{e2} + v_3, t_{e3z})\,\mathrm{rot}(\bar{z}, \omega_{e3z})\,\mathrm{rot}(\bar{y}, \omega_{e3y})\mathrm{rot}(\bar{x}, \omega_{e3x}) \\
&= \mathrm{tran}(0, 24.84120, 0),
\end{aligned}
\tag{3.22c}
$$

$$
\begin{aligned}
{}^{0}\bar{A}_{e4} = {} & \mathrm{tran}(t_{e4x}, v_1 + q_{e1} + v_2 + q_{e2} + v_3 + q_{e3} + v_4 + 1/\kappa_7, t_{e4z})\,\mathrm{rot}(\bar{z}, \omega_{e4z}) \\
& \mathrm{rot}(\bar{y}, \omega_{e4y})\ \mathrm{rot}(\bar{x}, \omega_{e4x}) = \mathrm{tran}(0, -2.93370, 0),
\end{aligned}
\tag{3.22d}
$$

$$
\begin{aligned}
{}^{0}\bar{A}_{e5} = {} & \mathrm{tran}(t_{e5x}, v_1 + q_{e1} + v_2 + q_{e2} + v_3 + q_{e3} + v_4 + q_{e4} + v_5 + 1/\kappa_9, t_{e5z}) \\
& \mathrm{rot}(\bar{z}, \omega_{e5z})\mathrm{rot}(\bar{y}, \omega_{e5y})\mathrm{rot}(\bar{x}, \omega_{e5x}) = \mathrm{tran}(0, 112.86490, 0),
\end{aligned}
\tag{3.22e}
$$

$$
\begin{aligned}
{}^{0}\bar{A}_{e6} = {} & \mathrm{tran}(t_{e6x}, v_1 + q_{e1} + v_2 + q_{e2} + v_3 + q_{e3} + v_4 + q_{e4} + v_5 + q_{e5} + v_6, t_{e6z}) \\
& \mathrm{rot}(\bar{z}, \omega_{e6z})\mathrm{rot}(\bar{y}, \omega_{e6y})\mathrm{rot}(\bar{x}, \omega_{e6x}) = \mathrm{tran}(0, 105.02900, 0).
\end{aligned}
\tag{3.22f}
$$

*Example 3.3* The following are the pose matrices of the two boundary frames, $(xyz)_{2j-1}$ and $(xyz)_{2j}$, with respect to each element frame $(xyz)_{ej}$ for $j = 1$ to $j = 6$ of Fig. 2.1 (also see Table 2.1):

$$ {}^{e1}\bar{A}_1 = I_{4\times4}, \tag{3.23a} $$

$$ {}^{e1}\bar{A}_2 = \mathrm{tran}(0, -1/\kappa_1 + q_{e1} + 1/\kappa_2, 0) = \mathrm{tran}(0, -78.45800, 0), \tag{3.23b} $$

$$ {}^{e2}\bar{A}_3 = I_{4\times4}, \tag{3.23c} $$

$$ {}^{e2}\bar{A}_4 = \mathrm{tran}(0, -1/\kappa_3 + q_{e2} + 1/\kappa_4, 0) = \mathrm{tran}(0, -528.62731, 0), \tag{3.23d} $$

$$ {}^{e3}\bar{A}_5 = \bar{I}_{4\times4}, \tag{3.23e} $$

$$ {}^{e3}\bar{A}_6 = \mathrm{tran}(0, q_{e3}, 0) = \mathrm{tran}(0, 7.62000, 0), \tag{3.23f} $$

$$ {}^{e4}\bar{A}_7 = \bar{I}_{4\times4}, \tag{3.23g} $$

$$ {}^{e4}\bar{A}_8 = \mathrm{tran}(0, -1/\kappa_7 + q_{e4} + 1/\kappa_8, 0) = \mathrm{tran}(0, 73.65490, 0), \tag{3.23h} $$

$$ {}^{e5}\bar{A}_9 = \bar{I}_{4\times4}, \tag{3.23i} $$

$$ {}^{e5}\bar{A}_{10} = \mathrm{tran}(0, -1/\kappa_9 + q_{e5} + 1/\kappa_{10}, 0) = \mathrm{tran}(0, -114.33300, 0), \tag{3.23j} $$

$$ {}^{e6}\bar{A}_{11} = \bar{I}_{4\times4}, \tag{3.23k} $$

*Example 3.4* Referring to the element pose matrix ${}^{0}\bar{A}_{ej}$ (j = 1 to j = 6) in Eqs. (3.22a–3.22f) and boundary pose matrices, ${}^{ej}\bar{A}_{2j-1}$ and ${}^{ej}\bar{A}_{2j}$ in Eqs. (3.23a–3.23k), one can have the following required pose matrices ${}^{0}\bar{A}_{i}$ (i = 1 to i = 11) of the boundary coordinate frames $(xyz)_i$ with respect to $(xyz)_0$:

$$\begin{aligned}{}^{0}\overline{A}_{1} &= {}^{0}\overline{A}_{e1}{}^{e1}\overline{A}_{1} \\ &= \mathrm{tran}(t_{e1x}, v_1 + 1/\kappa_1, t_{e1z})\,\mathrm{rot}(\overline{z}, \omega_{e1z})\,\mathrm{rot}(\overline{y}, \omega_{e1y})\,\mathrm{rot}(\overline{x}, \omega_{e1x}) \\ &= \mathrm{tran}(0, 38.22190, 0),\end{aligned} \tag{3.24a}$$

$$\begin{aligned}{}^{0}\overline{A}_{2} &= {}^{0}\overline{A}_{e1}{}^{e1}\overline{A}_{2} \\ &= \mathrm{tran}(t_{e1x}, v_1 + 1/\kappa_1, t_{e1z})\,\mathrm{rot}(\overline{z}, \omega_{e1z})\,\mathrm{rot}(\overline{y}, \omega_{e1y})\,\mathrm{rot}(\overline{x}, \omega_{e1x}) \\ &\quad \mathrm{tran}(0, -1/\kappa_1 + q_{e1} + 1/\kappa_2, 0) \\ &= \mathrm{tran}(0, 38.22190, 0)\,\mathrm{tran}(0, -78.45800, 0) = \mathrm{tran}(0, -40.23610, 0),\end{aligned} \tag{3.24b}$$

$$\begin{aligned}{}^{0}\overline{A}_{3} &= {}^{0}\overline{A}_{e2}{}^{e2}\overline{A}_{3} \\ &= \mathrm{tran}(t_{e2x}, v_1 + q_{e1} + v_2 + 1/\kappa_3, t_{e2z})\,\mathrm{rot}(\overline{z}, \omega_{e2z})\,\mathrm{rot}(\overline{y}, \omega_{e2y})\,\mathrm{rot}(\overline{x}, \omega_{e2x}) \\ &= \mathrm{tran}(0, -40.23610, 0),\end{aligned} \tag{3.24c}$$

$$\begin{aligned}{}^{0}\overline{A}_{4} &= {}^{0}\overline{A}_{e2}{}^{e2}\overline{A}_{4} \\ &= \mathrm{tran}(t_{e2x}, v_1 + q_{e1} + v_2 + 1/\kappa_3, t_{e2z})\,\mathrm{rot}(\overline{z}, \omega_{e2z})\,\mathrm{rot}(\overline{y}, \omega_{e2y})\,\mathrm{rot}(\overline{x}, \omega_{e2x}) \\ &\quad \mathrm{tran}(0, -1/\kappa_3 + q_{e2} + 1/\kappa_4, 0) \\ &= \mathrm{tran}(0, -40.23610, 0)\,\mathrm{tran}(0, -528.62731, 0) = \mathrm{tran}(0, -568.86341, 0),\end{aligned} \tag{3.24d}$$

$$\begin{aligned}{}^{0}\overline{A}_{5} &= {}^{0}\overline{A}_{e3}{}^{e3}\overline{A}_{5} \\ &= \mathrm{tran}(t_{e3x}, v_1 + q_{e1} + v_2 + q_{e2} + v_3, t_{e3z})\,\mathrm{rot}(\overline{z}, \omega_{e3z})\,\mathrm{rot}(\overline{y}, \omega_{e3y})\,\mathrm{rot}(\overline{x}, \omega_{e3x}) \\ &= \mathrm{tran}(0, 24.84120, 0),\end{aligned} \tag{3.24e}$$

$$\begin{aligned}{}^{0}\overline{A}_{6} &= {}^{0}\overline{A}_{e3}{}^{e3}\overline{A}_{6} \\ &= \mathrm{tran}(t_{e3x}, v_1 + q_{e1} + v_2 + q_{e2} + v_3, t_{e3z})\,\mathrm{rot}(\overline{z}, \omega_{e3z})\,\mathrm{rot}(\overline{y}, \omega_{e3y})\,\mathrm{rot}(\overline{x}, \omega_{e3x}) \\ &\quad \mathrm{tran}(0, q_{e3}, 0) \\ &= \mathrm{tran}(0, 24.84120, 0)\mathrm{tran}(0, 7.62000, 0) = \mathrm{tran}(0, 32.46120, 0),\end{aligned} \tag{3.24f}$$

$$
\begin{aligned}
{}^{0}\overline{A}_{7} &= {}^{0}\overline{A}_{e4}{}^{e4}\overline{A}_{7}\\
&= \mathrm{tran}(t_{e4x}, v_1 + q_{e1} + v_2 + q_{e2} + v_3 + q_{e3} + v_4 + 1/\kappa_7, t_{e4z})\,\mathrm{rot}(\overline{z}, \omega_{e4z})\\
&\quad \mathrm{rot}(\overline{y}, \omega_{e4y})\,\mathrm{rot}(\overline{x}, \omega_{e4x})\\
&= \mathrm{tran}(0, -2.93370, 0),
\end{aligned} \tag{3.24g}
$$

$$
\begin{aligned}
{}^{0}\overline{A}_{8} &= {}^{0}\overline{A}_{e4}{}^{e4}\overline{A}_{8}\\
&= \mathrm{tran}(t_{e4x}, v_1 + q_{e1} + v_2 + q_{e2} + v_3 + q_{e3} + v_4 + 1/\kappa_7, t_{e4z})\,\mathrm{rot}(\overline{z}, \omega_{e4z})\\
&\quad \mathrm{rot}(\overline{y}, \omega_{e4y})\,\mathrm{rot}(\overline{x}, \omega_{e4x})\,\mathrm{tran}(0, -1/\kappa_7 + q_{e4} + 1/\kappa_8, 0)\\
&= \mathrm{tran}(0, -2.93370, 0)\,\mathrm{tran}(0, 73.65490, 0) = \mathrm{tran}(0, 70.72120, 0),
\end{aligned} \tag{3.24h}
$$

$$
\begin{aligned}
{}^{0}\overline{A}_{9} &= {}^{0}\overline{A}_{e5}{}^{e5}\overline{A}_{9}\\
&= \mathrm{tran}(t_{e5x}, v_1 + q_{e1} + v_2 + q_{e2} + v_3 + q_{e3} + v_4 + q_{e4} + v_5 + 1/\kappa_9, t_{e5z})\\
&\quad \mathrm{rot}(\overline{z}, \omega_{e5z})\,\mathrm{rot}(\overline{y}, \omega_{e5y})\,\mathrm{rot}(\overline{x}, \omega_{e5x}) = \mathrm{tran}(0, 112.86490, 0),
\end{aligned} \tag{3.24i}
$$

$$
\begin{aligned}
{}^{0}\overline{A}_{10} &= {}^{0}\overline{A}_{e5}{}^{e5}\overline{A}_{10}\\
&= \mathrm{tran}(t_{e5x}, v_1 + q_{e1} + v_2 + q_{e2} + v_3 + q_{e3} + v_4 + q_{e4} + v_5 + 1/\kappa_9, t_{e5z})\\
&\quad \mathrm{rot}(\overline{z}, \omega_{e5z})\,\mathrm{rot}(\overline{y}, \omega_{e5y})\,\mathrm{rot}(\overline{x}, \omega_{e5x})\,\mathrm{tran}(0, -1/\kappa_9 + q_{e5} + 1/\kappa_{10}, 0)\\
&= \mathrm{tran}(0, 112.86490, 0)\,\mathrm{tran}(0, -114.33300, 0) = \mathrm{tran}(0, -1.46810, 0),
\end{aligned} \tag{3.24j}
$$

$$
\begin{aligned}
{}^{0}\bar{A}_{11} &= {}^{0}\bar{A}_{e6}{}^{e6}\bar{A}_{11}\\
&= \mathrm{tran}(t_{e6x}, v_1 + q_{e1} + v_2 + q_{e2} + v_3 + q_{e3} + v_4 + q_{e4} + v_5 + q_{e5} + v_6, t_{e6z})\\
&\quad \mathrm{rot}(\bar{z}, \omega_{e6z})\,\mathrm{rot}(\bar{y}, \omega_{e6y})\,\mathrm{rot}(\bar{x}, \omega_{e6x}) = \mathrm{tran}(0, 105.02900, 0).
\end{aligned} \tag{3.24k}
$$

Recall that when the pose matrix ${}^{0}\bar{A}_{i}$ of the $i^{th}$ boundary surface is expressed in terms of Eqs. (2.8) or (2.34), the parameters $t_{ix}$, $t_{iy}$, $t_{iz}$, $\omega_{iz}$, $\omega_{iy}$ and $\omega_{ix}$ are defined as the pose variables of the $i^{th}$ boundary surface. In the following chapters, the values of these variables are required to compute various quantities, such as the point spread function and modulation transfer function.

*Example 3.5* One can solve for the values of the six pose variables of ${}^{0}\bar{A}_{i}$ ($i = 1$ to $i = 11$) by using Eqs. (1.34)–(1.39), where ${}^{0}\bar{A}_{i}$ are given by Eqs. (3.24a)–(3.24k) for system shown in Fig. 2.1. The results are listed in Table 3.2. The values of $t_{ix} = t_{iz} = \omega_{iz} = \omega_{iy} = \omega_{iz} = 0$ shown in Table 3.2 indicate that this is an axis-symmetrical system.

*Example 3.6* According to Eq. (3.24a)–(3.24k), one can have the following element variable vectors for the system of Fig. 2.1:

$$\overline{X}_{e0} = [P_{0x} \quad P_{0y} \quad P_{0z} \quad \alpha_0 \quad \beta_0]^T, \tag{3.25a}$$

**Table 3.2** Values of boundary pose variables of Petzval lens system shown in Fig. 2.1

| | $t_{ix}$ | $t_{iy}$ | $t_{iz}$ | $\omega_{ix}$ | $\omega_{iy}$ | $\omega_{iz}$ |
|---|---|---|---|---|---|---|
| i = 1 | 0 | 38.2219 | 0 | 0° | 0° | 0° |
| i = 2 | 0 | −40.2361 | 0 | 0° | 0° | 0° |
| i = 3 | 0 | −40.2361 | 0 | 0° | 0° | 0° |
| i = 4 | 0 | −568.8634 | 0 | 0° | 0° | 0° |
| i = 5 | 0 | 24.8412 | 0 | 0° | 0° | 0° |
| i = 6 | 0 | 24.8412 | 0 | 0° | 0° | 0° |
| i = 7 | 0 | −2.9337 | 0 | 0° | 0° | 0° |
| i = 8 | 0 | 70.7212 | 0 | 0° | 0° | 0° |
| i = 9 | 0 | 112.8649 | 0 | 0° | 0° | 0° |
| i = 10 | 0 | −1.4681 | 0 | 0° | 0° | 0° |
| i = 11 | 0 | 105.029 | 0 | 0° | 0° | 0° |

$$\overline{X}_{e1} = [\, t_{e1x} \quad v_1 + 1/\kappa_1 \quad t_{e1z} \quad \omega_{e1x} \quad \omega_{e1y} \quad \omega_{e1z} \quad \xi_{air} \quad \xi_{e1} \quad \kappa_1 \quad q_{e1} \quad \kappa_2 \,]^T, \tag{3.25b}$$

$$\overline{X}_{e2} = [\, t_{e2x} \quad v_1 + q_{e1} + v_2 + 1/\kappa_3 \quad t_{e2z} \quad \omega_{e2x} \quad \omega_{e2y} \quad \omega_{e2z} \quad \xi_{air} \quad \xi_{e2} \quad \kappa_3 \quad q_{e2} \quad \kappa_4 \,]^T, \tag{3.25c}$$

$$\overline{X}_{e3} = [\, t_{e3x} \quad v_1 + q_{e1} + v_2 + q_{e2} + v_3 \quad t_{e3z} \quad \omega_{e3x} \quad \omega_{e3y} \quad \omega_{e3z} \quad \xi_{air} \quad q_{e3} \,]^T, \tag{3.25d}$$

$$\begin{aligned}\overline{X}_{e4} =& [\, t_{e4x} \quad v_1 + q_{e1} + v_2 + q_{e2} + v_3 + q_{e3} + v_4 + 1/\kappa_7 \quad t_{e4z} \quad \omega_{e4x} \quad \omega_{e4y} \quad \omega_{e4z} \\ & \xi_{air} \quad \xi_{e4} \quad \kappa_7 \quad q_{e4} \quad \kappa_8 \,]^T,\end{aligned} \tag{3.25e}$$

$$\begin{aligned}\overline{X}_{e5} = & [\, t_{e5x} \quad v_1 + q_{e1} + v_2 + q_{e2} + v_3 + q_{e3} + v_4 + q_{e4} + v_5 + 1/\kappa_9 \\ & t_{e5z} \quad \omega_{e5x} \quad \omega_{e5y} \quad \omega_{e5z} \quad \xi_{air} \quad \xi_{e5} \quad \kappa_9 \quad q_{e5} \quad \kappa_{10} \,]^T,\end{aligned} \tag{3.25f}$$

$$\begin{aligned}\overline{X}_{e6} = & [\, t_{e6x} \quad v_1 + q_{e1} + v_2 + q_{e2} + v_3 + q_{e3} + v_4 + q_{e4} + v_5 + q_{e5} + v_6 \\ & t_{e6z} \quad \omega_{e6x} \quad \omega_{e6y} \quad \omega_{e6z} \quad \xi_{air}]^T.\end{aligned} \tag{3.25g}$$

*Example 3.7* After deleting the repeated variables (e.g., $\xi_{air}$) in $\overline{X}_{ej}$ (j = 1 to j = 6), one obtains the following variable vector as the variable of this system:

$$
\begin{aligned}
\overline{X}_{sys} &= [x_v] \\
&= [P_{0x} \quad P_{0y} \quad P_{0z} \quad \beta_0 \quad \alpha_0 \\
&\quad t_{e1x} \quad v_1 \quad t_{e1z} \quad \omega_{e1x} \quad \omega_{e1y} \quad \omega_{e1z} \quad \xi_{air} \quad \xi_{e1} \quad q_{e1} \quad \kappa_1 \quad \kappa_2 \\
&\quad t_{e2x} \quad v_2 \quad t_{e2z} \quad \omega_{e2x} \quad \omega_{e2y} \quad \omega_{e2z} \quad \xi_{e2} \quad q_{e2} \quad \kappa_3 \quad \kappa_4 \\
&\quad t_{e3x} \quad v_3 \quad t_{e3z} \quad \omega_{e3x} \quad \omega_{e3y} \quad \omega_{e3z} \quad q_{e3} \\
&\quad t_{e4x} \quad v_4 \quad t_{e4z} \quad \omega_{e4x} \quad \omega_{e4y} \quad \omega_{e4z} \quad \xi_{e4} \quad q_{e4} \quad \kappa_7 \quad \kappa_8 \\
&\quad t_{e5x} \quad v_5 \quad t_{e5z} \quad \omega_{e5x} \quad \omega_{e5y} \quad \omega_{e5z} \quad \xi_{e5} \quad q_{e5} \quad \kappa_9 \quad \kappa_{10} \\
&\quad t_{e6x} \quad v_6 \quad t_{e6z} \quad \omega_{e6x} \quad \omega_{e6y} \quad \omega_{e6z}]^T.
\end{aligned}
\tag{3.26}
$$

*Example 3.8* The variable vector of the system shown in Fig. 2.9 is

$$
\overline{X}_{sys} = [P_{0x} \quad P_{0y} \quad P_{0z} \quad \alpha_0 \quad \beta_0 \quad t_{e1x} \quad v_1 \quad t_{e1z} \quad \omega_{e1x} \quad \omega_{e1y} \quad \omega_{e1z} \quad \xi_{air} \quad \xi_{e1} \quad q_{e1} \quad \kappa_1 \quad \kappa_2]^T.
\tag{3.27}
$$

*Example 3.9* The variable vector of the system shown in Fig. 2.14 is

$$
\overline{X}_{sys} = [P_{0x} \quad P_{0y} \quad P_{0z} \quad \alpha_0 \quad \beta_0 \quad t_{e1x} \quad v_1 \quad t_{e1z} \quad \omega_{e1x} \quad \omega_{e1y} \quad \omega_{e1z} \quad \xi_{air} \quad \xi_{e1} \quad q_{e1}]^T.
\tag{3.28}
$$

## 3.3 Non-Axially Symmetrical Optical Systems

Prisms are commonly employed in non-axially symmetrical optical systems because of their ability to output an image with a certain orientation or relocate the emergent rays in a given manner. In order to describe the pose of each element in 3-D space, we need an element coordinate frame $(xyz)_{ej}$ embedded at a convenient position and orientation in that element. Again, one can use Eq. (3.1) to define the pose matrix ${}^0\bar{A}_{ej}$ of $(xyz)_{ej}$ with respect to $(xyz)_0$. Then the methodology described in Sect. 1.5 can be employed to determine the pose matrix ${}^{ej}\bar{A}_i$ ($j = 1$ to $j = k$, $i = m_j - L_j + 1$ to $i = m_j$) of each boundary surface with respect to its element. An illustrative example containing $k = 4$ optical elements and $n = 13$ boundary surfaces is shown in Fig. 3.12. Table 3.3 lists the labels "i" of the first (i.e., $i = m_j - L_j + 1$) and the last (i.e., $i = m_j$) boundary surfaces of each element for the system in Fig. 3.12. Actually one can treat an axis-symmetrical system (e.g., Fig. 2.1) as a special case of a non-axially symmetrical system.

*Example 3.9* Referring Fig. 3.12, one has the following pose matrices ${}^0\bar{A}_{ej}$ ($j = 1$ to $j = 4$) to describe the position and orientation of an element coordinate frame $(xyz)_{ej}$ with respect to the world coordinate frame $(\mathbf{xyz})_0$:

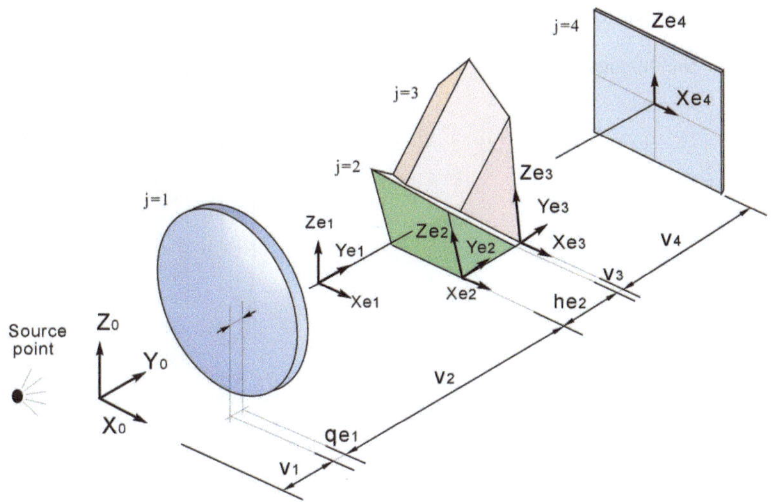


**Fig. 3.12** The illustrative system contains one double-convex lens, two prisms, and an image plane

**Table 3.3** Each element j of the system shown in Fig. 3.12 contains $L_j$ number of boundary surfaces labeled from $i = m_j - L_j + 1$ to $i = m_j$

| Element number | $j = 1$ | $j = 2$ | $j = 3$ | $j = k = 4$ |
|---|---|---|---|---|
| Lj (number of boundary surfaces in element j) | 2 | 4 | 6 | 1 |
| The first boundary surface of element j is $i = m_j - L_j + 1$ | 1 | 3 | 7 | 13 |
| The last boundary surface of element j is $i = m_j$ | 2 | 6 | 12 | 13 |

$$^{0}\bar{A}_{e1} = \text{tran}(t_{e1x}, v_1 + 1/\kappa_1, t_{e1z})\text{rot}(\bar{z}, \omega_{e1z})\text{rot}(\bar{y}, \omega_{e1y})\text{rot}(\bar{x}, \omega_{e1x}), \tag{3.29a}$$

$$^{0}\bar{A}_{e2} = \text{tran}(w_{e2}/2, v_1 + q_{e1} + v_2, -g_{e2})\text{rot}(\bar{z}, \omega_{e2z})\text{rot}(\bar{y}, \omega_{e2y})\text{rot}(\bar{x}, \omega_{e2x}), \tag{3.29b}$$

$$\begin{aligned} ^{0}\bar{A}_{e3} = \text{tran}(&w_{e3}/2, v_1 + q_{e1} + v_2 + h_{e2} \\ &+ v_3, -g_{e3})\text{rot}(\bar{z}, \omega_{e3z})\text{rot}(\bar{y}, \omega_{e3y})\text{rot}(\bar{x}, \omega_{e3x}), \end{aligned} \tag{3.29c}$$

$$\begin{aligned} ^{0}\bar{A}_{e4} = \text{tran}(&t_{e4x}, v_1 + q_{e1} + v_2 + h_{e2} + v_3 \\ &+ v_4, t_{e4z})\text{rot}(\bar{z}, \omega_{e4z})\text{rot}(\bar{y}, \omega_{e4y})\text{rot}(\bar{x}, \omega_{e4x}). \end{aligned} \tag{3.29d}$$

*Example 3.10* Referring to Figs. 3.13–3.16, one can have the following pose matrices $^{e1}\bar{A}_i$ (i = 1 to i = 2), $^{e2}\bar{A}_i$ (i = 3 to i = 6), $^{e3}\bar{A}_i$ (i = 7 to i = 12), and $^{e4}\bar{A}_{13}$, to describe the position and orientation of the boundary coordinate frame $(xyz)_i$ with respect to its element coordinate frame $(xyz)_{ej}$:

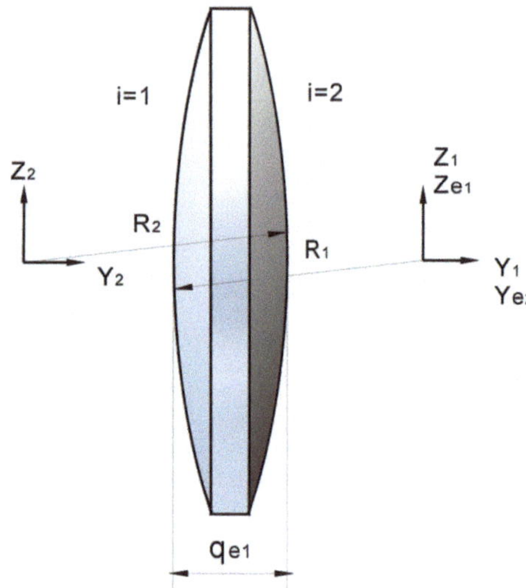


**Fig. 3.13** The element j = 1 of the system shown in Fig. 3.12 is a double-convex lens

$$^{e1}\bar{A}_1 = \bar{I}_{4\times4}, \tag{3.30a}$$

$$^{e1}\bar{A}_2 = \text{tran}(0, -1/\kappa_1 + q_{e1} + 1/\kappa_2, 0), \tag{3.30b}$$

$$^{e2}\bar{A}_3 = \text{tran}(0, 0, m_{e2}), \tag{3.30c}$$

$$^{e2}\bar{A}_4 = \text{tran}(0, 0, m_{e2})\text{rot}(\bar{x}, \psi_{e2}), \tag{3.30d}$$

$$^{e2}\bar{A}_5 = \text{rot}(\bar{x}, -\sigma_{e2} + 180°), \tag{3.30e}$$

$$^{e2}\bar{A}_6 = {}^{e2}\bar{A}_4, \tag{3.30f}$$

$$^{e3}\bar{A}_7 = \text{rot}(\bar{x}, \sigma_{e3}), \tag{3.30g}$$

$$^{e3}\bar{A}_8 = \bar{I}_{4\times4}, \tag{3.30h}$$

$$^{e3}\bar{A}_9 = \text{tran}(-w_{e3}, 0, h_{e3})\text{rot}(\bar{z}, -\psi_{e3})\text{rot}(\bar{x}, -45°), \tag{3.30i}$$

$$^{e3}\bar{A}_{10} = \text{tran}(0, 0, h_{e3})\text{rot}(\bar{z}, \psi_{e3})\text{rot}(\bar{x}, -45°), \tag{3.30j}$$

$$^{e3}\bar{A}_{11} = {}^{e3}\bar{A}_7, \tag{3.30k}$$

$$^{e3}\bar{A}_{12} = {}^{e3}\bar{A}_8, \tag{3.30l}$$

$$^{e4}\bar{A}_{13} = \bar{I}_{4\times4}. \tag{3.30m}$$

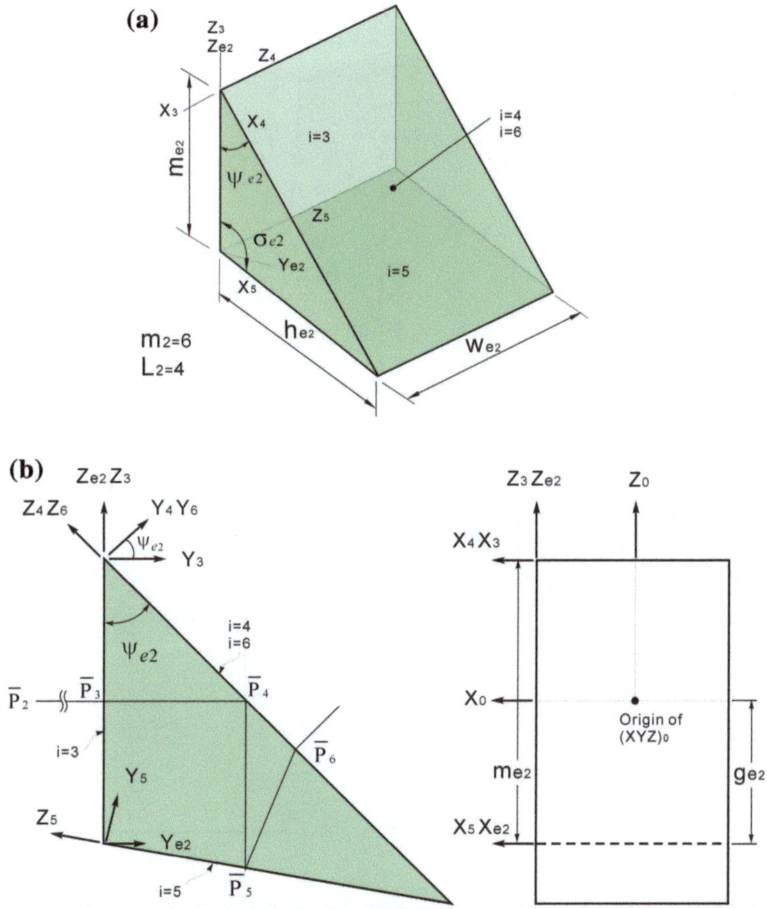


**Fig. 3.14** The element j = 2 of the system shown in Fig. 3.12 is a *right-angle* prism if $\sigma_{e2} = 90^{\circ}$. **a** Perspective view of element 2, **b** front and side views of element 2

*Example 3.11* As shown in Eqs. (2.8) and (2.34), one needs the pose matrices ${}^{0}\bar{A}_{i}$ (i = 1 to i = n = 13) for performing raytracing. Now they can be obtained by the following matrix concatenation:

$$\begin{aligned} {}^{0}\bar{A}_{1} &= {}^{0}\bar{A}_{e1}\,{}^{e1}\bar{A}_{1} \\ &= \text{tran}(t_{e1x}, v_{1} + 1/\kappa_{1}, t_{e1z})\,\text{rot}(\bar{z}, \omega_{e1z})\,\text{rot}(\bar{y}, \omega_{e1y})\,\text{rot}(\bar{x}, \omega_{e1x}), \end{aligned} \tag{3.31a}$$

$$\begin{aligned} {}^{0}\bar{A}_{2} &= {}^{0}\bar{A}_{e1}\,{}^{e1}\bar{A}_{2} \\ &= \text{tran}(t_{e1x}, v_{1} + 1/\kappa_{1}, t_{e1z})\,\text{rot}(\bar{z}, \omega_{e1z})\,\text{rot}(\bar{y}, \omega_{e1y})\,\text{rot}(\bar{x}, \omega_{e1x}) \\ &\quad \text{tran}(0, -1/\kappa_{1} + q_{e1} + 1/\kappa_{2}, 0), \end{aligned} \tag{3.31b}$$

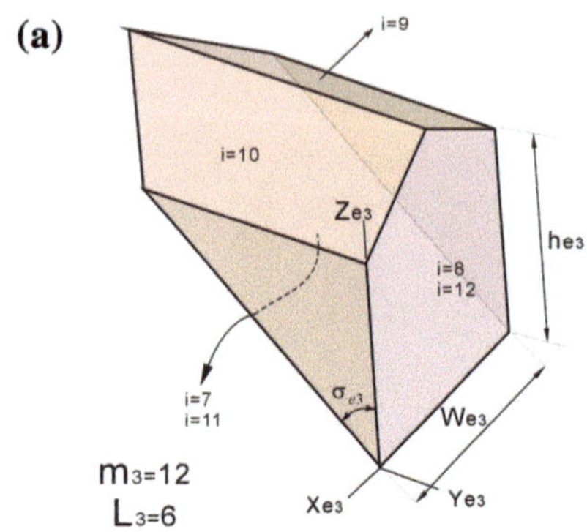


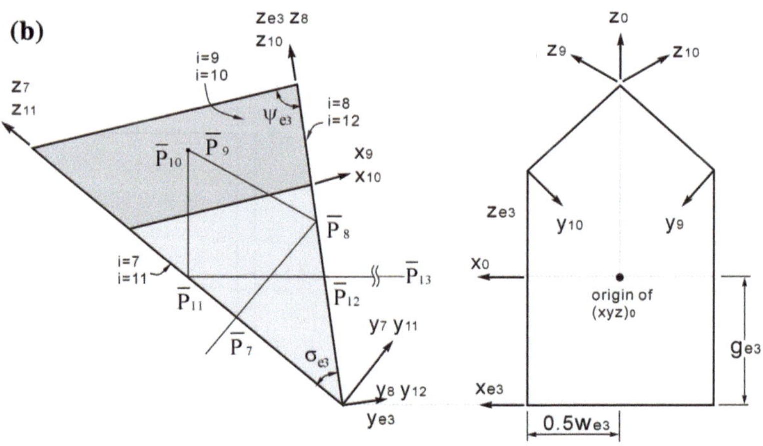


**Fig. 3.15** The element j = 3 of the system shown in Fig. 3.12 is a prism with a roof. **a** Perspective view of element 3 **b** Front and side views of element 3

**Fig. 3.16** The element j = 4 of the system shown in Fig. 3.12 is an image plane

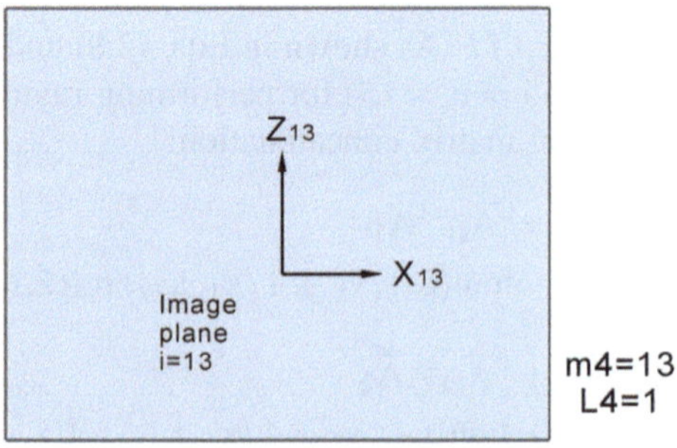

$$\begin{aligned}{}^{0}\overline{A}_{3} &= {}^{0}\overline{A}_{e2}{}^{e2}\overline{A}_{3}\\ &= \mathrm{tran}(w_{e2}/2, v_{1}+q_{e1}+v_{2}, -g_{e2})\,\mathrm{rot}(\bar{z}, \omega_{e2z})\,\mathrm{rot}(\bar{y}, \omega_{e2y})\,\mathrm{rot}(\bar{x}, \omega_{e2x})\\ &\quad \mathrm{tran}(0, 0, m_{e2}),\end{aligned} \tag{3.31c}$$

$$\begin{aligned}{}^{0}\overline{A}_{4} &= {}^{0}\overline{A}_{e2}{}^{e2}\overline{A}_{4}\\ &= \mathrm{tran}(w_{e2}/2, v_{1}+q_{e1}+v_{2}, -g_{e2})\,\mathrm{rot}(\bar{z}, \omega_{e2z})\,\mathrm{rot}(\bar{y}, \omega_{e2y})\,\mathrm{rot}(\bar{x}, \omega_{e2x})\\ &\quad \mathrm{tran}(0, 0, m_{e2})\,\mathrm{rot}(\bar{x}, \psi_{e2}),\end{aligned} \tag{3.31d}$$

$$\begin{aligned}{}^{0}\bar{A}_{5} &= {}^{0}\bar{A}_{e2}{}^{e2}\bar{A}_{5}\\ &= \mathrm{tran}(w_{e2}/2, v_{1}+q_{e1}+v_{2}, -g_{e2})\mathrm{rot}(\bar{z}, \omega_{e2z})\mathrm{rot}(\bar{y}, \omega_{e2y})\\ &\quad \mathrm{rot}(\bar{x}, \omega_{e2x})\mathrm{rot}(\bar{x}, -\sigma_{e2}+180^{\circ}),\end{aligned} \tag{3.31e}$$

$${}^{0}\bar{A}_{6} = {}^{0}\bar{A}_{4}, \tag{3.31f}$$

$$\begin{aligned}{}^{0}\bar{A}_{7} &= {}^{0}\bar{A}_{e3}{}^{e3}\bar{A}_{7}\\ &= \mathrm{tran}(w_{e3}/2, v_{1}+q_{e1}+v_{2}+h_{e2}+v_{3}, -g_{e3})\mathrm{rot}(\bar{z}, \omega_{e3z})\\ &\quad \mathrm{rot}(\bar{y}, \omega_{e3y})\mathrm{rot}(\bar{x}, \omega_{e3x})\mathrm{rot}(\bar{x}, \sigma_{e3}),\end{aligned} \tag{3.31g}$$

$$\begin{aligned}{}^{0}\bar{A}_{8} &= {}^{0}\bar{A}_{e3}{}^{e3}\bar{A}_{8}\\ &= \mathrm{tran}(w_{e3}/2, v_{1}+q_{e1}+v_{2}+h_{e2}+v_{3}, -g_{e3})\mathrm{rot}(\bar{z}, \omega_{e3z})\\ &\quad \mathrm{rot}\left(\bar{y}, \omega_{e3y}\right)\mathrm{rot}(\bar{x}, \omega_{e3x}),\end{aligned} \tag{3.31h}$$

$$\begin{aligned}{}^{0}\bar{A}_{9} &= {}^{0}\bar{A}_{e3}{}^{e3}\bar{A}_{9}\\ &= \mathrm{tran}(w_{e3}/2, v_{1}+q_{e1}+v_{2}+h_{e2}+v_{3}, -g_{e3})\mathrm{rot}(\bar{z}, \omega_{e3z})\mathrm{rot}(\bar{y}, \omega_{e3y})\mathrm{rot}(\bar{x}, \omega_{e3x})\\ &\quad \mathrm{tran}(-w_{e3}, 0, h_{e3})\mathrm{rot}(\bar{z}, -\psi_{e3})\mathrm{rot}(\bar{x}, -45^{\circ}),\end{aligned} \tag{3.31i}$$

$$\begin{aligned}{}^{0}\bar{A}_{10} &= {}^{0}\bar{A}_{e3}{}^{e3}\bar{A}_{10}\\ &= \mathrm{tran}(w_{e3}/2, v_{1}+q_{e1}+v_{2}+h_{e2}+v_{3}, -g_{e3})\mathrm{rot}(\bar{z}, \omega_{e3z})\mathrm{rot}(\bar{y}, \omega_{e3y})\mathrm{rot}(\bar{x}, \omega_{e3x})\\ &\quad \mathrm{tran}(0, 0, h_{e3})\mathrm{rot}(\bar{z}, \psi_{e3})\mathrm{rot}(\bar{x}, -45^{\circ}),\end{aligned} \tag{3.31j}$$

$${}^{0}\bar{A}_{11} = {}^{0}\bar{A}_{7}, \tag{3.31k}$$

$${}^{0}\bar{A}_{12} = {}^{0}\bar{A}_{8}, \tag{3.31l}$$

$$
\begin{aligned}
{}^{0}\bar{A}_{13} &= {}^{0}\bar{A}_{e4}{}^{e4}\bar{A}_{13} \\
&= \mathrm{tran}(t_{e4x}, v_1 + q_{e1} + v_2 + h_{e2} + v_3 + v_4, t_{e4z}) \\
&\quad \mathrm{rot}(\bar{z}, \omega_{e4z})\mathrm{rot}(\bar{y}, \omega_{e4y})\mathrm{rot}(\bar{x}, \omega_{e4x}).
\end{aligned} \tag{3.31m}
$$

Once having the desired pose matrix ${}^{0}\bar{A}_i$, one can apply the skew-ray tracing described in the Chap. 2 successively to trace rays traveling in this optical system to obtain the ray hits at the n[th] boundary surface, given as:

$$
\bar{P}_n = \begin{bmatrix} P_{nx} \\ P_{ny} \\ P_{nz} \\ 1 \end{bmatrix} = \begin{bmatrix} P_{n-1x} + \ell_{n-1x}\lambda_n \\ P_{n-1y} + \ell_{n-1y}\lambda_n \\ P_{n-1z} + \ell_{n-1z}\lambda_n \\ 1 \end{bmatrix}. \tag{3.31n}
$$

*Example 3.12* Referring to Fig. 3.13 and Eqs. (3.31a) and (3.31b), the variable vector $\bar{X}_{e1}$ of the 1[st] element in Fig. 3.12 has to include its six pose variables (i.e., $t_{e1x}$, $t_{e1y} = v_1 + 1/\kappa_1$, $t_{e1z}$, $\omega_{e1x}$, $\omega_{e1y}$ and $\omega_{e1z}$), the refractive index $\xi_{air}$ of air, the refractive index $\xi_{e1}$ of the element, the two curvatures, $\kappa_1$ and $\kappa_2$, of two spherical boundary surfaces, and element thickness $q_{e1}$.

$$
\bar{X}_{e1} = \begin{bmatrix} t_{e1x} & v_1 + 1/\kappa_1 & t_{e1z} & \omega_{e1x} & \omega_{e1y} & \omega_{e1z} & \xi_{e1} & \xi_{air} & \kappa_1 & q_{e1} & \kappa_2 \end{bmatrix}^T. \tag{3.32a}
$$

Referring to Fig. 3.14 and Eq. (3.31c) to Eq. (3.31f), the variable vector $\bar{X}_{e2}$ of the 2nd element in Fig. 3.12 has to contain its six pose variables, the refractive index $\xi_{air}$ of air, the element refractive index $\xi_{e2}$, and the characteristic dimensions $h_{e2}$, $w_{e2}$, $g_{e2}$, $m_{e2}$, $\psi_{e2}$, and $\sigma_{e2}$.

$$
\begin{aligned}
\bar{X}_{e2} = \big[ & w_{e2}/2 \quad v_1 + q_{e1} + v_2 \quad -g_{e2} \quad \omega_{e2x} \quad \omega_{e2y} \quad \omega_{e2z} \quad \xi_{e2} \quad \xi_{glue} \quad \xi_{air} \quad h_{e2} \quad w_{e2} \\
& g_{e2} \quad m_{e2} \quad \psi_{e2} \quad \sigma_{e2} \big]^T.
\end{aligned} \tag{3.32b}
$$

Referring to Fig. 3.15 and Eq. (3.31g) to Eq. (3.31l), the variable vector $\bar{X}_{e3}$ of the 3[rd] element in Fig. 3.12 should comprise its six pose variables with $v_3 = \sqrt{2}\mathrm{gap}_{glue}$, the refractive index $\xi_{air}$, the refractive index $\xi_{glue}$ of glue, the element refractive index $\xi_{e3}$, and the characteristic dimensions $h_{e3}$, $\psi_{e3}$, and $\sigma_{e3}$.

$$
\begin{aligned}
\bar{X}_{e3} = \Big[ & w_{e3}/2 \quad v_1 + q_{e1} + v_2 + h_{e2} + \sqrt{2}\,\mathrm{gap}_{glue} \quad -g_{e3} \quad \psi_{e2} - \sigma_{e3} \quad \omega_{e3y} \quad \omega_{e3z} \quad \xi_{air} \quad \xi_{e3} \\
& \xi_{glue} \quad h_{e3} \quad w_{e3} \quad g_{e2} \quad \psi_{e3} \quad \sigma_{e3} \Big]^T.
\end{aligned} \tag{3.32c}
$$

The variable vector $\bar{X}_{e4}$of the image plane shown in Fig. 3.12 (or see Fig. 3.16 and Eq. (3.31m)) has to include its six pose variables and the refractive index $\xi_{air}$ of air:

**Table 3.4** The boundary pose variables of the system shown in Fig. 3.12

| | tix | tiy | tiz | ωix | ωiy | ωiz |
|---|---|---|---|---|---|---|
| i = 1 | 0.00000 | 48.2219 | 0.00000 | 0° | 0° | 0° |
| i = 2 | 0.00000 | 41.0857 | 0.00000 | 0° | 0° | 0° |
| i = 3 | 5.00000 | 45.00000 | 5.00000 | 0° | 0° | 0° |
| i = 4 | 5.00000 | 45.00000 | 5.00000 | 45° | 0° | 0° |
| i = 5 | 5.00000 | 45.00000 | −5.00000 | 67.5° | 0° | 0° |
| i = 6 | 5.00000 | 45.00000 | 5.00000 | 45° | 0° | 0° |
| i = 7 | 5.00000 | 56.83848 | −5.00000 | 45° | 0° | 0° |
| i = 8 | 5.00000 | 56.83848 | −5.00000 | 0° | 0° | 0° |
| i = 9 | −5.00000 | 56.83848 | 9.14214 | 145° | 0° | 0° |
| i = 10 | 5.00000 | 56.83848 | 9.14214 | 145° | 0° | 0° |
| i = 11 | 5.00000 | 56.83848 | −5.00000 | 45° | 0° | 0° |
| i = 12 | 5.00000 | 56.83848 | 15.00000 | 0° | 0° | 0° |
| i = 13 | 0.00000 | 106.8385 | 0 | 0° | 0° | 0° |

$$\overline{X}_{e4} = \begin{bmatrix} t_{e4x} & v_1 + q_{e1} + v_2 + h_{e2} + \sqrt{2}gap_{glue} + v_4 & t_{e4z} & \omega_{e4x} & \omega_{e4y} & \omega_{e4z} & \xi_{air} \end{bmatrix}^T. \tag{3.32d}$$

*Example 3.13* One can find that some variables (e.g., $\xi_{air}$) appear repeatedly in $\overline{X}_{ej}$ (j = 1 to j = 4) in Example 3.12. We also can set $w_{e3} = w_{e2}$ and $g_{e3} = g_{e2}$ for simplifying the geometry of this system. After deleting these repeated variables, one obtains the following variable vector $\overline{X}_{sys}$ as system variable of the system in Fig. 3.12:

$$\begin{aligned} \overline{X}_{sys} = [P_{0x} \quad & P_{0y} \quad P_{0z} \quad \alpha_0 \quad \beta_0 \\ & t_{e1x} \quad v_1 \quad t_{e1z} \quad \omega_{e1x} \quad \omega_{e1y} \quad \omega_{e1z} \quad \xi_{air} \quad \xi_{e1} \quad q_{e1} \quad \kappa_1 \quad \kappa_2 \\ & v_2 \quad \omega_{e2x} \quad \omega_{e2y} \quad \omega_{e2z} \quad \xi_{glue} \quad \xi_{e2} \quad h_{e2} \quad w_{e2} \quad g_{e2} \quad m_{e2} \quad \psi_{e2} \quad \sigma_{e2}. \\ & \omega_{e3y} \quad \omega_{e3z} \quad \xi_{e3} \quad gap_{glue} \quad h_{e3} \quad \psi_{e3} \quad \sigma_{e3} \\ & t_{e4x} \quad v_4 \quad t_{e4z} \quad \omega_{e4x} \quad \omega_{e4y} \quad \omega_{e4z}]^T. \end{aligned} \tag{3.33}$$

*Example 3.14* Determine the six pose variables ${}^0\overline{A}_i$ (i = 1 to i = 13) for the system in Fig. 3.12 by using Eqs. (1.34, 1.35, .... ,1.39), when $\kappa_1 = 1/38.2219$, $\kappa_2 = -1/56.0587$, $q_{e1} = 5$, $h_{e2} = m_{e2} = w_{e2} = w_{e3} = 10$, $g_{e2} = 5$, $\psi_{e2} = 45°$, $\sigma_{e2} = 112.5°$, $gap_{glue} = 1.3$, $h_{e3} = 10.\sqrt{2}$, $\psi_{e3} = 90°$, $\sigma_{e3} = 45°$, $v_1 = 10$, $v_2 = 30$, $v_4 = 50$, $\xi_{air} = \xi_{glue} = 1$, $\xi_{e1} = 1.5$, $\xi_{e2} = \xi_{e3} = 1.3$, $t_{e1x} = t_{e1z} = \omega_{e1x} = \omega_{e1y} = \omega_{e1z} = \omega_{e2x} = \omega_{e2y} = \omega_{e2z} = \omega_{e3x} = \omega_{e3y} = \omega_{e3z} = t_{e4x} = t_{e4z} = \omega_{e4x} = \omega_{e4y} = \omega_{e4z} = 0$

The results are listed in Table 3.4.

There are many kinds of image quality metrics. They can be categorized into two distinct groups: geometric image quality metrics and diffractive image quality metrics. Geometric image quality metrics (examples are spot diagram, point spread function (PSF), and modulation transfer function (MTF)) can either be computed by ray-counting method or differential method. The former method is discussed in the following three sections, while the latter method will be addressed from Chap. 5–Chap. 7.

## 3.4 Evaluating the Spot Diagram by Raytracing

The spot diagram at each image is measured by two characteristic features: the spot position and the spot size [1–3]. The spot position is measured by the centroid $\begin{bmatrix} x_{n/centroid} & y_{n/centroid} & z_{n/centroid} & 1 \end{bmatrix}^T$. The spot size is measured by the root-mean-square (rms) value of the radius from the centroid of the rays at the spot. Prior to calculating the rms value of the radius, it is first necessary to determine the centroid of the image spot by averaging the available data that are obtained from the traces of a finite number (say, G) of rays distributed in the domain of $(\alpha_0, \beta_0)$ in some regular way, i.e.,

$$\begin{bmatrix} x_{n/centroid} & y_{n/centroid} & z_{n/centroid} & 1 \end{bmatrix}^T = \begin{bmatrix} \sum_1^G {}^nP_{nx} \Big/ G & 0 & \sum_1^G {}^nP_{nz} \Big/ G & 1 \end{bmatrix}^T, \tag{3.34}$$

where ${}^n\bar{P}_n = \begin{bmatrix} {}^nP_{nx} & 0 & {}^nP_{nz} & 1 \end{bmatrix}^T$ is the coordinates of the incidence point of a general ray on the image plane by ${}^n\bar{P}_n = {}^n\bar{A}_0\bar{P}_n = ({}^0\bar{A}_n)^{-1}\bar{P}_n$. Minimizing the deviations of the spot centroid from their ideal positions will reduce distortion. Minimizing the deviations of the colored spot centroid from each other at each image spot will reduce the lateral chromatism.

The rms value of the radius (denoted as rms) of the spot diagram on the image plane is calculated from

$$\begin{aligned} rms^2 &= \frac{1}{G}\sum_1^G \left[ ({}^nP_{nx} - x_{n/centroid})^2 + ({}^nP_{nz} - z_{n/centroid})^2 \right] \\ &= \frac{1}{G}\sum_1^G \left[ ({}^nP_{nx})^2 + ({}^nP_{nz})^2 \right] - \left( x^2_{n/centroid} + z^2_{n/centroid} \right). \end{aligned} \tag{3.35}$$

Minimizing the image spot sizes will reduce the following aberrations: spherical; coma; astigmatism; field curvature; and longitudinal color.

*Example 3.15* The centroid and rms radius of the source point $\bar{P}_0 = \begin{bmatrix} 0 & -507 & 170 & 1 \end{bmatrix}^T$ in Fig. 2.1 are $\begin{bmatrix} x_{11/centroid} & y_{11/centroid} & z_{11/centroid} & 1 \end{bmatrix}^T = \begin{bmatrix} 0 & 0 & -32.59679 & 1 \end{bmatrix}^T$ and 0.11571 mm, respectively, when 27860 rays are traced. Contrastingly, the centroid and rms radius are $\begin{bmatrix} 0 & 0 & -32.57903 & 1 \end{bmatrix}^T$ and 0.135 mm, respectively, if 296 rays are traced. It is observed that the accuracy of the centroid and the rms radius depends on the number of rays traced.

## 3.5 Determination of PSF by Raytracing

The PSF describes the impulse response of an optical system to a source point. Hereafter we use $B(x_n, z_n)$ to denote the distribution of PSF, since it is function of in-plane coordinates (i.e., $x_n$ and $z_n$) of image plane. It also represents the distribution of the ray density of spot diagrams. It is important to investigate the PSF of an optical system to gather all the available information, since it is the basic "brick" on which an image will be "built". However, real spread functions are rarely (if ever) represented by ordinary analytical functions in the literature (p. 372 in [4]). Therefore, Smith (p. 372 in [4]) determined the PSF of optical systems by counting the rays hitting the grids on the image plane (referred to hereafter as the ray-counting method, see Fig. 3.17). Again, the system in Fig. 2.1 is used as the illustrative example. Table 3.5 presents the ray density on the image plane by using $1/30 \times 1/30$ mmgrids when 4933 rays (uniformly distributed over the domain of $(\alpha_0, \beta_0)$) of the on-axis source point $\overline{P}_0 = \begin{bmatrix} 0 & -507 & 0 & 1 \end{bmatrix}^T$ are traced. Table 3.6 and Fig. 3.18 give the variation of normalized PSF by dividing the number of the rays in each grid by 291, the maximum ray density in Table 3.5. Figure 3.19 shows the normalized PSF of the same source point by using $1/200 \times 1/200$ mm grids on which 122397 rays are traced. Fig. 3.18 is overlaid in Fig. 3.19 for comparison purpose. Note that in this example, the PSF is an axis-symmetrical function since the system is an axis-symmetrical system and the source point lies on the optical axis. Therefore, these two figures, Figs. 3.18 and 3.19, only present cross-sectional views of the PSF distribution rather than the entire PSF distribution. Comparing these two figures, it is clear that the accuracy of the ray-counting method depends significantly on the number of traced rays and the size of the grids used to mesh the image plane. Table 3.7 illustrates the ray density of an off-axis source point $\overline{P}_0 = \begin{bmatrix} 0 & -507 & 170 & 1 \end{bmatrix}^T$ by tracing 27848 rays and using $0.05 \times 0.05$ mm grids. One can obtain the normalized PSF from Table 3.7 by taking the quotient of the number of rays hitting each grid divided by 1145, the maximum ray density in Table 3.7. It is observed from this table that the energy spread in sagittal direction is more intense than it is in meridional direction, indicating the existence of coma and astigmatism aberrations. Note that it is difficult to obtain an accurate estimation of the PSF distribution using the ray-counting method if the PSF distribution contains regions of rapid change. In addition, the ray-counting method requires the tracing of many rays in order to estimate the PSF over the entire image plane.

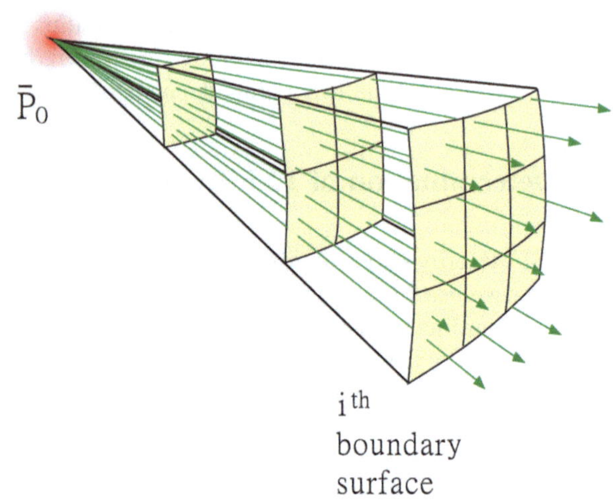


**Fig. 3.17** Ray-counting method determines the PSF of an optical system by counting the rays hitting the grids

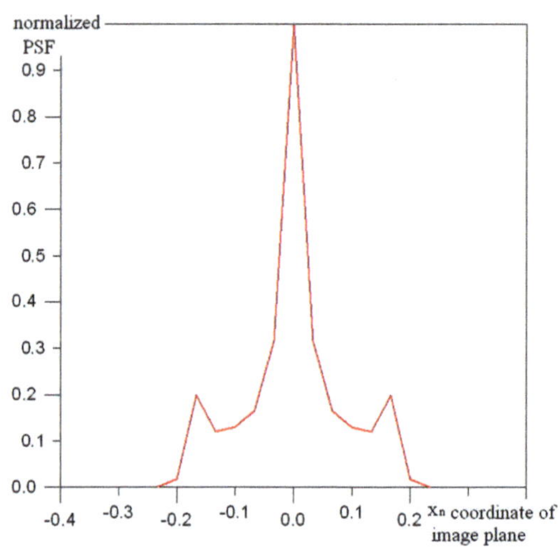


**Fig. 3.18** The cross-sectional view of the PSF distribution of Table 3.6

## 3.6 Estimation of MTF by Raytracing

The most accepted criterion for evaluating the imaging performance of an optical system is the MTF [5–14]. The MTF, which is both geometric and diffraction metrics, can be defined as the ratio of the contrast of the image to the contrast of the object. The MTF is also usually computed using the ray-counting method [4] by counting the number of rays hitting in each grid. In the following we first review this method using an on-axis source point as an illustrative example to

**Table 3.5** Ray density of the on-axis source point $\bar{P}_0 = [0 \quad -507 \quad 0 \quad 1]^T$ from ray-counting method on the image plane of the system shown in Fig. 2.1 with $1/30 \times 1/30$ mm grids and 4933 rays

$Z_{11}$

| | | | | | | | | | | | | | | |
|---|---|---|---|---|---|---|---|---|---|---|---|---|---|---|
| 0 | 0 | 0 | 0 | 0 | 0 | 0 | 0 | 0 | 0 | 0 | 0 | 0 | 0 | 0 |
| 0 | 0 | 0 | 0 | 0 | 0 | 0 | 4 | 0 | 0 | 0 | 0 | 0 | 0 | 0 |
| 0 | 0 | 0 | 0 | 21 | 48 | 59 | 53 | 59 | 48 | 21 | 0 | 0 | 0 | 0 |
| 0 | 0 | 0 | 33 | 46 | 37 | 34 | 37 | 34 | 37 | 46 | 33 | 0 | 0 | 0 |
| 0 | 0 | 18 | 50 | 37 | 35 | 35 | 38 | 35 | 35 | 37 | 50 | 18 | 0 | 0 |
| 0 | 0 | 45 | 35 | 34 | 41 | 44 | 48 | 44 | 41 | 34 | 35 | 45 | 0 | 0 |
| 0 | 0 | 57 | 35 | 38 | 42 | 70 | 93 | 70 | 42 | 38 | 35 | 57 | 0 | 0 |
| 0 | 5 | 58 | 35 | 38 | 48 | 92 | 291 | 92 | 48 | 38 | 35 | 58 | 5 | 0 |
| 0 | 0 | 52 | 34 | 39 | 50 | 68 | 100 | 68 | 50 | 39 | 34 | 52 | 0 | 0 |
| 0 | 0 | 43 | 39 | 34 | 38 | 43 | 46 | 43 | 38 | 34 | 39 | 43 | 0 | 0 |
| 0 | 0 | 18 | 46 | 33 | 34 | 35 | 35 | 35 | 34 | 33 | 46 | 18 | 0 | 0 |
| 0 | 0 | 0 | 31 | 50 | 33 | 33 | 34 | 33 | 33 | 50 | 31 | 0 | 0 | 0 |
| 0 | 0 | 0 | 0 | 20 | 49 | 56 | 53 | 56 | 49 | 20 | 0 | 0 | 0 | 0 |
| 0 | 0 | 0 | 0 | 0 | 0 | 0 | 4 | 0 | 0 | 0 | 0 | 0 | 0 | 0 |
| 0 | 0 | 0 | 0 | 0 | 0 | 0 | 0 | 0 | 0 | 0 | 0 | 0 | 0 | 0 |

$x_{11}$

$\Delta x_{11}$

understand it. Consider an on-axis source point $\bar{P}_0 = [0 \quad P_{0y} \quad 0 \quad 1]^T$ consisting of alternating light and dark bands with a luminous intensity (referred as Object Brightness Distribution Function (OBDF)) that varies according to a cosine function. The axis of OBDF can be either parallel to the direction leading away from the optical axis (sagittal) or perpendicular to this direction (meridional) (see Fig. 3.20). The OBDF in the sagittal direction can be expressed as

$$I(x_0) = b_0 + b_1 C(2\pi v x_0), \tag{3.36}$$

where v is the frequency of the brightness variation in cycles per unit length. $I_{max} = b_0 + b_1$ is the maximum brightness, $I_{min} = b_0 - b_1$ is the minimum brightness, and $x_0$ is the coordinate of the world coordinate frame $(xyz)_0$. The modulation of this OBDF is

$$M_0 = \frac{I_{max} - I_{min}}{I_{max} + I_{min}} = \frac{(b_0 + b_1) - (b_0 - b_1)}{(b_0 + b_1) + (b_0 - b_1)} = \frac{b_1}{b_0}. \tag{3.37}$$

To compute the MTF, one still needs line spread function $L(x_n)$, which can be computed by integration point spread function $B(x_n, z_n)$ along $z_n$ to give $L(x_n) = \int B(x_n, z_n) dz_n$. When this OBDF is imaged by an optical system, the energy intensity at $x_n$ on the image plane can be determined by the convolution of $I(x_0)$ with the line spread function $L(x_n)$:

**Table 3.6** The normalized PSF computed from Table 3.5

$Z_{11}$ (↑)

| | | | | | | | | | | | | | | |
|---|---|---|---|---|---|---|---|---|---|---|---|---|---|---|
| 0 | 0 | 0 | 0 | 0 | 0 | 0 | 0 | 0 | 0 | 0 | 0 | 0 | 0 | 0 |
| 0 | 0 | 0 | 0 | 0 | 0 | 0 | 0.01 | 0 | 0 | 0 | 0 | 0 | 0 | 0 |
| 0 | 0 | 0 | 0 | 0.07 | 0.16 | 0.02 | 0.18 | 0.2 | 0.16 | 0.07 | 0 | 0 | 0 | 0 |
| 0 | 0 | 0 | 0.11 | 0.16 | 0.13 | 0.12 | 0.13 | 0.12 | 0.13 | 0.16 | 0 | 0 | 0 | 0 |
| 0 | 0 | 0.06 | 0.17 | 0.13 | 0.02 | 0.12 | 0.13 | 0.12 | 0.12 | 0.13 | 0.17 | 0.06 | 0 | 0 |
| 0 | 0 | 0.15 | 0.12 | 0.12 | 0.14 | 0.15 | 0.16 | 0.15 | 0.14 | 0.12 | 0.12 | 0.15 | 0 | 0 |
| 0 | 0 | 0.20 | 0.12 | 0.13 | 0.14 | 0.24 | 0.32 | 0.24 | 0.14 | 0.13 | 0.12 | 0.20 | 0 | 0 |
| 0 | 0.02 | 0.20 | 0.12 | 0.13 | 0.16 | 0.32 | 1.00 | 0.32 | 0.16 | 0.13 | 0.12 | 0.20 | 0.02 | 0 |
| 0 | 0 | 0.18 | 0.12 | 0.13 | 0.17 | 0.23 | 0.34 | 0.23 | 0.17 | 0.13 | 0.12 | 0.18 | 0 | 0 |
| 0 | 0 | 0.15 | 0.13 | 0.12 | 0.13 | 0.15 | 0.16 | 0.15 | 0.13 | 0.12 | 0.13 | 0.15 | 0 | 0 |
| 0 | 0 | 0.06 | 0.16 | 0.11 | 0.12 | 0.12 | 0.12 | 0.12 | 0.12 | 0.11 | 0.16 | 0.06 | 0 | 0 |
| 0 | 0 | 0 | 0.11 | 0.17 | 0.11 | 0.11 | 0.12 | 0.11 | 0.11 | 0.17 | 0.11 | 0 | 0 | 0 |
| 0 | 0 | 0 | 0 | 0.07 | 0.17 | 0.19 | 0.18 | 0.19 | 0.17 | 0.07 | 0 | 0 | 0 | 0 |
| 0 | 0 | 0 | 0 | 0 | 0 | 0 | 0.01 | 0 | 0 | 0 | 0 | 0 | 0 | 0 |
| 0 | 0 | 0 | 0 | 0 | 0 | 0 | 0 | 0 | 0 | 0 | 0 | 0 | 0 | 0 |

$X_{11}$ (→)

$$
\begin{aligned}
I(x_n) &= \int L(\delta) I(x_n - \delta) d\delta \\
&= b_0 \int L(\delta) d\delta \, + \, b_1 C(2\pi v x_n) \int L(\delta) C(2\pi v \delta) d\delta \\
&\quad + b_1 S(2\pi v x_n) \int L(\delta) S(2\pi v \delta) d\delta.
\end{aligned}
\tag{3.38}
$$

One has to note that the PSF is usually normalized so that $\int L(\delta) d\delta = 1$, yielding that

$$
\begin{aligned}
I(x_n) &= b_0 + b_1 [\, L_c(v) \; C(2\pi v x_n) + L_s(v) \; S(2\pi v x_n)] \\
&= b_0 + b_1 G(v) C(2\pi v x_n - \varpi),
\end{aligned}
\tag{3.39}
$$

where

$$L_c(v) = \int L(\delta) \, C(2\pi v \delta) \, d\delta = \int L(x_n) \, C(2\pi v x_n) \, dx_n, \tag{3.40}$$

$$L_s(v) = \int L(\delta) \, S(2\pi v \delta) \, d\delta = \int L(x_n) \, S(2\pi v x_n) \, dx_n, \tag{3.41}$$

$$G(v) = \sqrt{L_c^2(v) + L_s^2(v)}, \tag{3.42}$$

**Table 3.7** Ray density of the off-axis source point $\overline{P}_0 = [0 \quad -507 \quad 170 \quad 1]^T$ from ray-counting method on the image plane with a 0.05 × 0.05 mm grids and 27848 rays for the system in Fig. 2.1. Coma and astigmatism aberration can be observed from this table

$Z_{11}$

| | | | | | | | | | | | | | | |
|---|---|---|---|---|---|---|---|---|---|---|---|---|---|---|
| 0 | 0 | 0 | 0 | 0 | 0 | 0 | 5 | 0 | 0 | 0 | 0 | 0 | 0 | 0 |
| 0 | 0 | 0 | 0 | 0 | 0 | 4 | 29 | 4 | 0 | 0 | 0 | 0 | 0 | 0 |
| 0 | 0 | 0 | 0 | 0 | 16 | 130 | 136 | 130 | 16 | 0 | 0 | 0 | 0 | 0 |
| 0 | 0 | 0 | 0 | 0 | 104 | 160 | 162 | 160 | 104 | 0 | 0 | 0 | 0 | 0 |
| 0 | 0 | 0 | 0 | 0 | 159 | 186 | 165 | 186 | 159 | 0 | 0 | 0 | 0 | 0 |
| 0 | 0 | 0 | 0 | 0 | 190 | 204 | 188 | 204 | 190 | 0 | 0 | 0 | 0 | 0 |
| 0 | 0 | 0 | 0 | 0 | 196 | 243 | 205 | 243 | 196 | 0 | 0 | 0 | 0 | 0 |
| 0 | 0 | 0 | 0 | 0 | 99 | 357 | 233 | 357 | 99 | 0 | 0 | 0 | 0 | 0 |
| 0 | 0 | 0 | 0 | 0 | 0 | 476 | 254 | 476 | 0 | 0 | 0 | 0 | 0 | 0 |
| 0 | 0 | 0 | 0 | 0 | 0 | 467 | 319 | 467 | 0 | 0 | 0 | 0 | 0 | 0 |
| 0 | 0 | 0 | 0 | 0 | 0 | 354 | 572 | 354 | 0 | 0 | 0 | 0 | 0 | 0 |
| 0 | 0 | 0 | 0 | 0 | 0 | 122 | 1067 | 122 | 0 | 0 | 0 | 0 | 0 | 0 |
| 0 | 0 | 0 | 0 | 0 | 0 | 83 | 1145 | 83 | 0 | 0 | 0 | 0 | 0 | 0 |
| 0 | 0 | 0 | 0 | 0 | 32 | 133 | 1008 | 133 | 32 | 0 | 0 | 0 | 0 | 0 |
| 0 | 0 | 0 | 0 | 7 | 87 | 154 | 836 | 154 | 87 | 7 | 0 | 0 | 0 | 0 |
| 0 | 0 | 0 | 1 | 52 | 100 | 172 | 690 | 172 | 100 | 52 | 1 | 0 | 0 | 0 |
| 0 | 0 | 0 | 27 | 78 | 106 | 185 | 542 | 185 | 106 | 78 | 27 | 0 | 0 | 0 |
| 0 | 0 | 9 | 64 | 81 | 111 | 191 | 422 | 191 | 111 | 81 | 64 | 9 | 0 | 0 |
| 0 | 2 | 49 | 65 | 84 | 113 | 189 | 322 | 189 | 113 | 84 | 65 | 49 | 2 | 0 |
| 0 | 31 | 56 | 70 | 83 | 108 | 179 | 273 | 179 | 108 | 83 | 70 | 56 | 31 | 0 |
| 0 | 28 | 59 | 67 | 82 | 112 | 162 | 208 | 162 | 112 | 82 | 67 | 59 | 28 | 0 |
| 0 | 19 | 58 | 69 | 79 | 107 | 151 | 189 | 151 | 107 | 79 | 69 | 58 | 19 | 0 |
| 0 | 9 | 51 | 66 | 81 | 104 | 133 | 152 | 133 | 104 | 81 | 66 | 51 | 9 | 0 |
| 0 | 0 | 37 | 64 | 79 | 92 | 122 | 133 | 122 | 92 | 79 | 64 | 37 | 0 | 0 |
| 0 | 0 | 20 | 61 | 75 | 89 | 111 | 121 | 111 | 89 | 75 | 61 | 20 | 0 | 0 |
| 0 | 0 | 0 | 42 | 76 | 82 | 101 | 107 | 101 | 82 | 76 | 42 | 0 | 0 | 0 |
| 0 | 0 | 0 | 10 | 50 | 81 | 91 | 95 | 91 | 81 | 50 | 10 | 0 | 0 | 0 |
| 0 | 0 | 0 | 0 | 13 | 54 | 79 | 81 | 79 | 54 | 13 | 0 | 0 | 0 | 0 |
| 0 | 0 | 0 | 0 | 0 | 7 | 29 | 47 | 29 | 7 | 0 | 0 | 0 | 0 | 0 |
| 0 | 0 | 0 | 0 | 0 | 0 | 0 | 0 | 0 | 0 | 0 | 0 | 0 | 0 | 0 |

$$\varpi(v) = \text{atan2}(L_s(v), L_c(v)), \tag{3.43}$$

where $\varpi(v)$ is the phase difference between $I(x_0)$ defined in Eq. (3.36) and the energy distribution function $I(x_n)$ given in Eq. (3.38). Note that a phase shift of 180°corresponds to a reversal of contrast, i.e. the image pattern is light where it should be dark and vice versa.

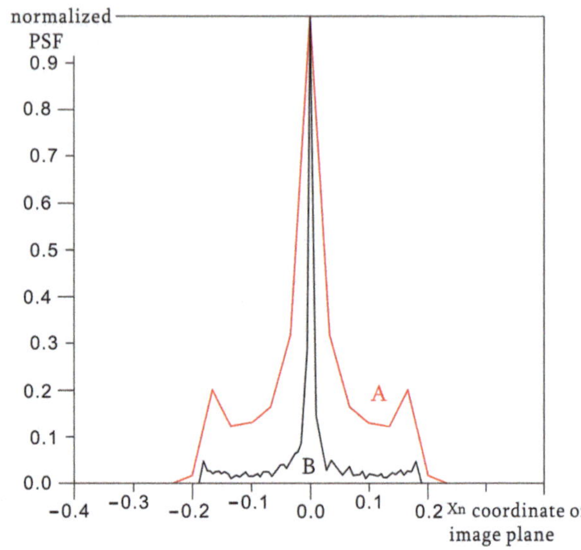


**Fig. 3.19** The cross-sectional view of the PSF distribution of $\overline{P}_0=[0 \quad -507 \quad 0 \quad 1]^T$ by tracing different numbers of rays and using different grid sizes for the system in Fig. 2.1 (A: $1/30 \times 1/30$ mm grids and 4933 rays; B: $1/200 \times 1/200$ mm grids and 122397 rays are traced)

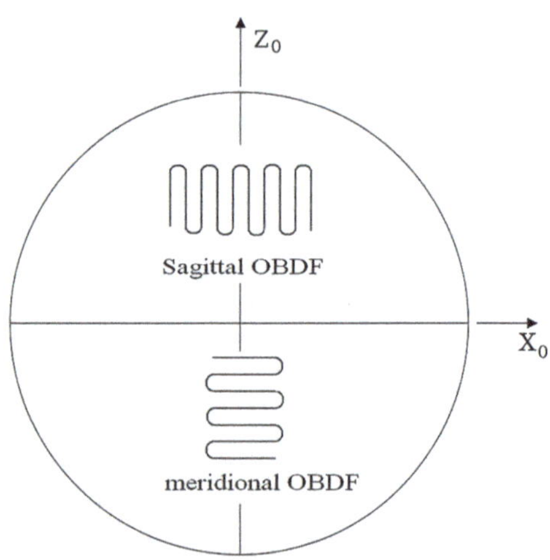


**Fig. 3.20** Sagittal and meridional directions of OBDF

The modulation of the intensity in the image is given by $M_n = b_1/b_0 G(\nu) = M_0 G(\nu)$. Therefore, $G(\nu)$ is the MTF, i.e.

$$\mathrm{MTF}(\nu) = M_n/M_0 = G(\nu) = \sqrt{L_c^2(\nu) + L_s^2(\nu)}. \tag{3.44}$$

Smith determined the line spread function $L(\delta)$by counting the rays that hit the image plane between the lines $\Delta x_n$ apart (e.g., Table 3.5). This yields the result that the summation of his line spread function is not equal to one (i.e.,

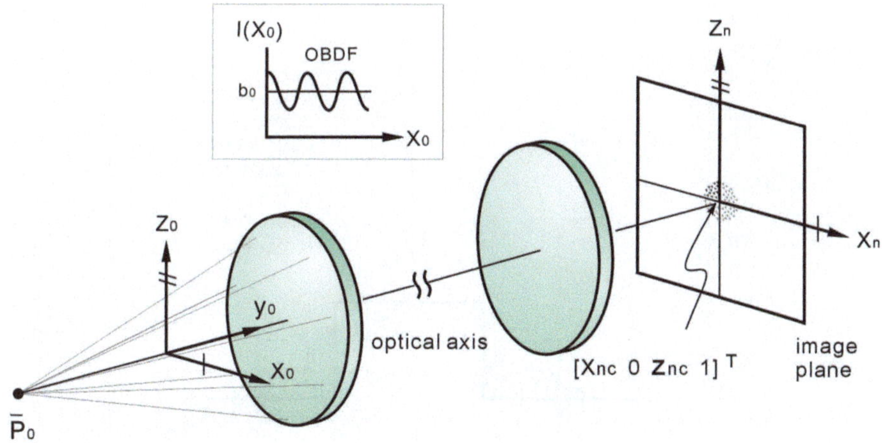


**Fig. 3.21** Radiation of light rays from an on-axis source point $\bar{P}_0$ onto the image plane in an axis–symmetrical optical system. Note that the MTF and line spread function are unchanged when the direction of the OBDF is changed

$\int L(x_n)dx_n \cong \sum L(x_n)\Delta x_n \neq 1$). In order to obtain the MTF, he divided Eq. (3.38) by $\sum L(x_n)\Delta x_n$, resulting in the different energy distribution equation at $x_n$:

$$\begin{aligned} I^*(x_n) &\cong \frac{I(x_n)}{\sum L(x_n)\Delta x_n} = b_0 + b_1\left[L_c^*(\nu)\, C(2\pi\nu x_n) + L_s^*(\nu)\, S(2\pi\nu x_n)\right] \\ &= b_0 + b_1 G^*(\nu) C(2\pi\nu x_n - \varpi^*). \end{aligned} \tag{3.45}$$

Here

$$L_c^*(\nu) \cong \sum L(x_n)\, C(2\pi\nu x_n)\, \Delta x_n \Big/ \sum L(x_n)\Delta x_n, \tag{3.46}$$

$$L_s^*(\nu) \cong \sum L(x_n)\, S(2\pi\nu x_n)\, \Delta x_n \Big/ \sum L(x_n)\Delta x_n, \tag{3.47}$$

$$G^*(\nu) = \sqrt{(L_c^*(\nu))^2 + (L_s^*(\nu))^2}, \tag{3.48}$$

$$\varpi^*(\nu) = \mathrm{atan2}(L_s^*(\nu), L_c^*(\nu)). \tag{3.49}$$

One has to note that $L_s^*(\nu) = 0$ is always true for an on-axis source point in an axis-symmetrical system, since $L_s^*(\nu)$ is an odd function with respect to $x_n$. His modulation in the image is given by $M_n^* = b_1/b_0 G^*(\nu) = M_0 G^*(\nu)$. Therefore, $G * (\nu)$ is his MTF:

$$\mathrm{MTF}^*(\nu) = M_n^*/M_0^* = G^*(\nu) = \sqrt{(L_c^*(\nu))^2 + (L_s^*(\nu))^2}. \tag{3.50}$$

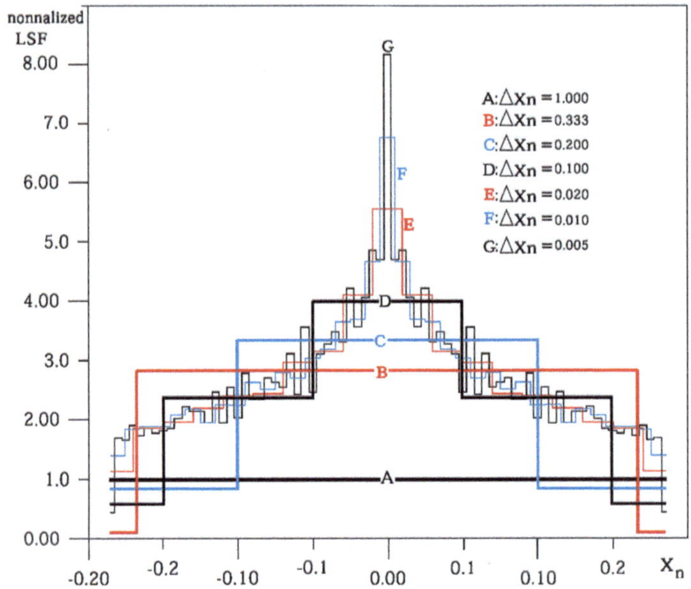


**Fig. 3.22** Variation of sagittal line spread function of $\overline{P}_0 = [0 \quad -507 \quad 0 \quad 1]^T$ under different grid sizes meshed on the image plane when 4933 rays are traced for the system of Fig. 2.1. Notably, the grid size significantly affects the line spread function resolution, which in turn affects the MTF of the ray-counting method. The proposed irradiance method of Chap. 6 does not need line spread function

The MTF curve can therefore be obtained by taking the frequency $\nu$ as the abscissa and MTF $*$ $(\nu)$ as the ordinate. The value of the MTF lies in the interval [0,1] and generally decreases with an increasing frequency $\nu$.

*Example 3.16* Figure 3.22 shows the variation of the line spread function of $\overline{P}_0 = [0 \quad -507 \quad 0 \quad 1]^T$ from ray-counting method with the change of the grid size meshed on the image plane of the system shown in Fig. 2.1. It shows the normalized line spread function, which is the quotient of the line spread function $L(x_n)$ divided by $\sum L(x_n)\Delta x_n$ (i.e. $L(x_n)/\sum L(x_n)\Delta x_n$). If the mesh on the image plane of the ray-counting method is large (e.g. curves A, B, C and D of Fig. 3.22), the line spread function resolution may be so poor that its value approaches unity (curve A is exactly the same as the aberration-free case), yielding over-estimated MTFs.

*Example 3.17* Figure 3.23 presents the MTF when the source point $\overline{P}_0 = [0 \quad -507 \quad 0 \quad 1]^T$ of the system shown in Fig. 2.1 is on-axis. Note that the sagittal and meridional MTFs of an on-axis source point are the same, and the phase shift $\varpi = 0$ for such an axis-symmetrical system.

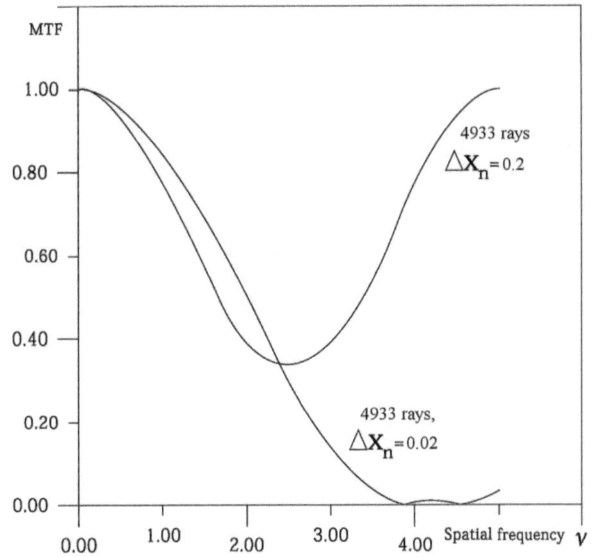


**Fig. 3.23** The MTF of an on-axis source point $\overline{P}_0 = [0 \quad -507 \quad 0 \quad 1]^T$ from ray-counting method for the system in Fig. 2.1. It is noted that the ray-counting method requires estimation of the line spread function, which relies on the number of rays traced and the meshing of the image plane and may have significant negative effect on the results

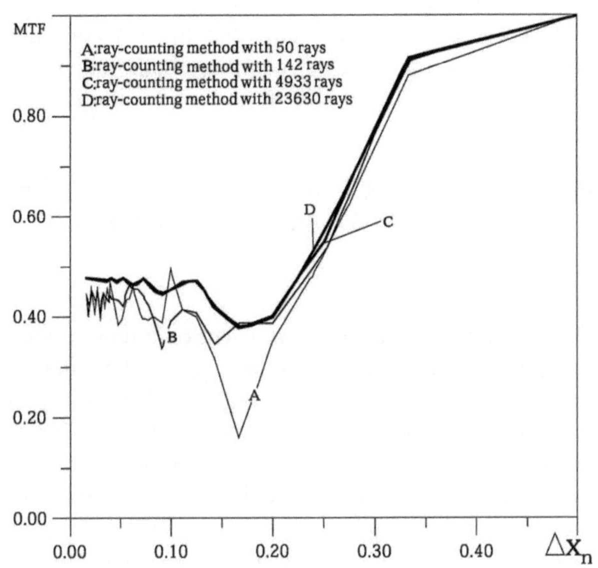


**Fig. 3.24** Variation of sagittal MTF of the on-axis source point $\overline{P}_0 = [0 \quad -507 \quad 0 \quad 1]^T$ for $v = 2$ (cycle per mm) by tracing different numbers of rays for the system in Fig. 2.1. Notably, the grid size meshed on the image plane significantly affects the MTF of the ray-counting method

*Example 3.18* Figure 3.24 shows the variation of MTF for the ray-counting method with the same source point $\overline{P}_0 = [0 \quad -507 \quad 0 \quad 1]^T$ but with various grid sizes on the image plane of system shown in Fig. 2.1. One can observe from this figure that the MTF from the ray-counting method depends on the mesh (i.e.$\Delta x_n$) of the image plane. If the mesh is too fine, the MTF may fluctuate dramatically.

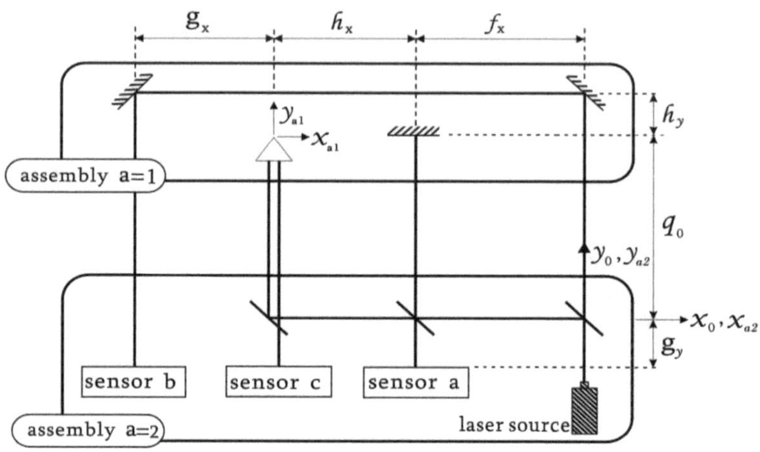


**Fig. 3.25** Schematic diagram of the motion measurement system used as a representative example in this study [15]

Furthermore, the MTF may be over-estimated if the mesh is too coarse, due to incorrect line spread function estimation.

Overall, one can conclude that the ray-counting method may fail in MTF computations, since the ray-counting method depends significantly on the number of rays traced and grid size on the image plane.

## 3.7 Determining Linear Equations of Position Sensitive Detector (PSD) for Small Motion Measurement Systems

High-accuracy laser-based optoelectronic motion measuring systems typically utilize light rays that travel from one optical boundary surface to another to perform motion measurements [15–21]. Figure 3.25 is the representative model of the motion measurement system reported by Ni and Wu [15]. This system consists of two optical assemblies: (1) assembly $a = 1$ is a measuring assembly attached rigidly to a moving table to be measured; (2) assembly $a = 2$ is a reference assembly mounted rigidly on the reference base. In Ni's model, assembly $a = 2$ contains a mixed group of optical elements, specifically two beam splitters and three PSDs (dual-axis lateral-effect photodetectors) to receive beams reflected from the measuring assembly. Assembly $a = 1$ also contains a mixed group of optical elements, specifically one corner-cube and three first-surface mirrors. The unit directional vector $\bar{\ell}_0 = \begin{bmatrix} 0 & 1 & 0 & 0 \end{bmatrix}^T$ of the laser source is aligned with the axis along which a machine tool table travels. Since assembly $a = 1$ is attached rigidly to the moving table, it emulates the motion (denoted as $\overline{V} = \begin{bmatrix} v_g \end{bmatrix} = \begin{bmatrix} t_{a1x} & t_{a1y} & t_{a1z} & \omega_{a1x} & \omega_{a1y} & \omega_{a1z} \end{bmatrix}^T$, $g = 1$ to $g = 6$) of the table. Two coordinate frames, $(xyz)_{a1}$ and $(xyz)_{a2}$, are respectively imbedded into assemblies a = 1 and a = 2 for modeling. The pose matrix ${}^0\bar{A}_{a2} = \bar{I}_{4times;4}$ of assembly a = 2 is the identity matrix, since $(xyz)_{a2}$ and $(xyz)_0$ are coinciding here for convenience. One should note that the origin of $(xyz)_{a1}$ must be placed at the point where its translational and rotational motions are to be measured. The pose matrix of $(xyz)_{a1}$ with respect to $(xyz)_0$ is defined by ${}^0\bar{A}_{a1} = \text{tran}(t_{a1z}, t_{a1y}, t_{a1x})$ $\text{rot}(\bar{z}, \omega_{a1z})\text{rot}(\bar{y}, \omega_{a1y})\text{rot}(\bar{x}, \omega_{a1x})$. If there is no error motion in the moving table, then the values are $t_{a1z} = \omega_{a1x} = \omega_{a1y} = \omega_{a1z} = 0$, $t_{a1x} = -(h_x + f_x)$ and $t_{a1y} = q_0$. When any error motion $\Delta\bar{V} = [\Delta v_g] = [\Delta t_{a1x} \quad \Delta t_{a1y} \quad \Delta t_{a1z} \quad \Delta\omega_{a1x} \quad \Delta\omega_{a1y} \quad \Delta\omega_{a1z}]^T$ exists in the motion of the table relative to its intended path, the laser beam reflected by the measuring assembly will shift its lateral position or change its angular orientation, depending on the type of error motion.

One application of the skew-ray tracing method is the modeling of an optical measurement system which measures small changes in a ray's final position (i.e., the incidence point $\bar{P}_n$, see Eq. (2.46) with g = n) to detect small motions. The sensor readings $[X \quad Z]^T$ can be obtained from ${}^n\bar{P}_n$ by the following transformation

$$\begin{bmatrix} X \\ 0 \\ Z \\ 1 \end{bmatrix} = \begin{bmatrix} {}^nP_{nx} \\ 0 \\ {}^nP_{nz} \\ 1 \end{bmatrix} = {}^n\bar{P}_n = {}^n\bar{A}_0\bar{P}_n$$

$$= \begin{bmatrix} I_{ix} & I_{iy} & I_{iz} & -(I_{ix}t_{ix} + I_{iy}t_{iy} + I_{iz}t_{iz}) \\ J_{ix} & J_{iy} & J_{iz} & -(J_{ix}t_{ix} + J_{iy}t_{iy} + J_{iz}t_{iz}) \\ K_{ix} & K_{iy} & K_{iz} & -(K_{ix}t_{ix} + K_{iy}t_{iy} + K_{iz}t_{iz}) \\ 0 & 0 & 0 & 1 \end{bmatrix} \begin{bmatrix} P_{nx} \\ P_{ny} \\ P_{nz} \\ 1 \end{bmatrix}. \tag{3.51}$$

Since PSD readings can obtain translational and rotational error motions $\Delta\bar{V}$, it is reasonable to express them as implicit nonlinear functions of $\bar{V} = [v_g]$, i.e. $[X \quad Z]^T = [X(\bar{V}) \quad Z(\bar{V})]^T$. Note that the translational and rotational error motions $\Delta\bar{V}$ are very small. Therefore we can use the first-order Taylor series expansion to expand $[X \quad Z]^T$ at $v_g = 0$ (g = 1 to g = 6) to obtain the PSD readings in linear form, as

$$\begin{bmatrix} X \\ Z \end{bmatrix} = \begin{bmatrix} X(\bar{0}) \\ Z(\bar{0}) \end{bmatrix} + \begin{bmatrix} \partial X/\partial\bar{V} \\ \partial Z/\partial\bar{V} \end{bmatrix}_{\Delta\bar{v}_g = \bar{0}_{1x6}} [\Delta v_g]. \tag{3.52}$$

Note that $[\partial X/\partial v_g \quad \partial X/\partial v_g]_{\Delta\bar{V} = \bar{0}_{1x6}}$ indicates that this matrix is evaluated at $\Delta\bar{V} = \bar{0}_{1x6}$. In the real world, it is impossible to exactly place an element at a specified position and orientation. Consequently, setting errors $[X(\bar{0}) \quad Z(\bar{0})]^T$ must be assumed to exist in any real system. In order to remove reading errors due to setting errors, this system has to be adjusted first to ensure the setting errors are as small as possible. The difference between the PSD readings and the setting errors, i.e. $[X \quad Z]^T - [X(\bar{0}) \quad Z(\bar{0})]^T = [\Delta X \quad \Delta Z]^T$ (designated as the "effective PSD readings"), at any single measurement will be used to determine the real positional/angular motion. Now the effective PSD readings can be expressed as $[\Delta X \quad \Delta Z]^T = [\partial X/\partial \bar{V} \quad \partial Z/\partial \bar{V}]^T_{\Delta v_g = 0}[\Delta v_g]$. If the translational and rotational motions are very small, we can express the sensor readings as

$$\begin{bmatrix} \Delta X \\ \Delta Z \end{bmatrix} = \begin{bmatrix} \partial X/\partial v_g \\ \partial X/\partial v_g \end{bmatrix}_{\Delta \bar{V} = \bar{0}_{1x6}} \Delta v_g$$

$$= \begin{bmatrix} c_{11g}g_y + c_{12g}q_0 + c_{13g}h_y + c_{14g}f_x + c_{15g}h_x + c_{16g}g_x \\ c_{21g}g_y + c_{22g}q_0 + c_{23g}h_y + c_{24g}f_x + c_{25g}h_x + c_{26g}g_x \end{bmatrix} \Delta v_g. \tag{3.53}$$

The leading coefficient, which represents the weight by which each translational and rotational motion contributes to the sensor readings, is the sensitivity of the corresponding translational and rotational motion. Now we can use finite difference methodology to determine the coefficients of each translational and rotational motion $\Delta v_g$ of Eq. (3.53). The following is the algorithm for obtaining the coefficients of Eq. (3.53) using the finite difference method.

(1) Label the boundary surfaces of the system sequentially from $i = 1$ to $i = n$.

(2) Assume there is only a small translational or rotational error motion, say $\Delta v_2 = \Delta t_{a1y} = 0.00001\text{mm}$, in the measuring assembly. Therefore, the pose matrix of the measuring assembly is defined by ${}^0\bar{A}_{a1} = \text{Trans}(0, q_0 + 0.00001, 0)$.

(3) To determine $[c_{122} \quad c_{222}]^T$, we must set $q_0 = 1$ while keeping $g_y = h_y = f_x = h_x = g_x = 0$. Calculate the coordinates of the incidence point $\bar{P}_n = [P_{nx} \quad P_{ny} \quad P_{nz} \quad 1]^T$ (Eq. (2.46) with $g = n$) and then $[X \quad 0 \quad Z \quad 1]^T$ (Eq. (3.51)) on the sensor surface by the skew-ray tracing methodology of Chap. 2. From Eq. (3.53) we have $[c_{122} \quad c_{222}]^T = [\partial X/\partial v_2 \quad \partial Z/\partial v_2]^T \approx [\Delta X/\Delta v_2 \quad \Delta Z/\Delta v_2]^T$. Similar procedure can be applied to obtain the other coefficients.

The illustrative system contains 3 separate sensors, sensors a-c in Fig. 3.25, in assembly $a = 2$. Each sensor interacts with a different set of the optical elements in assembly $a = 1$. Thus each sensor can be viewed as belonging to a unique subsystem. In the following example we will determine the linear equation of sensor b sub-system.

*Example 3.19* The sub-system of sensor b can be seen in Fig. 3.26. In this subsystem, two first-surface mirrors are located in assembly $a = 1$. Sensor b is located in assembly $a = 2$. The boundary coordinate frames $(xyz)_1$, $(xyz)_2$ and $(xyz)_3$ are established and defined by the pose matrices,

$${}^{a1}\bar{A}_1 = \text{tran}(f_x + h_x, h_y, 0)\text{rot}(\bar{z}, -45°),$$

$${}^{a1}\bar{A}_2 = \text{tran}(-g_x, h_y, 0)\text{rot}(\bar{z}, 45°),$$

and

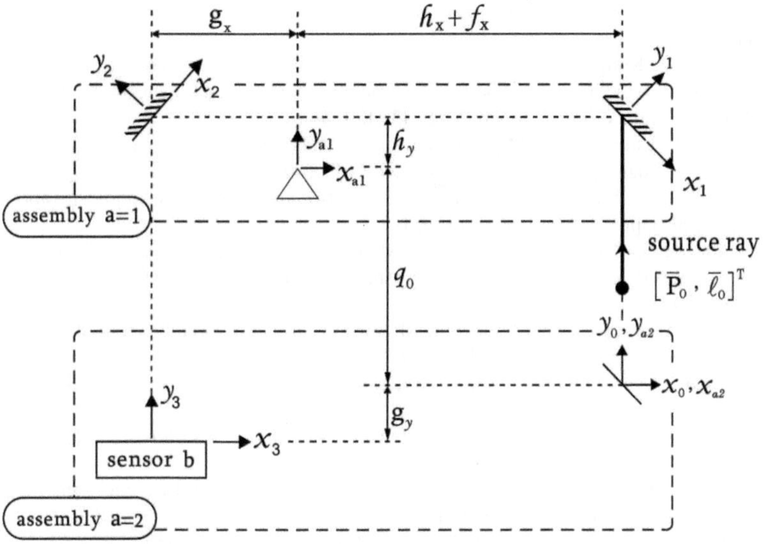


**Fig. 3.26** The model of sensor b of the representative motion measurement system

$^{a2}\bar{A}_3 = \text{tran}(-g_x - h_x - f_x, -g_y, 0)$.

The linear equations $[\Delta X_b \quad \Delta Z_b]^T$ are

$$\begin{bmatrix} \Delta X_b \\ \Delta Z_b \end{bmatrix} = \begin{bmatrix} \Delta^3 P_{3x} \\ \Delta^3 P_{3z} \end{bmatrix}$$

$$= \begin{bmatrix} -2\Delta t_{a1x} + (2h_y + f_x + h_x + g_x)\Delta\omega_{a1z} \\ [2(g_y + q_0 + h_y) + (f_x + h_x + g_x)]\Delta\omega_{a1x} - (f_x + h_x + g_x)\Delta\omega_{a1y} \end{bmatrix}. \tag{3.54}$$

The study of [15] gives the linear equation for sensor b as $\Delta Z_b = 2q_0\Delta\omega_{a1x} - (f_x + h_x + g_x)\Delta\omega_{a1y}$, and does not provide $\Delta X_b$. This is because [15] does not perform a strict skew-ray tracing of their measurement system, with the result that the effects of offset parameters $h_y$, $f_x$, $h_x$ and $g_x$ on the sensor readings are not accounted for.

## References


1. J.W. Foreman, Computation of RMS spot radii by ray tracing. Appl. Opt. **13**, 2585–2588 (1974)
2. T.B. Andersen, Evaluating RMS spot radii by ray tracing. Appl. Opt. **21**, 1241–1248 (1982)
3. B. Brixner, Lens design merit functions: rms image spot size and rms optical path difference. Appl. Opt. **17**, 715–716 (1978)

4. W.J. Smith, *Modern Optical Engineering*, 3rd edn. (Edmund Industrial Optics, Barrington, 2001)
5. K.H. Tseng, C. Kung, T.T. Liao, H.P. Chang, Calculation of modulation transfer function of an optical system by using skew ray tracing, Transactions of Canadian Society for Mechanical Engineering. J. Mech. Eng. **33**, 429–442 (2009)
6. S. Inoue, N. Tsumura, Y. Miyake, Measuring MTF of paper by sinusoidal test pattern projection. J. Imaging Sci. Technol. **41**, 657–661 (1997)
7. G.D. Boreman, S. Yang, Modulation transfer function measurement using three- and four-bar targets. Appl. Opt. **34**, 8050–8052 (1995)
8. D.N. Sitter, J.S. Goddard, R.K. Ferrell, Method for the measurement of the modulation transfer function of sampled imaging systems from bar-target patterns. Appl. Opt. **34**, 746–751 (1995)
9. R. Barakat, Determination of the optical transfer function directly from the edge spread function. J. Opt. Soc. Am. **55**, 1217–1221 (1965)
10. G.L. Rogers, Measurement of the modulation transfer function of paper. Appl. Opt. **37**, 7235–7240 (1998)
11. S.K. Park, R. Schowengerdt, M. Kaczynski, Modulation-transfer-function analysis for sampled image system. Appl. Opt. **23**, 2572–2582 (1984)
12. S. Inoue, N. Tsumura, Y. Miyake, Measuring MTF of paper by sinusoidal test pattern projection. J. Imaging Sci. Technol. **41**, 657–661 (1997)
13. K.H. Tseng, C. Kung, T.T. Liao, H.P. Chang, Calculation of modulation transfer function of an optical system by using skew ray tracing, Transactions of Canadian Society for Mechanical Engineering. J. Mech. Eng. **33**, 429–442 (2009)
14. E. Giakoumakis, M.C. Katsarioti, G.S. Panayiotakis, Modulation transfer function of thin transparent foils in radiographic cassettes. Appl. Phys. A Solids Surfaces **52**, 210–212 (1991)
15. J. Ni, S.M. Wu, An on-line measurement technique for machine volumetric error compensation. ASME J. Eng. Ind. **115**, 85–92 (1993)
16. K.H. Kim, K.F. Eman, S.M. Wu, Analysis alignment errors in a laser-based in-process cylindricity measurement system. J. Eng. Ind.-Trans ASME **109**, 321–329 (1987)
17. C.W. Park, K.F. Eman, S.M. Wu, An in-process flatness error measurement and compensatory control system. J. Eng. Ind.-Trans ASME **110**, 263–270 (1988)
18. E.H. Bokelberg, H.J. Sommer III, M.W. Tretheway, A six-degree-of-freedom laser vibrometer, part I: theoretical development. J. Sound Vibr. **178**, 643–654 (1994)
19. E.H. Bokelberg, H.J. Sommer III, M.W. Tretheway, A six-degree-of-freedom laser vibormeter, part II: experimental validation. J. Sound Vibr. **178**, 655–667 (1994)
20. P.D. Lin, K.F. Ehmann, Sensing of motion related errors in multi-axis machines. J. Dyn. Syst. Meas. Control-Trans ASME **118**, 425–433 (1996)
21. S.W. Lee, R. Mayor, J. Ni, Development of a six-degree-of-freedom geometric error measurement system for a meso-scale machine tool. J. Manuf. Sci. Eng.-Trans ASME **127**, 857–865 (2005)

# Chapter 4
# Paraxial Optics for Axis-Symmetrical Systems

Conventional paraxial optics, which is sometimes known as Gaussian optics or first-order optics, uses the $2 \times 2$ matrices as first estimates of meridional rays in early design stages of 2-D axis-symmetrical optical systems [1, 2]. In order to extend conventional paraxial optics for 3-D systems containing prisms, we have presented the $6 \times 6$ matrices obtained from the first-order Taylor series expansion to approximate skew-ray tracing equations [3]. We shall begin this chapter by modifying the $6 \times 6$ matrices of [3] to be applicable only to axis symmetrical systems. The obtained matrices are used to determine the cardinal points for an optical system. The imaging equations for multi-surface system are derived next.

## 4.1 Convention of Paraxial Optics

In paraxial optics no distinction is made between a spherical and an aspherical surface. This is due to the fact that in the paraxial region, spherical and aspherical surfaces are essentially indistinguishable if the vertex radius of curvature of that aspherical surface is used for the paraxial raytracing. It is also important to know that in paraxial optics rays are not reflected/or refracted at real boundary surfaces, but at the imaginary tangent planes (referred as paraxial surfaces, see Fig. 4.1) to real boundary surfaces at their vertices. Some symbols (e.g., $\bar{M}_i$ of Eq. (4.2)) of this chapter may be repeatedly being used in other chapters for different physical quantities. This will not cause any difficulty, since paraxial optics is always treated as an independent topic from the other in geometrical optics.

It is necessary to adopt a convention for the algebraic signs given to the various distances involved in paraxial optics. The following conventions are frequently used in paraxial optics:

1. The height above the optical axis is positive and that below the optical axis is negative. Thus h and h′ shown in Fig. 4.10 are positive and negative, respectively.

PD Lin, *New Computation Methods for Geometrical Optics*,
Springer Series in Optical Sciences 178, DOI: 10.1007/978-981-4451-79-6_4,

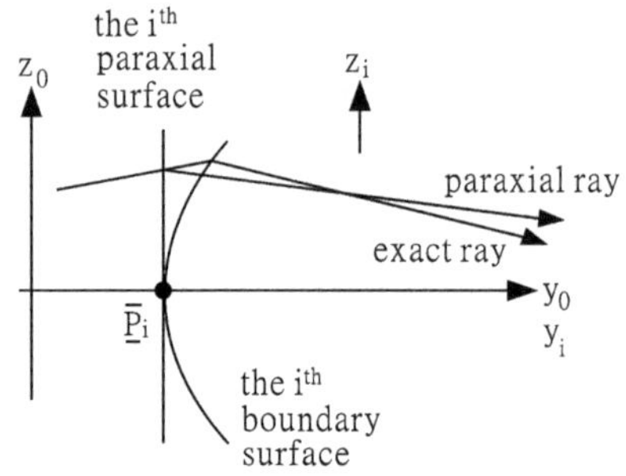


**Fig. 4.1** In paraxial optics rays are reflected/or refracted at paraxial surfaces, not at the real boundary surfaces

2. The distance to the left of a reference point is negative; and to the right positive. Thus the distances f and t of Fig. 4.10 are negative, while f′ and t′ of Fig. 4.10 are positive.
3. The focal length of a converging lens is positive and that of a diverging lens is negative.
4. Any refractive index $\xi_i$ of medium i is positive. The slope $\mu_i$ (which is the angle between the ray and optical axis, and measured in radians) of a ray is considered positive if a counterclockwise rotation turns the ray from the positive direction of $y_0$ axis to the direction of that ray.

## 4.2 Reflecting and Refracting Matrices of Paraxial Optics

To address paraxial optics for an axis symmetrical system, one first has to define the paraxial surfaces and differential rays. The ith paraxial surface is a virtual flat surface tangent to the real boundary surface at $\underline{\bar{P}}_i$, the vertex of the ith boundary surface (see Figs. 4.1 and 4.2). Shown in Fig. 4.2 are also a meridional ray $\begin{bmatrix} \bar{P}_i & \bar{\ell}_i \end{bmatrix}^T$ and an axial ray $\begin{bmatrix} \underline{\bar{P}}_i & \underline{\bar{\ell}}_i \end{bmatrix}^T$, where $\underline{\bar{P}}_i$ is the vertex of the ith boundary surface and $\underline{\bar{\ell}}_i$ is the unit directional vectors of the optical axis. The differential ray $\begin{bmatrix} \Delta\bar{P}_i & \Delta\bar{\ell}_i \end{bmatrix}^T$ (i = 0 to i = n) in an axis symmetrical system is the difference of $\begin{bmatrix} \bar{P}_i & \bar{\ell}_i \end{bmatrix}^T$ and $\begin{bmatrix} \underline{\bar{P}}_i & \underline{\bar{\ell}}_i \end{bmatrix}^T$, that is

$$\begin{bmatrix} \Delta\bar{P}_i \\ \Delta\bar{\ell}_i \end{bmatrix} = \begin{bmatrix} \bar{P}_i \\ \bar{\ell}_i \end{bmatrix} - \begin{bmatrix} \underline{\bar{P}}_i \\ \underline{\bar{\ell}}_i \end{bmatrix}. \tag{4.1}$$

Note that, as shown in Fig. 4.2, when the paraxial ray $\begin{bmatrix} \bar{P}_{i-1} & \bar{\ell}_{i-1} \end{bmatrix}^T$ travels from the (i − 1)th paraxial surface to the ith paraxial surface, it propagates along a straight-line path in medium i − 1 and then is reflected/or refracted by the ith paraxial surface. We can get an approximation for carrying the differential ray $\begin{bmatrix} \Delta\bar{P}_{i-1} & \Delta\bar{\ell}_{i-1} \end{bmatrix}^T$ from the (i − 1)th paraxial surface to the reflected/or refracted

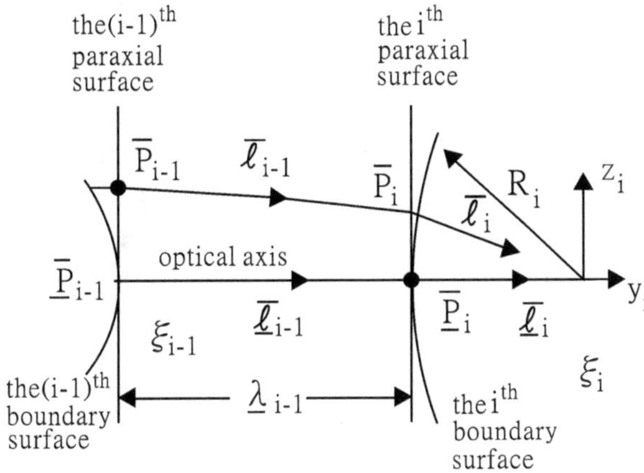


**Fig. 4.2** A paraxial ray propagates along a straight-line path and then is reflected/ or refracted by boundary surface

differential ray $\begin{bmatrix} \Delta\bar{P}_i & \Delta\bar{\ell}_i \end{bmatrix}^T$ at the ith paraxial surface by the following $6 \times 6$ matrix-based method:

$$\begin{bmatrix} \Delta\bar{P}_i \\ \Delta\bar{\ell}_i \end{bmatrix} = \bar{M}_i \bar{T}_{i-1} \begin{bmatrix} \Delta\bar{P}_{i-1} \\ \Delta\bar{\ell}_{i-1} \end{bmatrix}, \tag{4.2}$$

where $\bar{T}_{i-1}$ is the transfer matrix when ray travels along medium $i - 1$ and $\bar{M}_i$ is the reflection/or refraction matrix at the ith paraxial surface [3]. If $\underline{\lambda}_{i-1}$ is the length of the straight-line segment measured from $\underline{\bar{P}}_{i-1}$ to $\underline{\bar{P}}_i$ along optical axis, then we have

$$\bar{T}_{i-1} = \begin{bmatrix} 1 & 0 & 0 & \underline{\lambda}_{i-1} & 0 & 0 \\ 0 & 1 & 0 & 0 & \underline{\lambda}_{i-1} & 0 \\ 0 & 0 & 1 & 0 & 0 & \underline{\lambda}_{i-1} \\ 0 & 0 & 0 & 1 & 0 & 0 \\ 0 & 0 & 0 & 0 & 1 & 0 \\ 0 & 0 & 0 & 0 & 0 & 1 \end{bmatrix}. \tag{4.3}$$

The explicit expression of $\bar{M}_i$ depends not only on the boundary surface type, but also on whether the surface is reflecting or refracting one.

### *4.2.1 Reflecting and Refracting Matrices at a Flat Boundary Surface*

Many optical elements (e.g., beam-splitters) have reflecting and/or refracting flat boundary surfaces (Figs. 4.3 and 4.4). If the boundary coordinate frame $(xyz)_i$ is defined in such way that its $y_i z_i$ plane contains optical axis and its $y_i$-axis has the same sense as vector drawn from $\underline{\bar{P}}_{i-1}$ to $\underline{\bar{P}}_i$ (i.e., $\underline{\bar{\ell}}_{i-1} = \underline{\bar{\ell}}_i = \begin{bmatrix} 0 & 1 & 0 & 0 \end{bmatrix}^T$), then

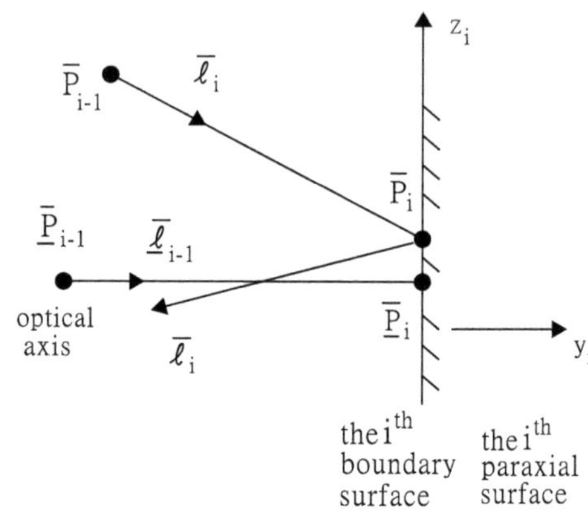


**Fig. 4.3** Flat first-surface mirror

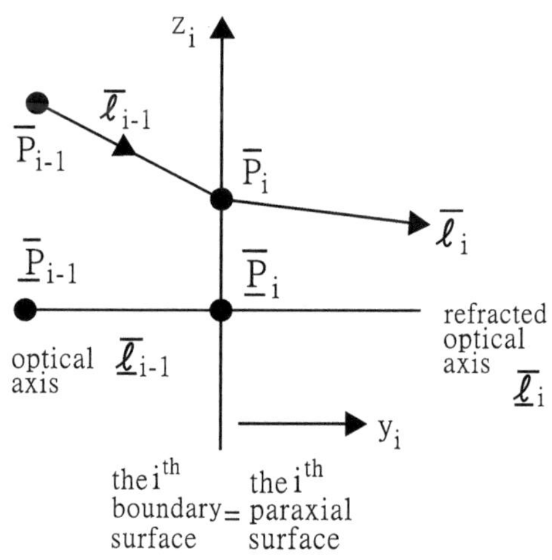


**Fig. 4.4** Flat refracting boundary surface

one has $s_i = 1$. The $6 \times 6$ refraction matrix $\bar{M}_i$ of such refracting flat boundary surface is given as

$$\bar{M}_i = \begin{bmatrix} 1 & 0 & 0 & 0 & 0 & 0 \\ 0 & 0 & 0 & 0 & 0 & 0 \\ 0 & 0 & 1 & 0 & 0 & 0 \\ 0 & 0 & 0 & N_i & 0 & 0 \\ 0 & 0 & 0 & 0 & N_i^2 & 0 \\ 0 & 0 & 0 & 0 & 0 & N_i \end{bmatrix}. \tag{4.4}$$

Similarly, if the boundary coordinate frame $(xyz)_i$ of a reflecting mirror is defined in the same way, then its $6 \times 6$ reflection matrix $\bar{M}_i$ is

$$\bar{M}_i = \begin{bmatrix} 1 & 0 & 0 & 0 & 0 & 0 \\ 0 & 0 & 0 & 0 & 0 & 0 \\ 0 & 0 & 1 & 0 & 0 & 0 \\ 0 & 0 & 0 & 1 & 0 & 0 \\ 0 & 0 & 0 & 0 & -1 & 0 \\ 0 & 0 & 0 & 0 & 0 & 1 \end{bmatrix}. \tag{4.5}$$

### *4.2.2 Reflecting and Refracting Matrices at a Spherical Boundary Surface*

A convex spherical mirror (Fig. 4.5) has a reflecting surface that curves outward, resembling a portion of the exterior of a sphere. A concave spherical mirror (Fig. 4.6) has a reflecting surface that curves inward, resembling a portion of the interior of a sphere. If the boundary coordinate frame $(xyz)_i$ is defined in such way that its $y_iz_i$ plane contains optical axis and its $y_i$-axis has the same sense as vector drawn from $\underline{\bar{P}}_{i-1}$ to $\underline{\bar{P}}_i$ (i.e., $\underline{\bar{\ell}}_{i-1} = \underline{\bar{\ell}}_i = [0 \quad 1 \quad 0 \quad 0]^T$), then one has $s_i = 1$. Furthermore, if a convex mirror has positive value of $R_i$ and a concave mirror has negative of $R_i$, then we have the following reflection matrix $\bar{M}_i$ valid for both convex and concave mirrors:

$$\bar{M}_i = \begin{bmatrix} 1 & 0 & 0 & 0 & 0 & 0 \\ 0 & 0 & 0 & 0 & 0 & 0 \\ 0 & 0 & 1 & 0 & 0 & 0 \\ 2/R_i & 0 & 0 & 1 & 0 & 0 \\ 0 & 0 & 0 & 0 & -1 & 0 \\ 0 & -2/R_i & 2/R_i & 0 & 0 & 1 \end{bmatrix}. \tag{4.6}$$

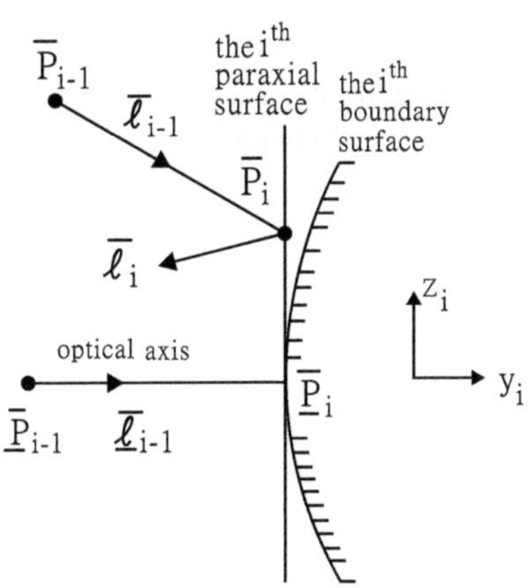


**Fig. 4.5** Convex spherical mirror

**Fig. 4.6** Concave spherical mirror

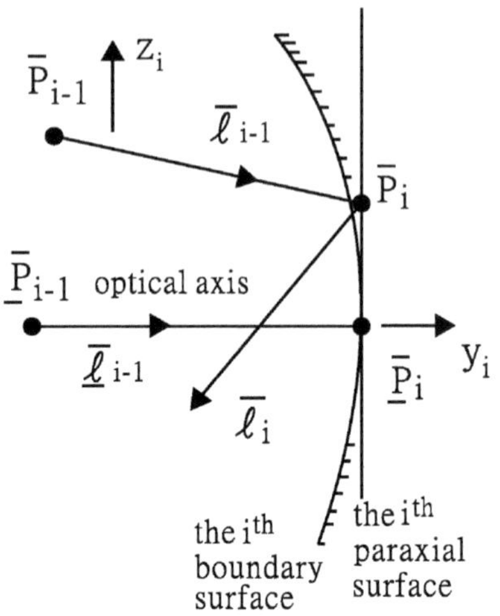


There are two types of spherical refracting boundary surfaces: concave and convex refracting spherical boundary surfaces. The optical axis of such a boundary surface (Figs. 4.7 and 4.8) is the straight-line segment passing through its geometrical center. Therefore, we can define the $y_i$ axis pointing from $\underline{\bar{P}}_{i-1}$ to $\underline{\bar{P}}_i$, leading $\underline{\bar{\ell}}_{i-1} = \underline{\bar{\ell}}_i = [0 \quad 1 \quad 0 \quad 0]^T$. If a convex boundary surface has positive value of $R_i$ and a concave boundary surface has negative of $R_i$, we have the following refraction matrix $\bar{M}_i$ valid for both convex and concave mirrors:

$$\bar{M}_i = \begin{bmatrix} 1 & 0 & 0 & 0 & 0 & 0 \\ 0 & 0 & 0 & 0 & 0 & 0 \\ 0 & 0 & 1 & 0 & 0 & 0 \\ (N_i - 1)/R_i & 0 & 0 & N_i & 0 & 0 \\ 0 & 0 & 0 & 0 & N_i^2 & 0 \\ 0 & 0 & (N_i - 1)/R_i & 0 & 0 & N_i \end{bmatrix}. \tag{4.7}$$

**Fig. 4.7** Refracting convex spherical boundary surface

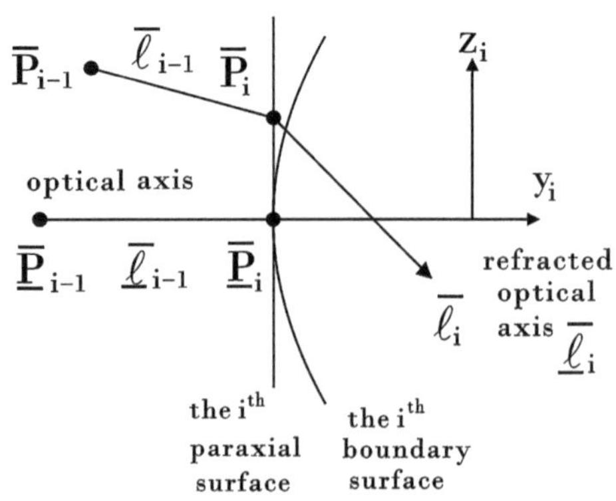

**Fig. 4.8** Refracting concave spherical boundary surface

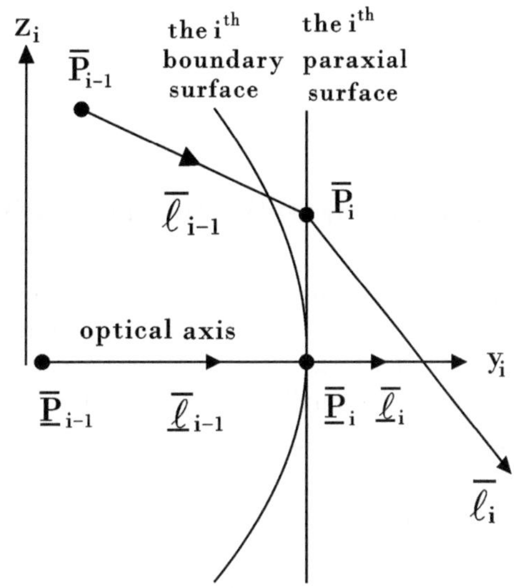


It is noted that Eqs. (4.4) and (4.5) can be deduced from Eqs. (4.6) and (4.7) by setting radius $R_i$ to be infinity. To trace a differential paraxial ray in an axis-symmetrical system possessing n paraxial surfaces, one should establish its paraxial surfaces and then label them sequentially from 1 to n. In an axis-symmetrical system the world coordinate frame $(xyz)_0$ and boundary coordinate frames $(xyz)_i$ ($i = 1$ to n) are parallel to each other, yielding that ${}^{i+1}\bar{B}_i$ ($i = 1$ to $n - 1$) of Eq. (42) in Ref. [3] is a unit matrix. One can apply Eq. (4.2) successively from $i = 0$ to $i = n$ to estimating the differential exit ray at the nth paraxial surface by

$$\begin{bmatrix} \Delta\overline{P}_n \\ \Delta\overline{\ell}_n \end{bmatrix} = (\overline{M}_n\overline{T}_{n-1})(\overline{M}_{n-1}\cdots\overline{T}_i)(\overline{M}_i\overline{T}_{i-1})\cdots(\overline{M}_2\overline{T}_1)(\overline{M}_1\overline{T}_0)\begin{bmatrix} \Delta\overline{P}_0 \\ \Delta\overline{\ell}_0 \end{bmatrix}. \quad (4.8)$$

Equation (4.8) can be further modified into the traditional $2 \times 2$ paraxial ray-tracing equation in textbooks [1] if rows 1, 2, 4, 5, columns 1, 2, 4, 5 of $\bar{M}_i$ and $\bar{T}_i$ are deleted, and $\begin{bmatrix} \Delta\overline{P}_i & \Delta\overline{\ell}_i \end{bmatrix}^T$ is replaced by the height-slope-angle matrix $\begin{bmatrix} z_i & \xi_i\mu_i \end{bmatrix}^T$ ($z_i$ is the ray height measured from optical axis, $\mu_i$ is the small angle in radian between the ray and optical axis with positive value when anticlockwise), yielding $\bar{M}_i$ and $\bar{T}_{i-1}$ ($i = 1$ to $i = n$) are given respectively as:

$$\bar{T}_{i-1} = \begin{bmatrix} 1 & \frac{\lambda_{i-1}}{\xi_{i-1}} \\ 0 & 1 \end{bmatrix}, \quad (4.9)$$

$$\bar{M}_i = \begin{bmatrix} 1 & 0 \\ \frac{-2\xi_i}{R_i} & -1 \end{bmatrix} \quad \text{for reflection,} \quad (4.10)$$

$$\bar{M}_i = \begin{bmatrix} 1 & 0 \\ \frac{\xi_{i-1}-\xi_i}{R_i} & 1 \end{bmatrix} \quad \text{for refraction.} \tag{4.11}$$

## 4.3 Paraxial Optics for Axis-Symmetrical Optical Systems

Paraxial optics uses the concepts of cardinal planes and cardinal points to trace paraxial rays for axis-symmetrical optical systems by using the $2 \times 2$ matrix method. To trace a paraxial ray, one has to: (1) establish and label its paraxial surfaces sequentially from $i = 1$ to $i = n$; and (2) mark the ray height-slope-angle matrix at each paraxial surface i by $[z_i \quad \xi_i\mu_i]^T$ ($i = 0$ to $i = n$) (see Fig. 4.9). One can then apply Eq. (4.8) to determining its height-slope-angle matrix $[z_n \quad \xi_n\mu_n]^T$ at the last paraxial surface by:

$$\begin{aligned} \begin{bmatrix} z_n \\ \xi_n\mu_n \end{bmatrix} &= (\bar{M}_n\bar{T}_n\bar{M}_{n-1}\bar{T}_{n-1}\cdots\bar{M}_i\bar{T}_i\cdots\bar{M}_2\bar{T}_1\bar{M}_1)\bar{T}_0 \begin{bmatrix} z_0 \\ \xi_0\mu_0 \end{bmatrix} \\ &= \begin{bmatrix} A & B \\ C & D \end{bmatrix} \begin{bmatrix} 1 & \frac{\lambda_0}{\xi_0} \\ 0 & 1 \end{bmatrix} \begin{bmatrix} z_0 \\ \xi_0\mu_0 \end{bmatrix} = \begin{bmatrix} A & B \\ C & D \end{bmatrix} \begin{bmatrix} z_1 \\ \xi_1\mu_1 \end{bmatrix}, \end{aligned} \tag{4.12}$$

where ABCD matrix describes the transformation between $[z_n \quad \xi_n\mu_n]^T$, the height-slope-angle matrix of a ray at the last paraxial surface, and $[z_1 \quad \xi_1\mu_1]^T$, the height-slope-angle matrix of that ray at the first paraxial surface.

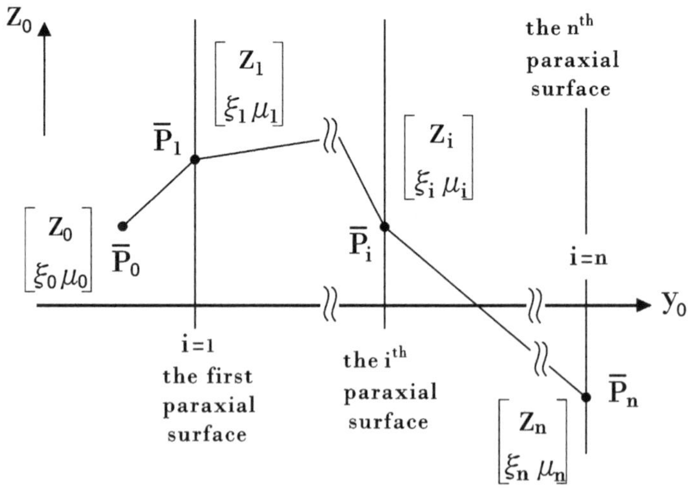


**Fig. 4.9** The ith paraxial surface and its height-slope-angle matrices

Recall that the components of ABCD matrix are such that the determinant, AD–BC, equals unity. Therefore, if we wish to trace a ray backwards, the matrix equation can be inverted to yield

$$\begin{bmatrix} z_1 \\ \xi_1 \mu_1 \end{bmatrix} = \begin{bmatrix} D & -B \\ -C & A \end{bmatrix} \begin{bmatrix} z_n \\ \xi_n \mu_n \end{bmatrix}. \tag{4.13}$$

One has to note that even though we use Eq. (4.13) to trace ray backward, the ray directions and the sign convention of rays and angles described in next section still have to be kept as they are in Eq. (4.12). Equations (4.12) and (4.13) allow one to determine the cardinal planes and cardinal points. They describe the behavior of the optical system by summarizing the effects of the combination of the individual optical boundary surfaces.

## 4.4 Cardinal Planes and Cardinal Points of an Optical System

To treat a well-corrected optical system (e.g., Fig. 2.1) as a "black box" in paraxial optics, we need to have some basic conception of cardinal planes and cardinal points (see Figs. 4.10 and 4.11). The back principal plane has the property that a ray emerging from the system appears to have crossed the back principal plane at the same distance from the optical axis that that ray appeared to have crossed the front principal plane. This means that, if the ray comes from left, the system can be treated as if all of the refraction happened at the back principal

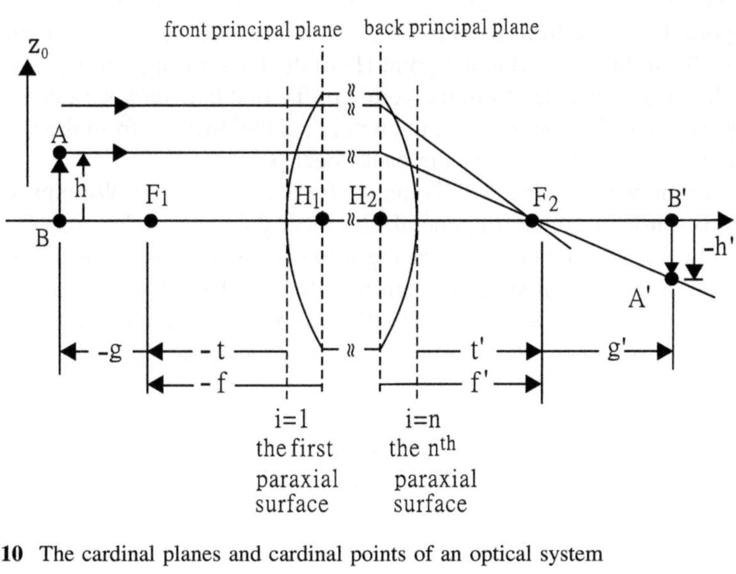


**Fig. 4.10** The cardinal planes and cardinal points of an optical system

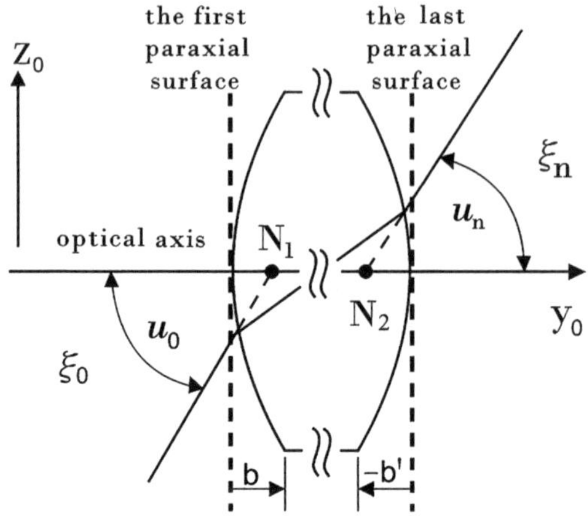


**Fig. 4.11** The node points of an optical system

plane. Similarly, the front principle plane is the plane at which the refraction can be considered to happen when a ray comes from right. The front and back principal points, $H_1$ and $H_2$, are the points where the front and back principal planes cross the optical axis of a system. The front and back focal points, $F_1$ and $F_2$, are the points at which rays parallel to the optical axis are brought to a common focus on the optical axis. The front and back focal planes are defined as the planes which are perpendicular to the optic axis and pass through the front and back focal points. Nodal points, $N_1$ and $N_2$, are two axial points such that a ray directed toward the front nodal point $N_1$ appears (after passing through the system) to emerge from the back nodal point $N_2$ parallel to its original direction. Nodal points are defined as those having unit angular magnification.

The effective front focal length, f, of a system is the distance from the front principal point $H_1$ to the front focal point $F_1$. The effective back focal length, $f'$, is the distance from the back principal point $H_2$ to the back focal point $F_2$. The front focal length, t, is the distance from the vertex of the first boundary surface $i = 1$ to the front focal point $F_1$. The back focal length, $t'$, is the distance from the vertex of the last boundary surface $i = n$ to the back focal point $F_2$.

Let us assume we have the ABCD matrix for a given system. We now seek to locate the six cardinal points. In general, the focal points $F_1$ and $F_2$ of an optical system can be easily calculated by tracing a ray parallel to the optical axis (i.e., $u_0 = 0$) through the optical system. Then the effective back focal length $f'$ is the ray height $z_0$ at the first paraxial surface divided by the negative ray slope angle (i.e., $f' = z_0/(-u_n)$) after the ray emerges from the back principal plane (see Fig. 4.12 and use Eq. (4.12)):

$$\begin{bmatrix} 0 \\ \xi_n \mu_n \end{bmatrix} = \begin{bmatrix} 1 & t'/\xi_n \\ 0 & 1 \end{bmatrix} \begin{bmatrix} A & B \\ C & D \end{bmatrix} \begin{bmatrix} z_0 \\ 0 \end{bmatrix}. \quad (4.14)$$

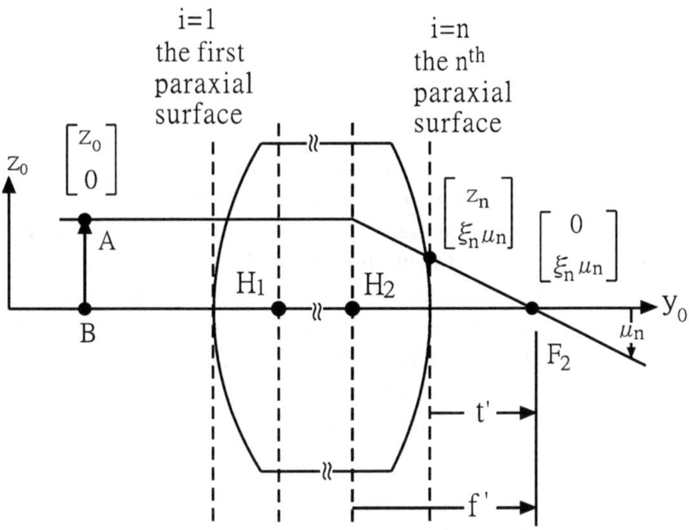


**Fig. 4.12** The effective back focal length f′ and back focal length t′ of an optical system with n boundary surfaces

One can obtain two equations, $t' = -A\xi_n/C$ and $\mu_n = Cz_0/\xi_n$, from the two components of Eq. (4.14), yielding

$$f' = \frac{z_0}{-\mu_n} = \frac{-\xi_n}{C}. \tag{4.15}$$

The back focal length t′ is the ray height $z_n$ at the last paraxial surface divided by the negative ray slope angle (i.e., $t' = z_n/(-u_n)$), yielding

$$t' = \frac{z_n}{-\mu_n} = \frac{-\xi_n A}{C}. \tag{4.16}$$

The distance from the last paraxial surface to the back principle point $H_2$ is just the difference between t′ and f′, i.e., $t' - f' = \xi_n(1 - A)/C$.

The front focal point $F_1$ and front principle point $H_1$ are found simply by tracing a parallel ray with height-slope-angle matrix $[z_n \quad \xi_n\mu_n]^T = [z_n \quad 0]^T$ traveling from right to left by using Eq. (4.13):

$$\begin{bmatrix} 0 \\ \xi_0\mu_0 \end{bmatrix} = \begin{bmatrix} 1 & -t/\xi_0 \\ 0 & 1 \end{bmatrix} \begin{bmatrix} D & -B \\ -C & A \end{bmatrix} \begin{bmatrix} z_n \\ 0 \end{bmatrix}. \tag{4.17}$$

One can obtain the following two equations from Eq. (4.17):

$$-t = \frac{z_0}{\mu_0} = \frac{-\xi_0 D}{C} \text{ or } t = \frac{\xi_0 D}{C}, \tag{4.18}$$

and

$$-f = \frac{z_n}{\mu_0} = -\frac{\xi_0}{C} \text{ or } f = \frac{\xi_0}{C}. \tag{4.19}$$

Finally, we wish to locate two nodal points $N_1$ and $N_2$ such that any ray entering the system directed towards $N_1$ appears on emergence as a ray coming from $N_2$ and making the same angle $\mu_0 = \mu_n$ with the optical axis (see Fig. 4.11). Let the distance of $N_1$ measured from the vertex of the first boundary surface $i = 1$ be denoted as b, and the distance of $N_2$ measured from the vertex of the last boundary surface $i = n$ be denoted as $b'$. It is noted that the ray path shown in Fig. 4.11 has been divided into three straight-line paths: one path from first paraxial surface to $N_1$, the second from $N_1$ to $N_2$, and the last path from $N_2$ to the last paraxial surface. Therefore, the ABCD matrix linking the first paraxial surface and last paraxial surface will be (pp. 56–57 of [1]).

$$\begin{bmatrix} A & B \\ C & D \end{bmatrix} = \begin{bmatrix} 1 & \frac{-b'}{\xi_n} \\ 0 & 1 \end{bmatrix} \begin{bmatrix} \phi_{11} & \phi_{12} \\ \phi_{21} & \phi_{22} \end{bmatrix} \begin{bmatrix} 1 & \frac{b}{\xi_0} \\ 0 & 1 \end{bmatrix}. \tag{4.20}$$

The second matrix of last equation is the matrix linking the second nodal point $N_2$ to the first nodal point $N_1$. According to the definition of nodal point, we then have

$$\begin{bmatrix} 0 \\ \xi_n \mu_n \end{bmatrix} = \begin{bmatrix} 0 \\ \xi_n \mu_0 \end{bmatrix} = \begin{bmatrix} \phi_{11} & \phi_{12} \\ \phi_{21} & \phi_{22} \end{bmatrix} \begin{bmatrix} 0 \\ \xi_0 \mu_0 \end{bmatrix}. \tag{4.21}$$

Equation (4.21) is true only if $\phi_{12} = 0$ and $\phi_{22} = \xi_n/\xi_0$. One can therefore obtain b and $b'$ by substituting $\phi_{12}$ and $\phi_{22}$ into Eq. (4.20), yielding:

$$b = \frac{D\xi_0 - \xi_n}{C}, \tag{4.22}$$

$$b' = \frac{A(D\xi_0 - \xi_n)}{C} - \xi_0 B. \tag{4.23}$$

For convenience we now recapitulate these results in Table 4.1. If the medium on both sides of the optical system is the same (e.g., air), then the front and back nodal points, $N_1$ and $N_2$, coincide with the front and back principal points, $H_1$ and $H_2$, respectively. One has to note that f is used to represent the effective back focal length in many text books as a result that the effective front focal length is not discussed.

## 4.5 Determination of Focal Points for a Thick Lens

Shown in Fig. 4.13 is a thick double convex lens with thickness $q_{e1}$ and refractive index $\xi_{e1}$. The lens is bounded by its two spherical boundary surfaces with radii $R_1$ and $R_2$, and surrounded by two different refractive media $\xi_0$ and $\xi_2$. Any incident

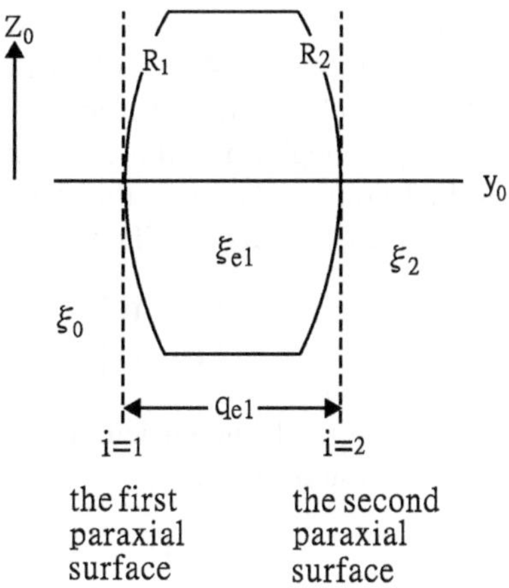


**Fig. 4.13** The thick double convex lens sounded by two different refractive media

**Table 4.1** The values of cardinal points

| Cardinal points | Measured from to | Value | Special case $\xi_n = \xi_0 = 1$ |
|---|---|---|---|
| f | From $H_1$ to $F_1$ | $\frac{\xi_0}{C}$ | $\frac{1}{C}$ |
| t | From the vertex of the first paraxial surface to $F_1$ | $\frac{\xi_0 D}{C}$ | $\frac{D}{C}$ |
| b | From the vertex of the first paraxial surface to $N_1$ | $\frac{D\xi_0-\xi_n}{C}$ | $\frac{D-1}{C}$ |
| f′ | From $H_2$ to $F_2$ | $\frac{-\xi_n}{C}$ | $\frac{-1}{C}$ |
| t′ | From the vertex of the last paraxial surface to $F_2$ | $\frac{-\xi_n A}{C}$ | $\frac{-A}{C}$ |
| b′ | From the vertex of the last paraxial surface to $N_2$ | $\frac{A(D\xi_0-\xi_n)}{C} - \xi_0 B = \frac{\xi_0 - A\xi_n}{C}$ | $\frac{1-A}{C}$ |

ray that passes through the lens will be refracted twice; one at each surface, and the other a straight-line path between the surfaces. The ABCD matrix can be obtained by

$$\begin{bmatrix} A & B \\ C & D \end{bmatrix} = \begin{bmatrix} 1 & 0 \\ \frac{\xi_{e1}-\xi_2}{R_2} & 1 \end{bmatrix} \begin{bmatrix} 1 & \frac{q_{e1}}{\xi_{e1}} \\ 0 & 1 \end{bmatrix} \begin{bmatrix} 1 & 0 \\ \frac{\xi_0-\xi_{e1}}{R_1} & 1 \end{bmatrix}$$
$$= \begin{bmatrix} 1 + \frac{q_{e1}}{\xi_{e1}} \frac{\xi_0-\xi_{e1}}{R_1} & \frac{q_{e1}}{\xi_{e1}} \\ \frac{\xi_0-\xi_{e1}}{R_1} + \frac{\xi_{e1}-\xi_2}{R_2}\left(1 + \frac{q_{e1}}{\xi_{e1}} \frac{\xi_0-\xi_{e1}}{R_1}\right) & \frac{\xi_{e1}-\xi_2}{R_2} \frac{q_{e1}}{\xi_{e1}} + 1 \end{bmatrix}, \tag{4.24}$$

Equation (4.24) is valid for biconvex, biconcave, and meniscus elements, provided that the sign notation stated in Sect. 4.1 is properly used on $R_1$, $R_2$, t′, and $\mu_2$. It is also applicable for plano-convex and plano-concave elements if $R_1$ or $R_2$

are assumed to be infinity. Now one can locate the cardinal points from Eqs. (4.15), (4.16), (4.18), (4.19), (4.22), and (4.23) for this thick lens.

If the thickness $q_{e1}$ of a lens is small enough so that its effect on calculation accuracy is negligible, the element is called a thin lens. We are interested in its cardinal points when this thin lens is surrounded by a single refractive medium, i.e., $\xi_0 = \xi_2 = \xi$. Its ABCD matrix can then be deduced from Eq. (4.24), given as:

$$\begin{bmatrix} A & B \\ C & D \end{bmatrix} = \begin{bmatrix} 1 & 0 \\ -(\xi_{e1} - \xi)\left(\frac{1}{R_1} - \frac{1}{R_2}\right) & 1 \end{bmatrix}. \tag{4.25}$$

Now the well known lens maker's formula of this thin lens is obtained from Eqs. (4.15) and (4.19):

$$\frac{1}{f} = \frac{-1}{f'} = \frac{-(\xi_{e1} - \xi)}{\xi}\left(\frac{1}{R_1} - \frac{1}{R_2}\right). \tag{4.26}$$

Equation (4.26) is one of the basic equations in optics. It states that the focal length of a thin lens is a function of the radii, $R_1$ and $R_2$, of lens surfaces and the lens refractive index $\xi_{e1}$. The reciprocal of $f'$ is defined as the optical power $\varphi$ of the lens, expressed in diopters ($m^{-1}$):

$$\varphi = \frac{1}{f'} = \frac{-1}{f} = \frac{(\xi_{e1} - \xi)}{\xi}\left(\frac{1}{R_1} - \frac{1}{R_2}\right). \tag{4.27}$$

The power $\varphi$ is a measurement of the ability to focus a parallel beam incident on a thin lens surrounded by a single medium. By substituting Eq. (4.27) into Eq. (4.25), one has the following simplified ABCD equation for this thin lens when it is surrounded by medium with refractive index $\xi$:

$$M_{lens} = \begin{bmatrix} 1 & 0 \\ -\xi/f' & 1 \end{bmatrix}. \tag{4.28}$$

## 4.6 Determination of the Focal Length of Curved Mirrors

We have investigated the cardinal points for systems on the assumption that a ray always travels from left to right. A curved mirror reverses the direction of propagation of a ray. Different types of curved mirrors are used in optical systems. Spherical, parabolic and hyperbolic mirrors are all common and may be used singly or in combination to produce the desired optical effect. A convex mirror, which is known as a diverging mirror, is a part of a spherical mirror whose reflecting surface is curved away from the centre of the sphere (Fig. 4.14). A concave mirror, which is also known as a converging mirror, is a part of a spherical mirror whose reflecting surface is curved towards the centre of the sphere (Fig. 4.15). It is evident that a mirror has only one focal point F, since the two

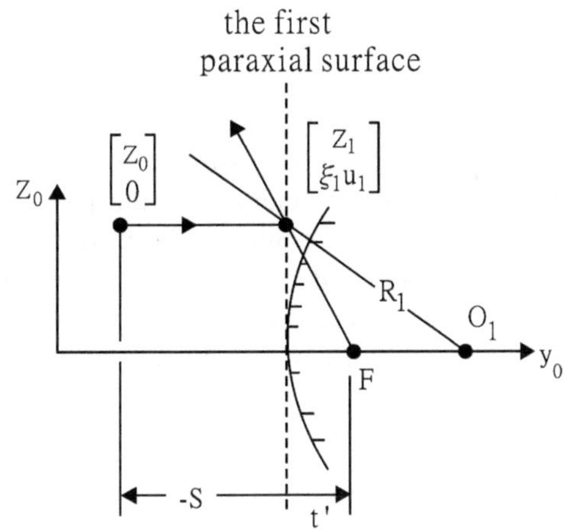


**Fig. 4.14** A convex mirror, whose center of curvature is located to the right of its surface, forms an erect virtual image

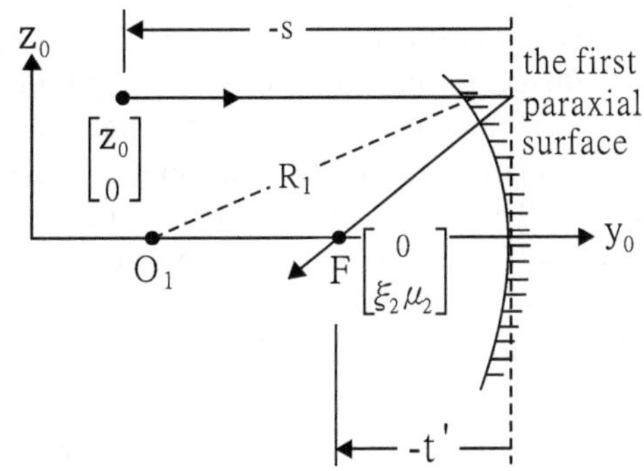


**Fig. 4.15** A concave mirror forms a real image

spaces are coincidental. The line joining the mirror vertex and its center $o_1$ is the optical axis. The ABCD matrix of both the convex and concave mirrors is

$$\begin{bmatrix} A & B \\ C & D \end{bmatrix} = \begin{bmatrix} 1 & 0 \\ \frac{2\xi_0}{-R_1} & -1 \end{bmatrix}, \tag{4.29}$$

where $R_1$ is positive for a convex mirror (Fig. 4.14) and negative for a concave mirror (Fig. 4.15). The focal length of the convex mirror can be determined by the following system equation:

$$\begin{bmatrix} z_1 \\ \xi_1\mu_1 \end{bmatrix} = \begin{bmatrix} z_1 \\ \xi_0\mu_1 \end{bmatrix} = \begin{bmatrix} 1 & 0 \\ \frac{2\xi_0}{-R_1} & -1 \end{bmatrix} \begin{bmatrix} 1 & \frac{-s}{\xi_0} \\ 0 & 1 \end{bmatrix} \begin{bmatrix} z_0 \\ 0 \end{bmatrix} = \begin{bmatrix} 1 \\ \frac{2\xi_0}{-R_1} \end{bmatrix} z_0. \tag{4.30}$$

Recall that the focal length of a converging lens is positive and the focal length of a diverging lens is negative. Therefore, the focal lengths of a convex mirror and a concave mirror are determined from $-\mu_1 = z_1/(-f')$ and $-\mu_1 = z_1/f'$, respectively. Both their effective back focal lengths are given by:

$$f' = \frac{z_1}{\mu_1} = \frac{-R_1}{2}. \tag{4.31}$$

As mentioned in Sect. 4.4 that t′ is the back focal length measured from the vertex of the boundary surface to the focal point $F_2$, one has to note that the principle plane coincides with the paraxial surface, leading $f' = -t'$ for both of convex and concave mirrors. From Eq. (4.31), we know the effective focal length of the concave and convex mirrors is half of the radius. The power $\varphi = 1/f' = -2/R_1$ of a concave mirror is positive, since $R_1$ is negative. Contrastingly, the power $\varphi = -2/R_1$ of a convex mirror remains negative as $R_1$ is now positive.

## 4.7 Image Position in an Optical System Using Cardinal Points

When the cardinal points of an optical system are known, the location and size of the image formed by the optical system can be readily determined. Shown in Fig. 4.16 are an off-axis source point A, an on-axis source point B, and their imaged points, A′ and B′. The length measured from the front principal point $H_1$ to B is s. The length measured from the back principal point $H_2$ to image B′ is s′, while g ($f = s + g$) is the length from the front focal point $F_1$ to B. The length from the back focal point $F_2$ to image point B′ is g′ ($f' = s' + g'$). The heights of the object and the image are h and h′, respectively. All above algebraic signs follow the sign convention stated in Sect. 4.1.

Generally, one can trace any two rays originating from the source point A with different slope angles to determine its image A′. But most researchers usually trace the following three rays:

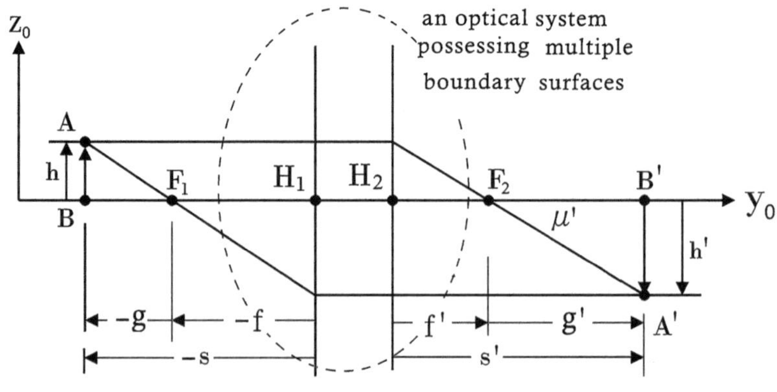


**Fig. 4.16** The image A′ can be determined by tracing different rays originated from its source point A

1. The parallel ray with height-slope-angle matrix $[z_0 \quad \xi_0\mu_0]^T = [h \quad 0]^T$. This ray will pass through the back focal point $F_2$; the refraction will appear to have occurred at the back principal plane.
2. The ray with height-slope-angle matrix $[z_0 \quad \xi_0\mu_0]^T = [h \quad \xi_0 h/g]^T$ ($\mu_0 = h/g$ is the negative slope angle). It passes through the front focal point $F_1$ and emerges from the system parallel to the optical axis.
3. A third ray could be constructed from A to the front nodal point $N_1$. This ray emerges from the back nodal point $N_2$ and is parallel to the entering ray.

Figure 4.17 shows an on-axis object B and its image B′ formed by an optical system having power $\varphi = 1/f'$ and being surrounded by media with refractive index $\xi$. One ray originating from B with slope angle $u$ propagates into the optical system. After refracting, the refracted ray intersects the optical axis with slope angle $u'$. The system matrix is

$$\begin{bmatrix} 0 \\ \xi\mu' \end{bmatrix} = \begin{bmatrix} 1 & \frac{s'}{\xi} \\ 0 & 1 \end{bmatrix} \begin{bmatrix} 1 & 0 \\ \frac{-\xi}{f'} & 1 \end{bmatrix} \begin{bmatrix} 1 & -\frac{s}{\xi} \\ 0 & 1 \end{bmatrix} \begin{bmatrix} 0 \\ \xi\mu \end{bmatrix} = \begin{bmatrix} -\frac{s}{\xi} + \frac{s'}{\xi}\left(\frac{s}{f'} + 1\right) \\ \frac{s}{f'} + 1 \end{bmatrix} \xi\mu. \quad (4.32)$$

The location of the image is determined by the first component of Eq. (4.32), given as:

$$\frac{1}{f'} = \frac{1}{s'} - \frac{1}{s}, \quad (4.33)$$

where $f'$ is the effective back focal length. Equation (4.33) is very useful in calculating the image locations of a simplified system surrounded by a single medium with refractive index $\xi$.

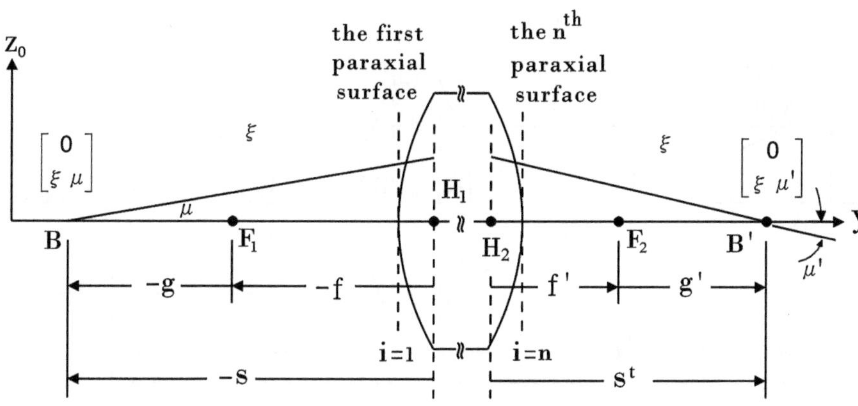


**Fig. 4.17** The object and image

## 4.8 Lateral Magnification

Optical magnification includes lateral magnification, longitudinal magnification, angular magnification, as well as other magnifications. Lateral magnification m is the ratio of some linear dimension $h'$, perpendicular to the optical axis, of an image formed by an optical system, to the corresponding linear dimension $h$ of the object (see Fig. 4.18). Figure 4.18 is a simplified optical system with power $\varphi = 1/\mathrm{f}$ and surrounded by a media with refractive index $\xi$. Also shown in this figure are the ray heights and slope angles at an object point A and its image A′. Now we have the following system equation:

$$\begin{bmatrix} \mathrm{h}' \\ \xi u' \end{bmatrix} = \begin{bmatrix} 1 & \frac{\mathrm{s}'}{\xi} \\ 0 & 1 \end{bmatrix} \begin{bmatrix} 1 & 0 \\ -\frac{\xi}{\mathrm{f}'} & 1 \end{bmatrix} \begin{bmatrix} 1 & \frac{-\mathrm{s}}{\xi} \\ 0 & 1 \end{bmatrix} \begin{bmatrix} \mathrm{h} \\ 0 \end{bmatrix} = \mathrm{h} \begin{bmatrix} 1 - \frac{\mathrm{s}'}{\mathrm{f}'} \\ -\frac{\xi}{\mathrm{f}'} \end{bmatrix}. \tag{4.34}$$

The lateral magnification m, by its definition, can be estimated from the first component of Eq. (4.34), given as:

$$\mathrm{m} = \frac{\mathrm{h}'}{\mathrm{h}} = 1 - \frac{\mathrm{s}'}{\mathrm{f}'} = \frac{\mathrm{f}'}{\mathrm{f}' + \mathrm{s}} = \frac{\mathrm{s}'}{\mathrm{s}}. \tag{4.35}$$

Note that Eq. (4.35) shows that the lateral magnification m is a negative number.

## 4.9 Longitudinal Magnification

Longitudinal magnification $\hat{\mathrm{m}}$ is the magnification along the optical axis, i.e., the magnification of the longitudinal thickness of the object or the magnification of a longitudinal motion along the optical axis. Mathematically $\hat{\mathrm{m}}$ of a simplified

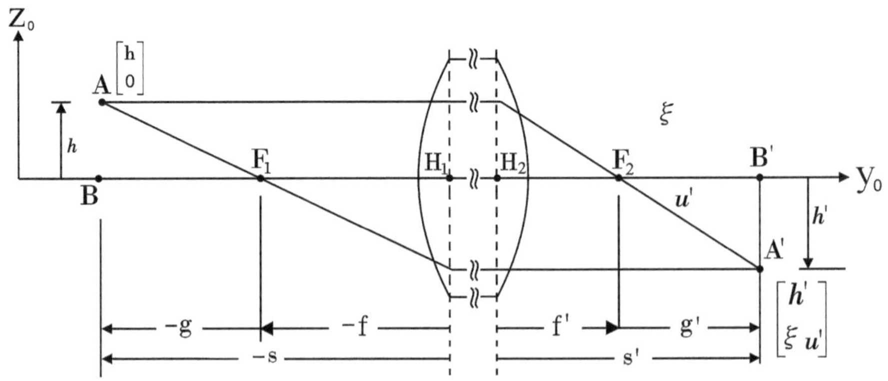


**Fig. 4.18** Lateral magnification m is the ratio of image height $h'$ to object height h

system with power $\varphi = 1/f'$ and surrounded by single medium can be estimated directly from Eq. (4.33), given as:

$$\hat{m} = \lim_{\Delta s \to 0} \frac{\Delta s'}{\Delta s} = \frac{ds'}{ds} = \frac{f'^2}{(f' + s)^2} = m^2. \tag{4.36}$$

One also can estimate longitudinal magnification $\hat{m}$ using the finite difference method,

$$\hat{m} = \frac{\Delta s'}{\Delta s} = \frac{s_2' - s_1'}{s_2 - s_1}, \tag{4.37}$$

where $s_1$ and $s_2$ denote the distances from the front principle point $H_1$ to the front and back edges of an object, respectively (see Fig. 4.19). $s_1'$ and $s_2'$ are the distances from the back principle point $H_2$ to the corresponding edges of the image, respectively. We have $s_1' = s_1 f'/(s_1 + f')$ and $s_2' = s_2 f'/(s_2 + f')$ from Eq. (4.33). After substituting these two equations into Eq. (4.37), one also obtains Eq. (4.36). One has to note that $\hat{m}$ is ordinarily positive, indicating that the object and the image always move in the same direction. Furthermore, since lateral magnification m varies with position $y_0$, longitudinal magnification $\hat{m}$ is also a function of $y_0$.

## 4.10 Focal Lengths of Two-Element Systems Surrounded by Air

We now consider a combination of two separated elements surrounded by air with a reflective index $\xi_{air} = 1$. Assume, as shown in Fig. 4.20, there are two elements spaced with a distance $v_2$, their powers $\phi_a$ and $\phi_b$ and back focus length $f'$. The ABCD matrix of this system is given by:

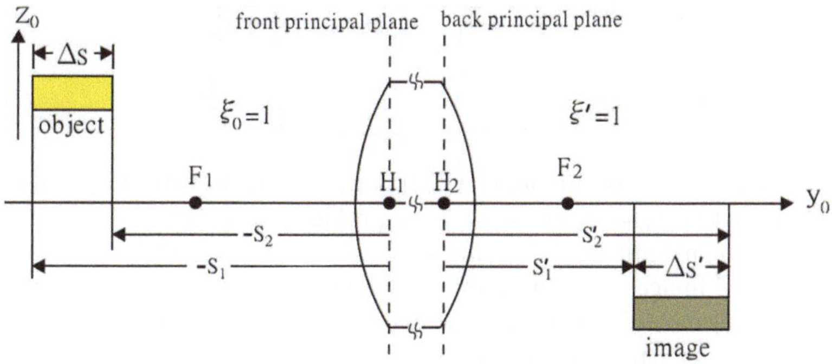


**Fig. 4.19** The longitudinal magnification can be estimated by finite difference method

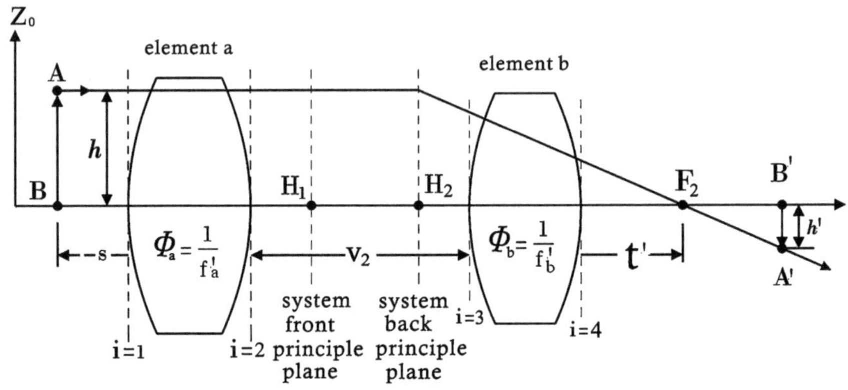


**Fig. 4.20** Two-component systems

$$\begin{bmatrix} A & B \\ C & D \end{bmatrix} = \begin{bmatrix} 1 & 0 \\ -\phi_b & 1 \end{bmatrix} \begin{bmatrix} 1 & v_2 \\ 0 & 1 \end{bmatrix} \begin{bmatrix} 1 & 0 \\ -\phi_a & 1 \end{bmatrix}$$

$$= \begin{bmatrix} 1 - v_2\phi_a & v_2 \\ -(\phi_a + \phi_b - \phi_a\phi_b v_2) & -\phi_b v_2 + 1 \end{bmatrix}. \quad (4.38)$$

With Eq. (4.38), the power $\phi_{ab}$ and effective back focal length $f'_{ab}$ can be estimated as:

$$\phi_{ab} = \phi_a + \phi_b - \phi_a\phi_b v_2 = \frac{1}{f'_a} + \frac{1}{f'_b} - \frac{v_2}{f'_a f'_b} = \frac{1}{f'_{ab}}. \quad (4.39)$$

The front and back focal lengths (i.e., $t_{ab}$ and $t'_{ab}$), the effective front and back focal lengths (i.e., $f_{ab}$ and $f'_{ab}$) can be estimated by Eqs. (4.15), (4.16), (4.18), and (4.19), yielding

$$t'_{ab} = -t_{ab} = \frac{1 - \phi_a v_2}{\phi_a + \phi_b - \phi_a\phi_b v_2} = \frac{f'_b(f'_a - v_2)}{f'_a + f'_b - v_2}, \quad (4.40)$$

$$f'_{ab} = -f_{ab} = \frac{1}{\phi_a + \phi_b - \phi_a\phi_b v_2} = \frac{f'_a f'_b}{f'_a + f'_b - v_2}. \quad (4.41)$$

There are two types of problems which occur in a two-element optical system. With reference to Fig. 4.21, the first type of problem occurs when we are given the required system magnification m, the space $v_2$ of the two elements, the object position $-s$ and image position $s'$, and neglecting the thicknesses of the elements. We wish to determine the powers of the two elements. This problem can be solved by tracing two rays whose height-slope-angle matrices are $[z_0 \quad \xi_0\mu_0]^T = [h \quad 0]^T$ and $[z_0 \quad \xi_0\mu_0]^T = [0 \quad \mu]^T$. One can use the following two system equations from these two rays:

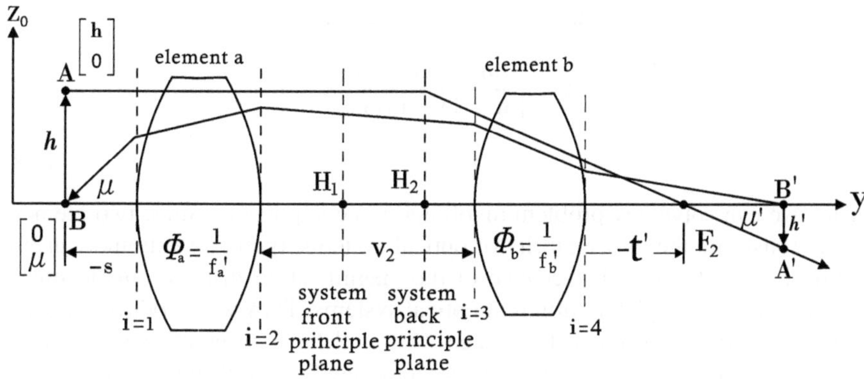


**Fig. 4.21** A two-element system operating at finite conjugate

$$\begin{bmatrix} h' \\ \mu' \end{bmatrix} = \begin{bmatrix} 1 & s' \\ 0 & 1 \end{bmatrix} \begin{bmatrix} 1 & 0 \\ -\phi_b & 1 \end{bmatrix} \begin{bmatrix} 1 & v_2 \\ 0 & 1 \end{bmatrix} \begin{bmatrix} 1 & 0 \\ -\phi_a & 1 \end{bmatrix} \begin{bmatrix} 1 & -s \\ 0 & 1 \end{bmatrix} \begin{bmatrix} h \\ 0 \end{bmatrix}$$

$$= \begin{bmatrix} (1-\phi_b s')(1-\phi_a v_2) - \phi_a s' \\ -\phi_b(1-\phi_a v_2) - \phi_a \end{bmatrix} h, \tag{4.42}$$

$$\begin{bmatrix} 0 \\ \mu'' \end{bmatrix} = \begin{bmatrix} 1 & s' \\ 0 & 1 \end{bmatrix} \begin{bmatrix} 1 & 0 \\ -\phi_b & 1 \end{bmatrix} \begin{bmatrix} 1 & v_2 \\ 0 & 1 \end{bmatrix} \begin{bmatrix} 1 & 0 \\ -\phi_a & 1 \end{bmatrix} \begin{bmatrix} 1 & -s \\ 0 & 1 \end{bmatrix} \begin{bmatrix} 0 \\ \mu \end{bmatrix}$$

$$= \begin{bmatrix} -s + v_2 + s v_2 \phi_a + s'[s(\phi_b + \phi_a - \phi_a\phi_b v_2) - \phi_b v_2 + 1] \\ s(\phi_b + \phi_a - \phi_a \phi_b v_2) - \phi_b v_2 + 1 \end{bmatrix} \mu. \tag{4.43}$$

One can solve for the required powers $\phi_a$ and $\phi_b$ from the first components of Eqs. (4.42) and (4.43) with $m = h'/h$:

$$\phi_a = \frac{(m v_2 - m v_2 - s')}{m s v_2}, \tag{4.44}$$

$$\phi_b = \frac{(v_2 - ms + s')}{v_2 s'}. \tag{4.45}$$

For the second type of problem, we are faced with the inverse case, in which we are given the component powers $\phi_a$ and $\phi_b$ (or the effective back focal lengths $f'_a$ and $f'_b$), the desired object-to-image distance $T = -s + v_2 + s'$, and the magnification $m$. We can determine the locations for the two elements by substituting the first component of Eq. (4.42) into the first component of Eq. (4.43) which results in the following quadratic equation:

$$0 = v_2^2 - v_2 T + T(f'_a + f'_b) + \frac{(m-1)^2 f'_a f'_b}{m}. \tag{4.46}$$

Then $s$ and $s'$ are easily determined from the valid solution of Eq. (4.46):

$$s = \frac{(m-1)v_2 + T}{(m-1) - mv_2\phi_a}, \tag{4.47}$$

$$s' = T + s - v_2. \tag{4.48}$$

Then we can solve any problem involving two components. Since two-element systems constitute the vast majority of optical systems, these are extremely useful equations. Note that the sign change of the magnification m from plus to minus will result in two completely different optical systems. They will produce the same enlargement (or reduction) of the image. One will have an erect image, and the other an inverted one. Nevertheless one system may be significantly more suitable than the other in specific applications.

## 4.11 The Optical Invariant in Paraxial Optics

The linearity of the paraxial raytracing equations results in the existence of an optical invariant, which is often denoted by $\Phi$. Figure 4.22 shows that two paraxial rays in a medium of refractive index $\xi$ propagate into a spherical boundary surface with radius R and then refract. After refracting, the two rays propagate in the element with refractive index $\xi'$. The optical invariant can be obtained by tracing two paraxial rays shown in Fig. 4.22:

$$\begin{bmatrix} z_a \\ \xi_a \mu_a \end{bmatrix} = \begin{bmatrix} z_a \\ \xi \mu_a \end{bmatrix} = \begin{bmatrix} 1 & 0 \\ \frac{\xi' - \xi}{R} & 1 \end{bmatrix} \begin{bmatrix} z'_a \\ \xi' \mu'_a \end{bmatrix} \tag{4.49}$$

$$\begin{bmatrix} z_b \\ \xi_b \mu_b \end{bmatrix} = \begin{bmatrix} z_b \\ \xi \mu_b \end{bmatrix} = \begin{bmatrix} 1 & 0 \\ \frac{\xi' - \xi}{R} & 1 \end{bmatrix} \begin{bmatrix} z'_b \\ \xi' \mu'_b \end{bmatrix} \tag{4.50}$$

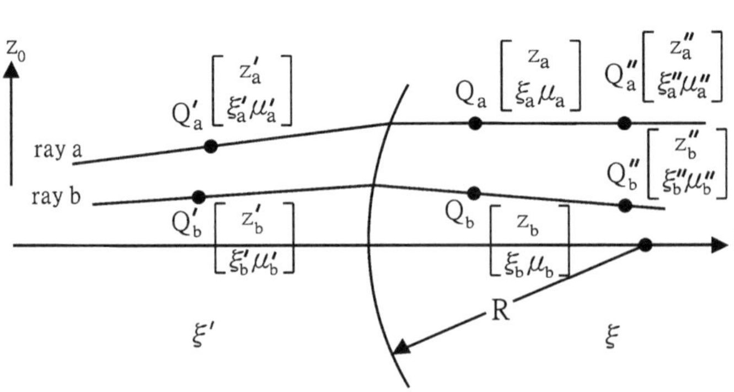


**Fig. 4.22** The two paraxial rays refracted by a spherical surface and then travelling in straight-line paths can determine the paraxial invariant

From Eqs. (4.49) and (4.50), we can get the following result:

$$\frac{\xi' - \xi}{R} = \frac{\xi \mu_a - \xi' \mu'_a}{z_a} = \frac{\xi \mu_b - \xi' \mu'_b}{z_b}. \quad (4.51)$$

Now we can obtain the optical invariant from Eq. (4.51) with $z_a = z'_a$ and $z_b = z'_b$:

$$\Phi = \xi(\mu_a z_b - \mu_b z_a) = \xi'(\mu'_a z'_b - \mu'_b z'_a). \quad (4.52)$$

The invariant in the straight-line traveling process (Fig. 4.22) can also be derived by continuously tracing these two rays travelling along their straight-line paths:

$$\begin{bmatrix} z''_a \\ \xi''_a \mu''_a \end{bmatrix} = \begin{bmatrix} z''_a \\ \xi \mu_a \end{bmatrix} = \begin{bmatrix} 1 & \frac{t}{\xi} \\ 0 & 1 \end{bmatrix} \begin{bmatrix} z_a \\ \xi_a \mu_a \end{bmatrix} = \begin{bmatrix} 1 & \frac{t}{\xi} \\ 0 & 1 \end{bmatrix} \begin{bmatrix} z_a \\ \xi \mu_a \end{bmatrix}, \quad (4.53)$$

$$\begin{bmatrix} z''_b \\ \xi''_b \mu''_b \end{bmatrix} = \begin{bmatrix} z''_b \\ \xi \mu_b \end{bmatrix} = \begin{bmatrix} 1 & \frac{t}{\xi} \\ 0 & 1 \end{bmatrix} \begin{bmatrix} z_b \\ \xi_b \mu_b \end{bmatrix} = \begin{bmatrix} 1 & \frac{t}{\xi} \\ 0 & 1 \end{bmatrix} \begin{bmatrix} z_b \\ \xi \mu_b \end{bmatrix}. \quad (4.54)$$

Thus, one has $(z'_a - z_a)/(\xi \mu_a) = t/\xi = (z'_b - z_b)/(\xi \mu_b)$ from Eqs. (4.53) and (4.54), and the invariant defined in Eq. (4.52) is also valid in this path. Through a similar series of operations, one can show that $\Phi = \xi(\mu_a z_b - \mu_b z_a)$ for a given surface is equal to $\Phi = \xi(\mu_a z_b - \mu_b z_a)$ for the next surface. Thus this term is invariant not only across the surface but also across the space between surfaces. It is therefore invariant throughout the entire optical system or any continuous part of the system.

The numerical value of invariant $\Phi$ may be calculated in any of several ways, and the invariant may then be used to arrive at the value of other quantities without the necessity of certain intermediate operations or raytracing calculations which would otherwise be required. The optical invariant is one of the most useful tools available to optical engineers. It can be used in developing optical layouts and system concepts, and providing guidance in optical design. The optical invariant is a "must" in an optical engineer's toolbox.

### *4.11.1 The Invariant and Magnification*

Once we have the optical invariant $\Phi$, we can estimate the magnifications, image height, and telescopic magnification data of a third ray from two traced rays, for an optical system. Figure 4.23 shows an object producing an image. By Eq. (4.52), we have the following equation:

$$\Phi = \xi_0(\mu_a \cdot 0 - \mu_b h) = \xi_2(\mu'_a \cdot 0 - \mu'_b h') = -\xi_0 \mu_b h = -\xi_2 \mu'_b h'. \quad (4.55)$$

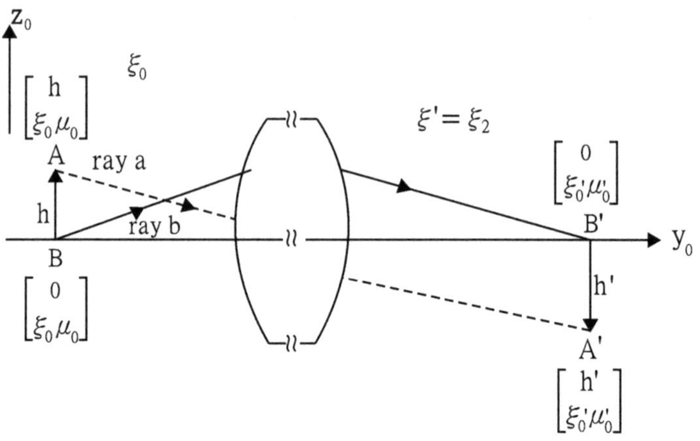


**Fig. 4.23** The invariant of a system can estimate the lateral magnification by only tracing an on-axis source point

Equation (4.55) can be rearranged to give a generalized expression for the magnification m of the optical system,

$$m = \frac{h'}{h} = \frac{\xi_0 \mu_b}{\xi_2 \mu_b'}. \tag{4.56}$$

This equation is useful for systems where the object and image are not in the air but in two media with different refractive indices. Equation (4.56) indicates that one can trace the ray b originating from B, an on-axis source point, with height-slope-angle matrix $[0 \quad \xi_0 \mu]^T$ to obtain the height-slope-angle matrix $[0 \quad \xi_0 \mu]^T$ at point B′, the image of B, and to estimate the magnification m of this system. If both object and image are in air, we have $m = \mu_b / \mu_b'$, indicating that the lateral magnification is the ratio of slope angles, $\mu_b$ and $\mu_b'$, of an incoming ray from B and its exit ray.

### 4.11.2 Image Height for Object at Infinity

Another useful application of the optical invariant Φ is the estimation of the image height of an optical system when its object is at infinity (see Fig. 4.24). Using Eq. (4.52), we can get the image height for an object at infinity from:

$$\Phi = \xi_0(\mu_a \cdot z_b - 0 \cdot z_a) = \xi_2(\mu_a' \cdot 0 - \mu_b' h') = \xi_0 \mu_a z_b = -\xi_2 \mu_b' h'. \tag{4.57}$$

Then we can have the image height

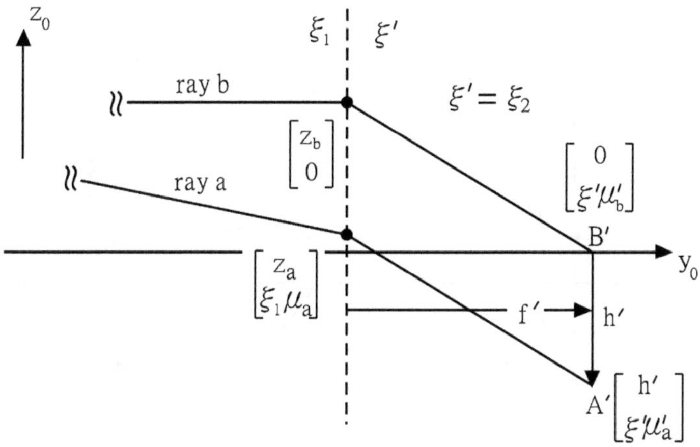


**Fig. 4.24** The image height of an optical system with its object at infinity can be estimated by using the optical invariant

$$h' = -\frac{\xi_0 \mu_a}{\xi_2 \mu'_b} z_b \,. \tag{4.58}$$

If both object and image are in air, we have $h' = -z_b \mu_a / \mu'_b$. Recall that we obtain the back focal length $t' = z_n/(-u_n) = z_b/(-u'_b)$ from Eq. (4.16). Therefore, Eq. (4.58) can be further simplified as:

$$h' = \mu_a t'. \tag{4.59}$$

Equation (4.59) shows that the image height is the product of the slope angle $\mu_a$ and the back focal length $t'$ of the optical system, when its object is at infinity.

### *4.11.3 Data of a Third Ray from Two Traced Rays*

Figure 4.25 shows that three rays propagate form a medium of refractive index $\xi$ to a medium of refractive index $\xi'$. If we know the height-slope-angle matrices of rays a and b, we can use the optical invariant to determine the ray height-slope-angle matrix of the ray c. In other words, a paraxial system is completely described by the ray data of any two unrelated rays. From the optical invariant Eq. (4.52), we can have the following two equations:

$$\xi(\mu_c z_a - \mu_a z_c) = \xi' z'_a u'_c - \xi' \mu'_a z'_c, \tag{4.60}$$

$$\xi(\mu_c z_b - \mu_b z_c) = \xi' z'_b \mu'_c - \xi' \mu'_b z'_c. \tag{4.61}$$

And then we can get $z'_c$ and $\mu'_c$ from Eqs. (4.60) and (4.61).

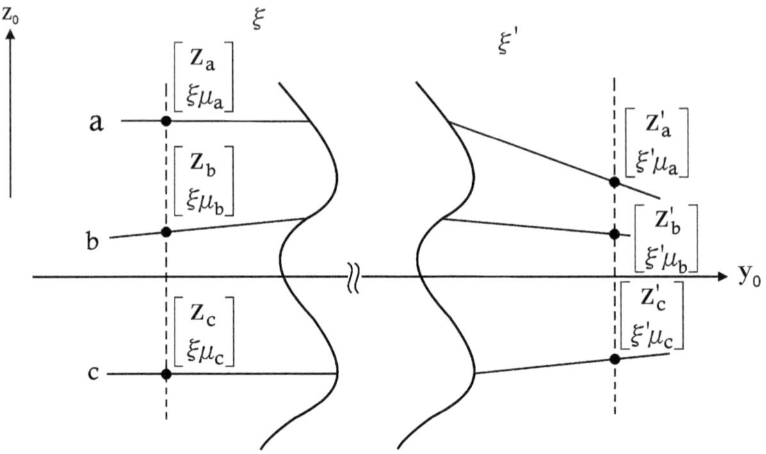


**Fig. 4.25** The height-slope-angle matrix of the third ray can be determined from those of the first and the second rays

$$\mu'_c = \frac{\begin{vmatrix} \xi(\mu_c z_a - \mu_a z_c) & -\xi'\mu'_a \\ \xi(\mu_c z_b - \mu_b z_c) & -\xi'\mu'_b \end{vmatrix}}{\begin{vmatrix} \xi' z' & -\xi'\mu'_a \\ \xi' z'_b & -\xi'\mu'_b \end{vmatrix}} = \frac{\xi(\mu_c z_b - \mu_b z_c)}{\xi'(\mu'_a z'_b - \mu'_b z'_a)}\mu'_a + \frac{\xi(\mu_a z_c - z_a\mu_c)}{\xi'(\mu'_a z'_b - \mu'_b z'_a)}\mu'_b, \tag{4.62}$$

$$z'_c = \frac{\begin{vmatrix} \xi' z'_a & \xi(\mu_c z_a - \mu_a z_c) \\ \xi' z'_b & \xi(\mu_c z_b - \mu_b z_c) \end{vmatrix}}{\begin{vmatrix} \xi' z' & -\xi'\mu'_a \\ \xi' z'_b & -\xi'\mu'_b \end{vmatrix}} = \frac{\xi(\mu_c z_b - \mu_b z_c)}{\xi'(\mu'_a z'_b - \mu'_b z'_a)}z'_a + \frac{\xi(\mu_a z_c - z_a\mu_c)}{\xi'(\mu'_a z'_b - \mu'_b z'_a)}z'_b. \tag{4.63}$$

Equations (4.62) and (4.63) show that we can determine the ray data of a third ray without further raytracing, by using:

$$u'_c = A\mu'_a + B\mu'_b, \tag{4.64}$$

$$z'_c = Az'_a + Bz'_b, \tag{4.65}$$

where $A = \frac{\xi(\mu_c Z_b - \mu_b Z_c)}{\xi'(\mu'_a Z'_b - \mu'_b Z'_a)}$ and $B = \frac{\xi(\mu_a Z_c - Z_a \mu_c)}{\xi'(\mu'_a Z'_b - \mu'_b Z'_a)}$.

## *4.11.4 Focal Length Determination*

Another application of the optical invariant is the determination of the back focal length $t'$ and the effective back focal length $f'$. The height $z'_c$ and angle $\mu'_c$ of a parallel ray, ray c in Fig. 4.26, can be determined by using Eqs. (4.66) and (4.67):

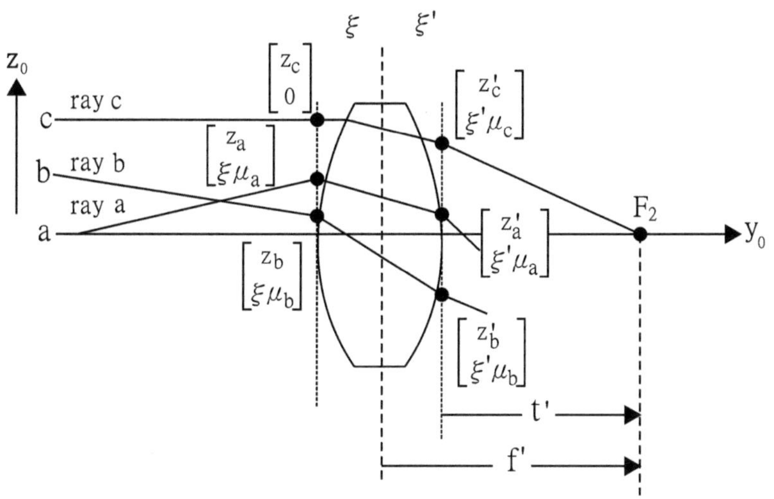


**Fig. 4.26** blf and eblf can be determined from the optical invariant

$$\mu'_c = A\mu'_a + B\mu'_b, \tag{4.66}$$

$$z'_c = Az'_a + Bz'_b, \tag{4.67}$$

where $A = \frac{\xi(-\mu_b z_c)}{\xi'(\mu'_a z'_b - \mu'_b z'_a)}$ and $B = \frac{\xi(\mu_a z_c)}{\xi'(\mu'_a z'_b - \mu'_b z'_a)}$.

Then we can get the effective back focal length f′ and the back focal length t′ by using Eqs. (4.15) and (4.16):

$$t' = -\frac{z'_c}{\mu'_c} = -\frac{-\xi\,\mu_b z'_a + \xi\,\mu_a z'_b}{-\xi\,\mu_b \mu'_a + \xi\,\mu_a \mu'_b}, \tag{4.68}$$

$$f' = -\frac{z_c}{\mu'_c} = -\frac{z_c}{A\,\mu'_a + B\,\mu'_b} = \frac{\xi'(\mu'_b z'_a - \mu'_a z'_b)}{\xi(\mu_a \mu'_b - \mu_b \mu'_a)}. \tag{4.69}$$

Most optical computer programs make use of Eqs. (4.68) and (4.69) to calculate the focal lengths. Such programs usually put an object at a remote, but finite distance and thus cannot directly calculate the focal lengths without a special calculation.

## References


1. A. Gerrard, J.M. Burch, *Introduction to Matrix Methods in Optics* (Wiley, New York, 1975)
2. A.E. Attard, Matrix optical analysis of skew rays in mixed systems of spherical and orthogonal cylindrical lenses. Appl. Opt. **23**, 2706–2709 (1984)
3. P.D. Lin, C.K. Sung, Matrix-based paraxial skew ray-tracing in 3D systems with non-coplanar optical axis. OPTIK—Int. J. Light Electron Opt. **117**, 329–340 (2006)

# Chapter 5
# The Jacobian Matrix of a Ray with Respect to System Variable Vector

The automation in optical design work has made variational raytracing to estimate the Jacobian matrix with respect to system variables by using finite difference method [1–14]. This means that each variable (e.g., $\xi_{e1}$ of Fig. 2.1) of the system is varied by a small amount, $\Delta\xi_{e1}$, and that the corresponding changes of optical quantities (e.g., the incident point $\overline{P}_{11}$ of Fig. 2.1) are calculated by normal raytracing. The differences of the obtained optical quantities $\Delta\overline{P}_{11}$ are then divided by the difference of that variable $\Delta\xi_{e1}$ so that the components of the Jacobian matrix $\partial\overline{P}_{11}/\partial\xi_{e1}$ are estimated by $\Delta\overline{P}_{11}/\Delta\xi_{e1}$. The finite difference method is straightforward and is easily implemented in computer codes, at the cost of additional raytracing steps. However, the accuracy of the results is reliant upon an appropriate choice of the $\Delta\xi_{e1}$. For example, an excessive $\Delta\xi_{e1}$ violates the assumption of local linearity, while an overly-small $\Delta\xi_{e1}$ reduces the difference between the original and the perturbed solutions and leads to high rounding errors. Instead of finite difference method the differential method can be used for determination of the Jacobian matrix. This fact is known in geometrical optics since many years and there exist several publications on the subject [15–22]. Unfortunately, only parts of system variables are considered in these publications, yielding that they are only valid for axis-symmetrical systems. Comparatively, in this chapter the Jacobian matrix $\partial\overline{R}_i/\partial\overline{X}_{sys}$ is determined by differential method for optical system containing flat and spherical boundary surfaces. This methodology is also applicable for systems with aspherical boundary surface if one follows the algorithm of Ref. [23].

## 5.1 The First-Order Derivative of a Merit Function

It is known from Chap. 2 that the ray $\overline{R}_i$ is not only function of the incoming ray $\overline{R}_{i-1}$ (Figs. 5.1, 5.2, 5.3 and 5.4), but also of boundary variable vector $\overline{X}_i$ given in Eq. (2.28) (Fig. 5.5). In mathematical language, ray $\overline{R}_i$ is a recursive function with the given function $\overline{R}_0 = \overline{R}_0(\overline{X}_0)$, where $\overline{X}_0$ is given in Eq. (2.3), as

PD Lin, *New Computation Methods for Geometrical Optics*,
Springer Series in Optical Sciences 178, DOI: 10.1007/978-981-4451-79-6_5,

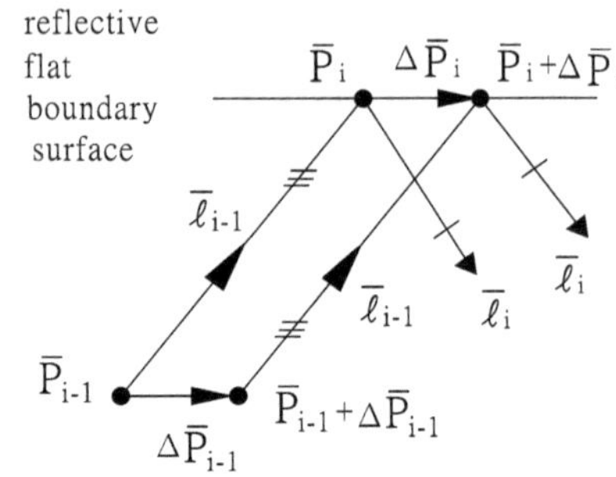


**Fig. 5.1** The variations of incidence point and its unit directional vector, $\Delta\overline{P}_i$ and $\Delta\overline{\ell}_i$, due to the change of $\overline{P}_{i-1}$ at a reflective flat boundary surface

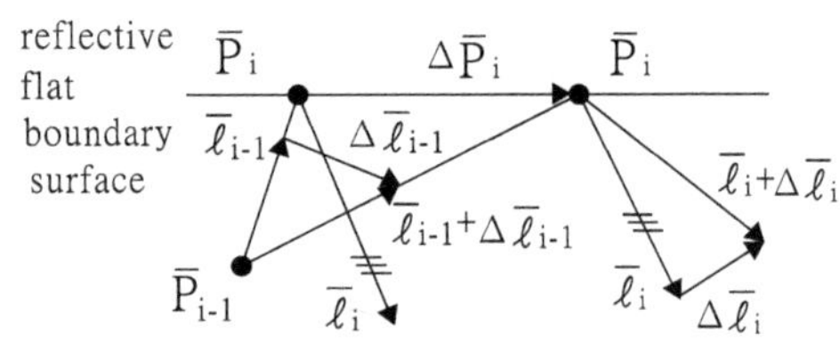


**Fig. 5.2** The variations of incidence point and its unit directional vector, $\Delta\overline{P}_i$ and $\Delta\overline{\ell}_i$, due to the change of $\overline{\ell}_{i-1}$ at a reflective flat boundary surface

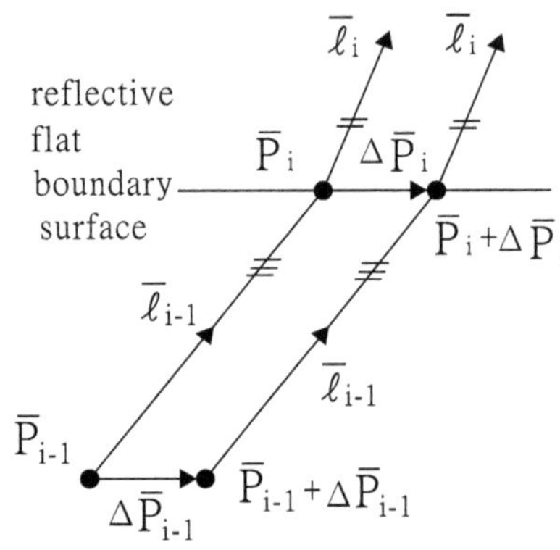


**Fig. 5.3** The variations of incidence point and its unit directional vector, $\Delta\overline{P}_i$ and $\Delta\overline{\ell}_i$, due to the change of $\overline{P}_{i-1}$ at a refractive flat boundary surface

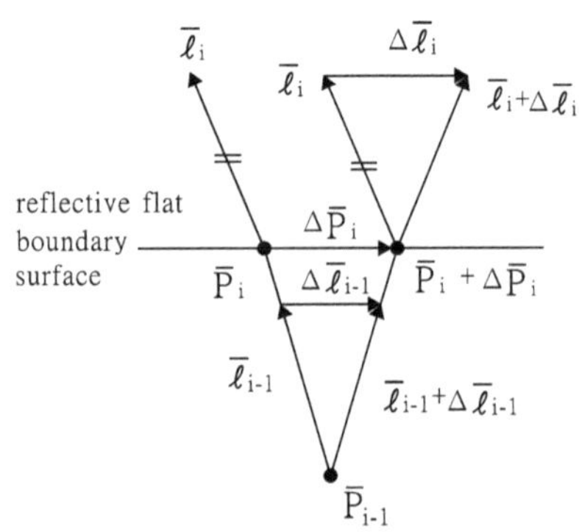


**Fig. 5.4** The variations of incidence point and its unit directional vector, $\Delta\overline{P}_i$ and $\Delta\overline{\ell}_i$, due to the change of $\overline{\ell}_{i-1}$ at a refractive flat boundary surface

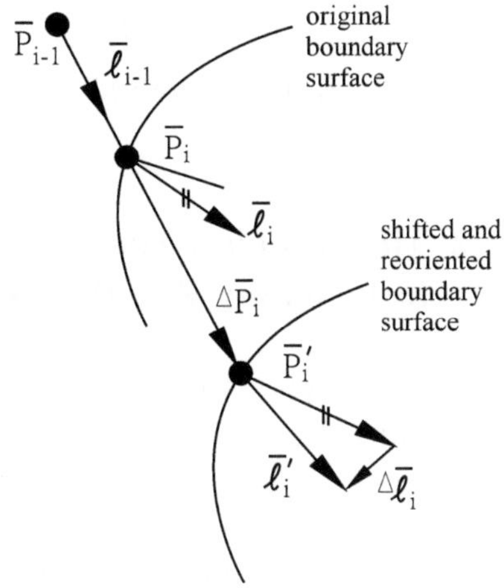


**Fig. 5.5** Change in refracted/reflected ray $\Delta\bar{R}_i = \begin{bmatrix}\Delta\bar{P}_i & \Delta\bar{\ell}_i\end{bmatrix}^T$ as result of change in boundary variables, $\Delta\bar{X}_i$

$$\bar{R}_i = \bar{R}_i\left(\bar{R}_{i-1}, \bar{X}_i\right). \tag{5.1}$$

If a merit function $\Phi$ is defined in terms of the ray $\bar{R}_n$ on the image plane, its Jacobian matrix $\partial\Phi/\partial\bar{X}_{sys}$ has the form

$$\begin{aligned}
\frac{\partial\Phi}{\partial\bar{X}_{sys}} &= \frac{\partial\Phi}{\partial\bar{R}_n}\left(\frac{\partial\bar{R}_n}{\partial\bar{X}_n}\frac{\partial\bar{X}_n}{\partial\bar{X}_{sys}} + \frac{\partial\bar{R}_n}{\partial\bar{R}_{n-1}}\frac{\partial\bar{R}_{n-1}}{\partial\bar{X}_{sys}}\right) \\
&= \frac{\partial\Phi}{\partial\bar{R}_n}\frac{\partial\bar{R}_n}{\partial\bar{X}_n}\frac{\partial\bar{X}_n}{\partial\bar{X}_{sys}} + \frac{\partial\Phi}{\partial\bar{R}_n}\frac{\partial\bar{R}_n}{\partial\bar{R}_{n-1}}\left(\frac{\partial\bar{R}_{n-1}}{\partial\bar{X}_{n-1}}\frac{\partial\bar{X}_{n-1}}{\partial\bar{X}_{sys}} + \frac{\partial\bar{R}_{n-1}}{\partial\bar{R}_{n-2}}\frac{\partial\bar{R}_{n-2}}{\partial\bar{X}_{sys}}\right) + \ldots \\
&= \frac{\partial\Phi}{\partial\bar{R}_n}\frac{\partial\bar{R}_n}{\partial\bar{X}_n}\frac{\partial\bar{X}_n}{\partial\bar{X}_{sys}} + \frac{\partial\Phi}{\partial\bar{R}_n}\frac{\partial\bar{R}_n}{\partial\bar{R}_{n-1}}\frac{\partial\bar{R}_{n-1}}{\partial\bar{X}_{n-1}}\frac{\partial\bar{X}_{n-1}}{\partial\bar{X}_{sys}} + \ldots + \frac{\partial\Phi}{\partial\bar{R}_n}\frac{\partial\bar{R}_n}{\partial\bar{R}_{n-1}}\frac{\partial\bar{R}_{n-1}}{\partial\bar{R}_{n-2}} \\
&\quad \cdots\frac{\partial\bar{R}_2}{\partial\bar{R}_1}\frac{\partial\bar{R}_1}{\partial\bar{X}_1}\frac{\partial\bar{X}_1}{\partial\bar{X}_{sys}} + \frac{\partial\Phi}{\partial\bar{R}_n}\frac{\partial\bar{R}_n}{\partial\bar{R}_{n-1}}\frac{\partial\bar{R}_{n-1}}{\partial\bar{R}_{n-2}}\cdots\frac{\partial\bar{R}_i}{\partial\bar{R}_{i-1}}\cdots\frac{\partial\bar{R}_1}{\partial\bar{R}_0}\frac{\partial\bar{R}_0}{\partial\bar{X}_0}\frac{\partial\bar{X}_0}{\partial\bar{X}_{sys}}.
\end{aligned} \tag{5.2}$$

In general, Eq. (5.2) indicates that the derivative matrices $\partial\bar{R}_i/\partial\bar{R}_{i-1}$(i = 1 to i = n), $\partial\bar{R}_i/\partial\bar{X}_i$ (i = 0 to i = n), and $\partial\bar{X}_i/\partial\bar{X}_{sys}$(i = 0 to i = n) are required in advance to determine the Jacobian matrix of the merit function $\Phi$. Therefore, this chapter is conceptually divided into the following sections. Sections 5.2 and 5.3 investigate the Jacobian matrix $\partial\bar{R}_i/\partial\bar{R}_{i-1}$ of ray $\bar{R}_i$ with respect to its incoming ray $\bar{R}_{i-1}$ for flat and spherical boundary surfaces. Sections 5.4 and 5.5 provide the Jacobian matrix $\partial\bar{R}_i/\partial\bar{X}_i$ of ray $\bar{R}_i$ with respect to boundary variable vector $\bar{X}_i$ for these two boundary surfaces. Section 5.6 addresses the Jacobian matrix of a ray $\bar{R}_i$ with respect to the system variable vector $\bar{X}_{sys}$ of an optical system. In Sections 5.7 and 5.8, the first- and second-order derivative matrices (i.e., $\partial\bar{X}_i/\partial\bar{X}_{sys}$ and

$\partial^2\overline{X}_i/\partial\overline{X}^2_{sys}$) of the boundary variable vector $\overline{X}_i$ with respect to the system variable vector $\overline{X}_{sys}$ will be addressed.

## 5.2 The Jacobian Matrix of Ray $\overline{R}_i$ with Respect to Incoming Ray $\overline{R}_{i-1}$ at a Flat Boundary Surface

As shown in Figs. 5.1 and 5.2 (or Figs. 5.3 and 5.4), any change in the incidence point, $\Delta\overline{P}_{i-1}$, and/or unit directional vector, $\Delta\overline{\ell}_{i-1}$, of the incoming ray from last boundary surface $\bar{r}_{i-1}$ causes a corresponding change in the refracted/or reflected ray, $\Delta\overline{R}_i$, at the present boundary surface $\bar{r}_i$. The ith ray $\overline{R}_i$ includes the incidence point $\overline{P}_i$ and its reflected/or refracted unit directional vector $\overline{\ell}_i$. Therefore, the Jacobian matrix $\partial\overline{R}_i/\partial\overline{R}_{i-1}$comprises two components, namely $\partial\overline{P}_i/\partial\overline{R}_{i-1}$ and $\partial\overline{\ell}_i/\partial\overline{R}_{i-1}$, as discussed in the following.

(a) $\partial\overline{P}_i/\partial\overline{R}_{i-1} = \left[\partial\overline{P}_i/\partial x_v\right]$, $x_v \in \overline{R}_{i-1}$, at a flat boundary surface can be obtained by differentiating Eq. (2.36) to give

$$\frac{\partial\overline{P}_i}{\partial\overline{R}_{i-1}} = \left[\frac{\partial\overline{P}_i}{\partial x_v}\right] = \begin{bmatrix} \partial P_{ix}/\partial x_v \\ \partial P_{iy}/\partial x_v \\ \partial P_{iz}/\partial x_v \end{bmatrix} = \begin{bmatrix} 1 & 0 & 0 & \lambda_i & 0 & 0 \\ 0 & 1 & 0 & 0 & \lambda_i & 0 \\ 0 & 0 & 1 & 0 & 0 & \lambda_i \end{bmatrix} + \begin{bmatrix} \ell_{i-1x} \\ \ell_{i-1y} \\ \ell_{i-1z} \end{bmatrix} \left[\frac{\partial\lambda_i}{\partial x_v}\right], \tag{5.3}$$

where $[\partial\lambda_i/\partial x_v]$ is given as

$$\frac{\partial\lambda_i}{\partial\overline{R}_{i-1}} = \left[\frac{\partial\lambda_i}{\partial x_v}\right] = -\frac{1}{E_i}\begin{bmatrix} J_{ix} & J_{iy} & J_{iz} & 0 & 0 & 0 \end{bmatrix} + \frac{D_i}{E_i^2}\begin{bmatrix} 0 & 0 & 0 & J_{ix} & J_{iy} & J_{iz} \end{bmatrix}, \tag{5.4}$$

and $D_i$ and $E_i$ are defined in Eqs. (2.39) and (2.40).

(b) When $\overline{\ell}_i$ is the unit directional vector of the reflected ray at a flat boundary surface, $\partial\overline{\ell}_i/\partial\overline{R}_{i-1} = \left[\partial\overline{\ell}_i/\partial x_v\right]$, $x_v \in \overline{R}_{i-1}$, can be obtained by differentiating Eq. (2.44) to give

$$\frac{\partial\overline{\ell}_i}{\partial\overline{R}_{i-1}} = \left[\frac{\partial\overline{\ell}_i}{\partial x_v}\right] = \begin{bmatrix} \partial\ell_{ix}/\partial x_v \\ \partial\ell_{iy}/\partial x_v \\ \partial\ell_{iz}/\partial x_v \end{bmatrix}$$
$$= \begin{bmatrix} 0 & 0 & 0 & 1-2J_{ix}J_{ix} & -2J_{ix}J_{iy} & -2J_{ix}J_{iz} \\ 0 & 0 & 0 & -2J_{iy}J_{ix} & 1-2J_{iy}J_{iy} & -2J_{iy}J_{iz} \\ 0 & 0 & 0 & -2J_{iz}J_{ix} & -2J_{iz}J_{iy} & 1-2J_{iz}J_{iz} \end{bmatrix}. \tag{5.5}$$

(c) When $\bar{\ell}_i$ is the unit directional vector of the refracted ray at a flat boundary surface, $\partial\bar{\ell}_i/\partial\bar{R}_{i-1} = [\partial\bar{\ell}_i/\partial x_v]$, $x_v \in \bar{R}_{i-1}$, can be obtained by differentiating (2.45) to give

$$\frac{\partial\bar{\ell}_i}{\partial\bar{R}_{i-1}} = \left[\frac{\partial\bar{\ell}_i}{\partial x_v}\right] = \begin{bmatrix} \partial\ell_{ix}/\partial x_v \\ \partial\ell_{iy}/\partial x_v \\ \partial\ell_{iz}/\partial x_v \end{bmatrix} = \begin{bmatrix} 0 & 0 & 0 & N_i & 0 & 0 \\ 0 & 0 & 0 & 0 & N_i & 0 \\ 0 & 0 & 0 & 0 & 0 & N_i \end{bmatrix}$$
$$+ \left( \frac{s_i E_i N_i^2}{\sqrt{1 - N_i^2 + (N_i E_i)^2}} - N_i \right) \begin{bmatrix} 0 & 0 & 0 & J_{ix}J_{ix} & J_{ix}J_{iy} & J_{ix}J_{iz} \\ 0 & 0 & 0 & J_{iy}J_{ix} & J_{iy}J_{iy} & J_{iy}J_{iz} \\ 0 & 0 & 0 & J_{iz}J_{ix} & J_{iz}J_{iy} & J_{iz}J_{iz} \end{bmatrix}. \tag{5.6}$$

As indicated by Eqs. (5.5) and (5.6), the change of $\bar{P}_{i-1}$ doesn't cause any change of $\bar{\ell}_i$ when a ray is reflected/or refracted by a flat boundary surface. This fact also is shown by Figs. 5.1 and 5.3.

Combining Eqs. (5.3) and (5.5) for a reflection process, or Eqs. (5.3) and (5.6) for a refraction process, one obtains the Jacobian matrix $\partial\bar{R}_i/\partial\bar{R}_{i-1}$ of the ith ray $\bar{R}_i$ with respect to its incoming ray $\bar{R}_{i-1}$, as

$$\frac{\partial\bar{R}_i}{\partial\bar{R}_{i-1}} = \bar{M}_i = \begin{bmatrix} \frac{\partial\bar{P}_i}{\partial\bar{P}_{i-1}} & \frac{\partial\bar{P}_i}{\partial\bar{\ell}_{i-1}} \\ \bar{0}_{3\times3} & \frac{\partial\bar{\ell}_i}{\partial\bar{\ell}_{i-1}} \end{bmatrix}. \tag{5.7}$$

Equation (5.7) can be used to determine the change of refracted/or reflected ray, $\Delta\bar{R}_i$, in terms of the product of the Jacobian matrix $\bar{M}_i$ and the change of incoming ray, $\Delta\bar{R}_{i-1}$:

$$\Delta\bar{R}_i = \bar{M}_i\,\Delta\bar{R}_{i-1} = \begin{bmatrix} \frac{\partial\bar{P}_i}{\partial\bar{P}_{i-1}} & \frac{\partial\bar{P}_i}{\partial\bar{\ell}_{i-1}} \\ \bar{0}_{3\times3} & \frac{\partial\bar{\ell}_i}{\partial\bar{\ell}_{i-1}} \end{bmatrix} \begin{bmatrix} \Delta\bar{P}_{i-1} \\ \Delta\bar{\ell}_{i-1} \end{bmatrix}. \tag{5.8}$$

If all the boundary surfaces of an optical system are flat ones, by successive applications of Eq. (5.8) one can express the change of any intermediate ray, $\Delta\bar{R}_g$, at the gth boundary surface in terms of the changes of its source ray, $\begin{bmatrix} \Delta\bar{P}_0 & \Delta\bar{\ell}_0 \end{bmatrix}^T$, as

$$\Delta\bar{R}_g = \bar{M}_g\bar{M}_{g-1}\ldots\ldots\bar{M}_2\bar{M}_1\,\Delta\bar{R}_0 = \begin{bmatrix} \frac{\partial\bar{P}_g}{\partial\bar{P}_0} & \frac{\partial\bar{P}_g}{\partial\bar{\ell}_0} \\ \bar{0}_{3\times3} & \frac{\partial\bar{\ell}_g}{\partial\bar{\ell}_0} \end{bmatrix} \begin{bmatrix} \Delta\bar{P}_0 \\ \Delta\bar{\ell}_0 \end{bmatrix}. \tag{5.9}$$

The determination of $\Delta\overline{R}_g$ from Eq. (5.9) needs the change of the source ray, $\Delta\overline{R}_0$, which can be obtained by directly differentiating the source ray $\overline{R}_0 = \begin{bmatrix}\overline{P}_0 & \overline{\ell}_0\end{bmatrix}^T$ (see Eqs. (2.1) and (2.2)), resulting in $\Delta\overline{R}_0 = \partial\overline{R}_0/\partial\overline{X}_0\Delta\overline{X}_0 = \overline{S}_0\,\Delta\overline{X}_0$. Here $\Delta\overline{X}_0$ is the change of the variable vector $\overline{X}_0$ [Eq. (2.3)] of the source ray $\overline{R}_0$, and

$$\frac{\partial\overline{R}_0}{\partial\overline{X}_0} = \overline{S}_0 = \begin{bmatrix} \partial P_{0x}/\partial\overline{X}_0 \\ \partial P_{0y}/\partial\overline{X}_0 \\ \partial P_{0z}/\partial\overline{X}_0 \\ \partial\ell_{0x}/\partial\overline{X}_0 \\ \partial\ell_{0y}/\partial\overline{X}_0 \\ \partial\ell_{0z}/\partial\overline{X}_0 \end{bmatrix} = \begin{bmatrix} 1 & 0 & 0 & 0 & 0 \\ 0 & 1 & 0 & 0 & 0 \\ 0 & 0 & 1 & 0 & 0 \\ 0 & 0 & 0 & -C\beta_0 S(90^\circ+\alpha_0) & -S\beta_0 C(90^\circ+\alpha_0) \\ 0 & 0 & 0 & C\beta_0 C(90^\circ+\alpha_0) & -S\beta_0 S(90^\circ+\alpha_0) \\ 0 & 0 & 0 & 0 & C\beta_0 \end{bmatrix}. \tag{5.10}$$

Note that the matrix multiplication of Eq. (5.9) shows that $\partial\overline{\ell}_g/\partial\overline{P}_0 = \overline{0}_{3\times3}$, indicating that the directional vector $\overline{\ell}_g$ of any ray crossing flat boundary surfaces is independent of its source point $\overline{P}_0$. Furthermore, from Eqs. (5.8), and (5.9) one has

$$\frac{\partial\overline{\ell}_g}{\partial\overline{\ell}_0} = \frac{\partial\overline{\ell}_g}{\partial\overline{\ell}_{g-1}}\frac{\partial\overline{\ell}_{g-1}}{\partial\overline{\ell}_{g-2}}\cdots\frac{\partial\overline{\ell}_i}{\partial\overline{\ell}_{i-1}}\cdots\frac{\partial\overline{\ell}_2}{\partial\overline{\ell}_1}\frac{\partial\overline{\ell}_1}{\partial\overline{\ell}_0}. \tag{5.11}$$

Here $\partial\overline{\ell}_i/\partial\overline{\ell}_{i-1}$ for reflection and refraction processes can be respectively obtained from Eqs. (5.5) and (5.6) after the unit normal vector $\overline{n}_i$ of Eq. (2.35) is used:

$$\begin{aligned}\frac{\partial\overline{\ell}_i}{\partial\overline{\ell}_{i-1}} = \frac{\partial\overline{\ell}_i(\overline{n}_i)}{\partial\overline{\ell}_{i-1}} &= \begin{bmatrix} 1-2J_{ix}J_{ix} & -2J_{ix}J_{iy} & -2J_{ix}J_{iz} \\ -2J_{iy}J_{ix} & 1-2J_{iy}J_{iy} & -2J_{iy}J_{iz} \\ -2J_{iz}J_{ix} & -2J_{iz}J_{iy} & 1-2J_{iz}J_{iz} \end{bmatrix} \\ &= \begin{bmatrix} 1-2n_{ix}n_{ix} & -2n_{ix}n_{iy} & -2n_{ix}n_{iz} \\ -2n_{iy}n_{ix} & 1-2n_{iy}n_{iy} & -2n_{iy}n_{iz} \\ -2n_{iz}n_{ix} & -2n_{iz}n_{iy} & 1-2n_{iz}n_{iz} \end{bmatrix} \\ &= \overline{I}_{3\times3} - 2\overline{n}_i\overline{n}_i^T,\end{aligned} \tag{5.12}$$

and

$$\frac{\partial\bar{\ell}_i}{\partial\bar{\ell}_{i-1}} = \frac{\partial\bar{\ell}_i(\bar{n}_i, N_i)}{\partial\bar{\ell}_{i-1}} = N_i\begin{bmatrix}1 & 0 & 0\\0 & 1 & 0\\0 & 0 & 1\end{bmatrix}$$
$$+\left(\frac{s_iE_iN_i^2}{\sqrt{1-N_i^2+(N_iE_i)^2}} - N_i\right)\begin{bmatrix}J_{ix}J_{ix} & J_{ix}J_{iy} & J_{ix}J_{iz}\\J_{iy}J_{ix} & J_{iy}J_{iy} & J_{iy}J_{iz}\\J_{iz}J_{ix} & J_{iz}J_{iy} & J_{iz}J_{iz}\end{bmatrix}$$
$$= N_i\begin{bmatrix}1 & 0 & 0\\0 & 1 & 0\\0 & 0 & 1\end{bmatrix} + \left(\frac{s_iE_iN_i^2}{\sqrt{1-N_i^2+(N_iE_i)^2}} - N_i\right)\begin{bmatrix}n_{ix}n_{ix} & n_{ix}n_{iy} & n_{ix}n_{iz}\\n_{iy}n_{ix} & n_{iy}n_{iy} & n_{iy}n_{iz}\\n_{iz}n_{ix} & n_{iz}n_{iy} & n_{iz}n_{iz}\end{bmatrix}$$
$$= N_i\bar{I}_{3\times3} + (H_i - N_i)\bar{n}_i\bar{n}_i^T. \tag{5.13}$$

Equation (5.12) indicates that in reflection process the Jacobian matrix $\partial\bar{\ell}_i/\partial\bar{\ell}_{i-1}$ is function of the unit normal vector $\bar{n}_i$ of the ith boundary surface. Contrastingly, in the refraction case $\partial\bar{\ell}_i/\partial\bar{\ell}_{i-1}$ is not only function of $\bar{n}_i$, but also of $N_i = \xi_{i-1}/\xi_i$ and $E_i$.

*Example 5.1* Prisms are common optical elements containing only flat boundary surfaces. One of the major functions of prisms is reorientation of an image. By using Eq. (5.11), a merit function can be defined to specify the image orientation change after the image is refracted/or reflected by a prism. To achieve this, it is first necessary to establish a world coordinate frame $(xyz)_0$ to mark the orientation of an object before the prism (e.g., Fig. 5.6). Note that, for simplicity, coordinate

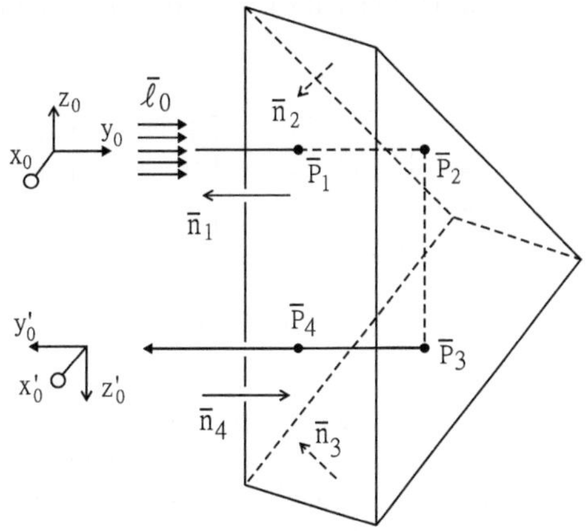


**Fig. 5.6** A right-angle prism can invert the image from top to bottom or right to left if the rays enter and exit this prism perpendicularly

frame $(xyz)_0$ always emits light rays in the $y_0$ direction and then enter into the first flat boundary surface. The object's image $(xyz)'_0$ will indicate the orientation of the image after the light rays pass through the prism. An example is the right-angle prism of Fig. 5.6 with $\bar{n}_1 = [0 \quad -1 \quad 0 \quad 0]^T, \bar{n}_2 = \left[0 \quad -\sqrt{1/2} \quad -\sqrt{1/2} \quad 0\right]^T$, $\bar{n}_3 = \left[0 \quad -\sqrt{1/2} \quad \sqrt{1/2} \quad 0\right]^T$ and $\bar{n}_4 = [0 \quad 1 \quad 0 \quad 0]^T$. The image orientation change can be obtained from Eq. (5.11) with g = 4, if the rays are perpendicular to the first and the last flat boundary surfaces, given as

$$\begin{aligned}\frac{\partial\bar{\ell}_4}{\partial\bar{\ell}_0} &= \frac{\partial\bar{\ell}_4(\bar{n}_4, N_4)}{\partial\bar{\ell}_3}\frac{\partial\bar{\ell}_3}{\partial\bar{\ell}_2}\frac{\partial\bar{\ell}_2}{\partial\bar{\ell}_1}\frac{\partial\bar{\ell}_1(\bar{n}_1, N_1)}{\partial\bar{\ell}_0} \\ &= \begin{bmatrix} 1/N_1 & 0 & 0 \\ 0 & 1/N_1^2 & 0 \\ 0 & 0 & 1/N_1 \end{bmatrix}\begin{bmatrix} 1 & 0 & 0 \\ 0 & 0 & 1 \\ 0 & 1 & 0 \end{bmatrix}\begin{bmatrix} 1 & 0 & 0 \\ 0 & 0 & -1 \\ 0 & -1 & 0 \end{bmatrix}\begin{bmatrix} N_1 & 0 & 0 \\ 0 & N_1^2 & 0 \\ 0 & 0 & N_1 \end{bmatrix} \\ &= \begin{bmatrix} 1 & 0 & 0 \\ 0 & -1 & 0 \\ 0 & 0 & -1 \end{bmatrix}.\end{aligned} \tag{5.14}$$

This equation indicates that a right-angle prism can invert the image from top to bottom or right to left if the rays enter and exit this prism perpendicularly.

*Example 5.2* The retroreflective property of the corner-cube retroreflectors has been put to use in safety reflectors and the laser ranging targets. Here the corner-cube mirror (Fig. 5.7), which is made of three-front-surface mirrors, will be used as illustrative example. The unit normal vectors of its three mirrors are $\bar{n}_1 = [0 \quad 0 \quad 1 \quad 0]^T$, $\bar{n}_2 = [0 \quad 1 \quad 0 \quad 0]^T$, and $\bar{n}_3 = [1 \quad 0 \quad 0 \quad 0]^T$. We can apply Eq. (5.11) with g = 3 to prove this retroreflective property by

$$\begin{aligned}\frac{\partial\bar{\ell}_3}{\partial\bar{\ell}_0} = \frac{\partial\bar{\ell}_3}{\partial\bar{\ell}_2}\frac{\partial\bar{\ell}_2}{\partial\bar{\ell}_1}\frac{\partial\bar{\ell}_1}{\partial\bar{\ell}_0} &= \begin{bmatrix} 1 & 0 & 0 \\ 0 & 1 & 0 \\ 0 & 0 & -1 \end{bmatrix}\begin{bmatrix} 1 & 0 & 0 \\ 0 & -1 & 0 \\ 0 & 0 & 1 \end{bmatrix}\begin{bmatrix} -1 & 0 & 0 \\ 0 & 1 & 0 \\ 0 & 0 & 1 \end{bmatrix} \\ &= \begin{bmatrix} -1 & 0 & 0 \\ 0 & -1 & 0 \\ 0 & 0 & -1 \end{bmatrix}.\end{aligned} \tag{5.15}$$

*Example 5.3* Another type of corner-cube retroreflectors is solid glass corner-cube (Fig. 5.8), which may be a glass-filled half-cube, cut along the diagonal, for enhanced dimensional stability under temperature excursions. In addition, this construction is thought to be advantageous for minimizing the errors in the retroreflector assembly to within the required angular precision. One can see that

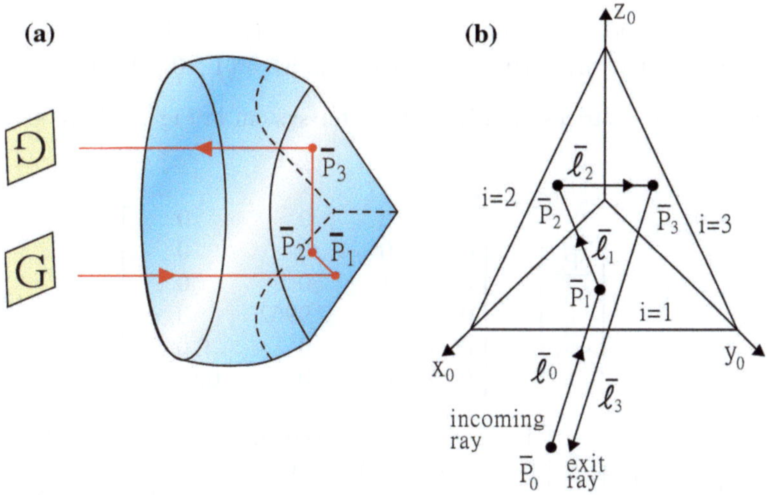


**Fig. 5.7** A perfect corner-cube mirror can reflect a beam parallel to the incident beam independent of beam/corner-cube alignment

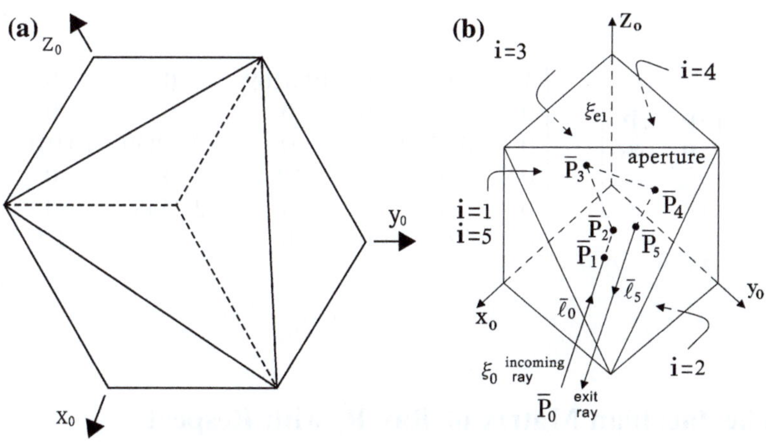


**Fig. 5.8** A solid glass corner-cube is a glass-filled half-cube, cut along the diagonal, for enhanced dimensional stability under temperature excursions

mathematically the corner-cube mirror is a special case of solid glass corner-cube with $N_1 = \xi_0/\xi_{e1} = N_5 = \xi_{e1}/\xi_0 = 1$, where $\xi_{e1}$ is the refractive index of the solid glass corner-cube. Its retroreflective property can be proved numerically by:

$$\frac{\partial\bar{\ell}_5}{\partial\bar{\ell}_0} = \frac{\partial\bar{\ell}_4(\bar{n}_5, N_5)}{\partial\bar{\ell}_4}\frac{\partial\bar{\ell}_4}{\partial\bar{\ell}_3}\frac{\partial\bar{\ell}_3}{\partial\bar{\ell}_2}\frac{\partial\bar{\ell}_2}{\partial\bar{\ell}_1}\frac{\partial\bar{\ell}_1(\bar{n}_1, N_1)}{\partial\bar{\ell}_0}, \tag{5.16}$$

where the unit normal vectors are: $\bar{n}_1 = \bar{n}_5 = \begin{bmatrix} 1/\sqrt{3} & 1/\sqrt{3} & 1/\sqrt{3} & 0 \end{bmatrix}^T$, $\bar{n}_2 = \begin{bmatrix} 0 & 0 & 1 & 0 \end{bmatrix}^T$, $\bar{n}_3 = \begin{bmatrix} 0 & 1 & 0 & 0 \end{bmatrix}^T$, and $\bar{n}_4 = \begin{bmatrix} 1 & 0 & 0 & 0 \end{bmatrix}^T$.

*Example 5.4* The matrices $\bar{S}_0$, $\bar{M}_1$, and $\bar{M}_2$ for system in Fig. 2.14 are given by

$$\bar{S}_0 = \begin{bmatrix} \frac{\partial \bar{P}_0}{\partial \bar{X}_0} \\ \frac{\partial \bar{\ell}_0}{\partial \bar{X}_0} \end{bmatrix} = \begin{bmatrix} 1 & 0 & 0 & 0 & 0 \\ 0 & 1 & 0 & 0 & 0 \\ 0 & 0 & 1 & 0 & 0 \\ 0 & 0 & 0 & -0.9962 & 0 \\ 0 & 0 & 0 & 0 & -0.0872 \\ 0 & 0 & 0 & 0 & 0.9962 \end{bmatrix}, \tag{5.17a}$$

$$\bar{M}_1 = \begin{bmatrix} \frac{\partial \bar{P}_1}{\partial \bar{P}_0} & \frac{\partial \bar{P}_1}{\partial \bar{\ell}_0} \\ \frac{\partial \bar{\ell}_1}{\partial \bar{P}_0} & \frac{\partial \bar{\ell}_1}{\partial \bar{\ell}_0} \end{bmatrix} = \begin{bmatrix} 1 & 0 & 0 & 10.0382 & 0 & 0 \\ 0 & 0 & 0 & 0 & 0 & 0 \\ 0 & -0.0875 & 1 & 0 & -0.8782 & 10.0382 \\ 0 & 0 & 0 & 0.6667 & 0 & 0 \\ 0 & 0 & 0 & 0 & 0.4435 & 0 \\ 0 & 0 & 0 & 0 & 0 & 0.6667 \end{bmatrix}, \tag{5.17b}$$

$$\bar{M}_2 = \begin{bmatrix} \frac{\partial \bar{P}_2}{\partial \bar{P}_1} & \frac{\partial \bar{P}_2}{\partial \bar{\ell}_1} \\ \frac{\partial \bar{\ell}_2}{\partial \bar{P}_1} & \frac{\partial \bar{\ell}_2}{\partial \bar{\ell}_1} \end{bmatrix} = \begin{bmatrix} 1 & 0 & 0 & 10.0169 & 0 & 0 \\ 0 & 0 & 0 & 0 & 0 & 0 \\ 0 & -0.0582 & 1 & 0 & -0.5830 & 10.0169 \\ 0 & 0 & 0 & 1.5 & 0 & 0 \\ 0 & 0 & 0 & 0 & 2.2548 & 0 \\ 0 & 0 & 0 & 0 & 0 & 1.5 \end{bmatrix}. \tag{5.17c}$$

## 5.3 The Jacobian Matrix of Ray $\bar{R}_i$ with Respect to Incoming Ray $\bar{R}_{i-1}$ at a Spherical Boundary Surface

The derivation of the Jacobian matrix $\bar{M}_i = \partial \bar{R}_i / \partial \bar{R}_{i-1} = \begin{bmatrix} \partial \bar{P}_i / \partial \bar{R}_{i-1} & \partial \bar{\ell}_i / \partial \bar{R}_{i-1} \end{bmatrix}^T$ of a spherical boundary surface is more lengthy and cumbersome than its counterparts of a flat boundary surface, since raytracing equations of a spherical boundary surface are comparatively complicated. Again, we will divide the determination of $\bar{M}_i$ into two parts, $\partial \bar{P}_i / \partial \bar{R}_{i-1}$ and $\partial \bar{\ell}_i / \partial \bar{R}_{i-1}$, as in the following.

(a) $\partial \bar{P}_i / \partial \bar{R}_{i-1} = \begin{bmatrix} \partial \bar{P}_i / \partial x_v \end{bmatrix}$, $x_v \in \bar{R}_{i-1}$, can be obtained by differentiating Eq. (2.10) to give (see Figs. 5.9 and 5.10):

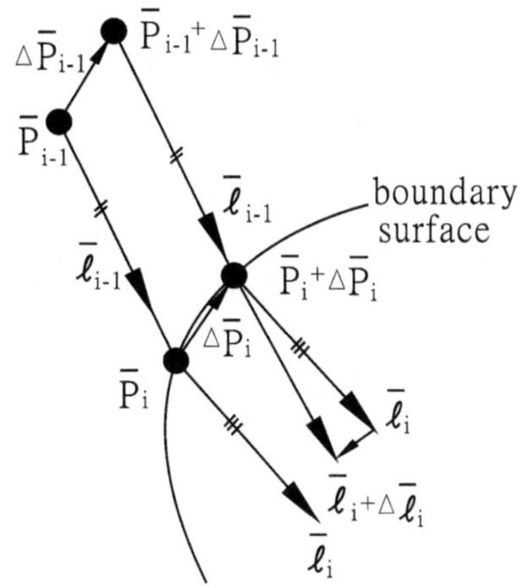


**Fig. 5.9** The variations of $\overline{P}_i$ and $\overline{\ell}_i$ due to the change of $\overline{P}_{i-1}$ at a spherical boundary surface

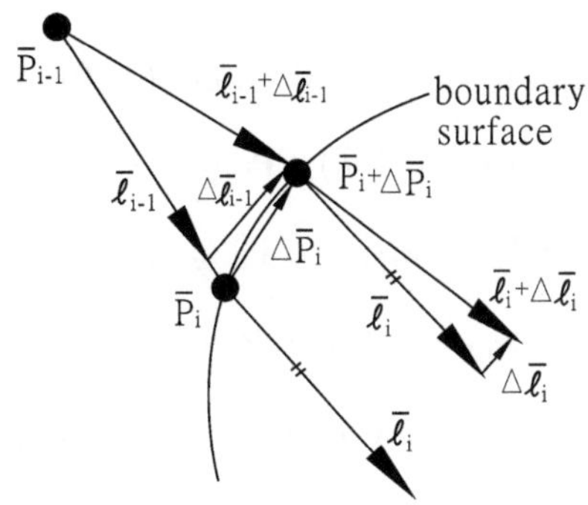


**Fig. 5.10** The variations of $\overline{P}_i$ and $\overline{\ell}_i$ due to the change of $\overline{\ell}_{i-1}$ at a spherical boundary surface

$$\frac{\partial\overline{P}_i}{\partial\overline{R}_{i-1}} = \left[\frac{\partial\overline{P}_i}{\partial x_v}\right] = \begin{bmatrix} \partial P_{ix}/\partial x_v \\ \partial P_{iy}/\partial x_v \\ \partial P_{iz}/\partial x_v \end{bmatrix} = \begin{bmatrix} 1 & 0 & 0 & \lambda_i & 0 & 0 \\ 0 & 1 & 0 & 0 & \lambda_i & 0 \\ 0 & 0 & 1 & 0 & 0 & \lambda_i \end{bmatrix} + \begin{bmatrix} \ell_{i-1x} \\ \ell_{i-1y} \\ \ell_{i-1z} \end{bmatrix} \left[\frac{\partial\lambda_i}{\partial x_v}\right]. \tag{5.18}$$

$\lambda_i$ is given in Eq. (2.15), and

$$\frac{\partial\lambda_i}{\partial\overline{R}_{i-1}} = \left[\frac{\partial\lambda_i}{\partial x_v}\right] = \left[-\frac{\partial D_i}{\partial x_v}\right] \pm \frac{1}{2\sqrt{D_i^2 - E_i}}\left(2D_i\left[\frac{\partial D_i}{\partial x_v}\right] - \left[\frac{\partial E_i}{\partial x_v}\right]\right), \tag{5.19}$$

$$\left[\frac{\partial D_i}{\partial x_v}\right] = [\,\ell_{i-1x} \quad \ell_{i-1y} \quad \ell_{i-1z} \quad P_{i-1x} - t_{ix} \quad P_{i-1y} - t_{iy} \quad P_{i-1z} - t_{iz}\,], \tag{5.20}$$

$$\left[\frac{\partial E_i}{\partial x_v}\right] = \left[2(P_{i-1x} - t_{ix}) \quad 2(P_{i-1y} - t_{iy}) \quad 2(P_{i-1z} - t_{iz}) \quad 0 \quad 0 \quad 0\right]. \tag{5.21}$$

(b) When $\bar{\ell}_i$ is the unit directional vector of the refracted ray at a spherical boundary surface, $\partial\bar{\ell}_i/\partial\bar{R}_{i-1} = \left[\partial\bar{\ell}_i/\partial x_v\right]$, $x_v \in \bar{R}_{i-1}$, can be obtained by differentiating Eq. (2.27) to give

$$\frac{\partial\bar{\ell}_i}{\partial\bar{R}_{i-1}} = \left[\frac{\partial\bar{\ell}_i}{\partial x_v}\right] = \begin{bmatrix} \partial\ell_{ix}/\partial x_v \\ \partial\ell_{iy}/\partial x_v \\ \partial\ell_{iz}/\partial x_v \end{bmatrix} = \left(\frac{-N_i^2 C\theta_i}{\sqrt{1-N_i^2+(N_i C\theta_i)^2}} + N_i\right)\begin{bmatrix} n_{ix} \\ n_{iy} \\ n_{iz} \end{bmatrix}\left[\frac{\partial(C\theta_i)}{\partial x_v}\right]$$
$$+ N_i \begin{bmatrix} 0 & 0 & 0 & 1 & 0 & 0 \\ 0 & 0 & 0 & 0 & 1 & 0 \\ 0 & 0 & 0 & 0 & 0 & 1 \end{bmatrix} + \left(-\sqrt{1-N_i^2+(N_i C\theta_i)^2} + N_i C\theta_i\right)\begin{bmatrix} \partial n_{ix}/\partial x_v \\ \partial n_{iy}/\partial x_v \\ \partial n_{iz}/\partial x_v \end{bmatrix}, \tag{5.22}$$

where

$$\left[\frac{\partial(C\theta_i)}{\partial x_v}\right] = \left[C\theta_{i,1} \quad C\theta_{i,2} \quad C\theta_{i,3} \quad C\theta_{i,4} \quad C\theta_{i,5} \quad C\theta_{i,6}\right] \tag{5.23}$$

can be obtained from Eq. (2.20) to give

$$C\theta_{i,1} = \partial(C\theta_i)/\partial P_{i-1x} = -s_i\left(\ell_{i-1x}\, n_{ix,1} + \ell_{i-1y}\, n_{iy,1} + \ell_{i-1z}\, n_{iz,1}\right), \tag{5.23a}$$

$$C\theta_{i,2} = \partial(C\theta_i)/\partial P_{i-1y} = -s_i\left(\ell_{i-1x}\, n_{ix,2} + \ell_{i-1y}\, n_{iy,2} + \ell_{i-1z}\, n_{iz,2}\right), \tag{5.23b}$$

$$C\theta_{i,3} = \partial(C\theta_i)/\partial P_{i-1z} = -s_i\left(\ell_{i-1x}\, n_{ix,3} + \ell_{i-1y}\, n_{iy,3} + \ell_{i-1z}\, n_{iz,3}\right), \tag{5.23c}$$

$$C\theta_{i,4} = \partial(C\theta_i)/\partial \ell_{i-1x} = -s_i\left(\ell_{i-1x}\, n_{ix,4} + \ell_{i-1y}\, n_{iy,4} + \ell_{i-1z}\, n_{iz,4}\right) - s_i n_{ix}, \tag{5.23d}$$

$$C\theta_{i,5} = \partial(C\theta_i)/\partial \ell_{i-1y} = -s_i\left(\ell_{i-1x}\, n_{ix,5} + \ell_{i-1y}\, n_{iy,5} + \ell_{i-1z}\, n_{iz,5}\right) - s_i n_{iy}, \tag{5.23e}$$

$$C\theta_{i,6} = \partial(C\theta_i)/\partial \ell_{i-1z} = -s_i\left(\ell_{i-1x}\, n_{ix,6} + \ell_{i-1y}\, n_{iy,6} + \ell_{i-1z}\, n_{iz,6}\right) - s_i n_{iz}. \tag{5.23f}$$

The last term of Eq. (5.22) can be obtained from Eq. (2.9) to give

$$\begin{bmatrix} \partial n_{ix}/\partial x_v \\ \partial n_{iy}/\partial x_v \\ \partial n_{iz}/\partial x_v \end{bmatrix} = \begin{bmatrix} n_{ix,1} & n_{ix,2} & n_{ix,3} & n_{ix,4} & n_{ix,5} & n_{ix,6} \\ n_{iy,1} & n_{iy,2} & n_{iy,3} & n_{iy,4} & n_{iy,5} & n_{iy,6} \\ n_{iz,1} & n_{iz,2} & n_{iz,3} & n_{iz,4} & n_{iz,5} & n_{iz,6} \end{bmatrix}$$
$$= s_i \begin{bmatrix} I_{ix} & J_{ix} & K_{ix} \\ I_{iy} & J_{iy} & K_{iy} \\ I_{iz} & J_{iz} & K_{iz} \end{bmatrix}\left(\begin{bmatrix} -S\beta_i C\alpha_i \\ -S\beta_i S\alpha_i \\ C\beta_i \end{bmatrix}\left[\frac{\partial\beta_i}{\partial x_v}\right] + \begin{bmatrix} -C\beta_i S\alpha_i \\ C\beta_i C\alpha_i \\ 0 \end{bmatrix}\left[\frac{\partial\alpha_i}{\partial x_v}\right]\right). \tag{5.24}$$

The required term $[\partial\alpha_i/\partial x_v]$ in Eq. (5.24) can be obtained by differentiating Eq. (2.18), that is

$$\left[\frac{\partial \alpha_i}{\partial x_v}\right] = \frac{1}{\sigma_i^2+\rho_i^2}\left(\sigma_i\left[\frac{\partial \rho_i}{\partial x_v}\right] - \rho_i\left[\frac{\partial \sigma_i}{\partial x_v}\right]\right), \tag{5.25}$$

where $[\partial \rho_i/\partial x_v]$ and $[\partial \sigma_i/\partial x_v]$ are obtained from Eqs. (2.13) and (2.12), respectively, as follows:

$$[\partial \rho_i/\partial x_v] = \begin{bmatrix}\rho_{i,1} & \rho_{i,2} & \rho_{i,3} & \rho_{i,4} & \rho_{i,5} & \rho_{i,6}\end{bmatrix}, \tag{5.26}$$

$$\rho_{i,1} = \partial \rho_i/\partial P_{i-1x} = (J_{ix}\ell_{i-1x} + J_{iy}\ell_{i-1y} + J_{iz}\ell_{i-1z})\lambda_{i,1} + J_{ix}, \tag{5.26a}$$

$$\rho_{i,2} = \partial \rho_i/\partial P_{i-1y} = (J_{ix}\ell_{i-1x} + J_{iy}\ell_{i-1y} + J_{iz}\ell_{i-1z})\lambda_{i,2} + J_{iy}, \tag{5.26b}$$

$$\rho_{i,3} = \partial \rho_i/\partial P_{i-1z} = (J_{ix}\ell_{i-1x} + J_{iy}\ell_{i-1y} + J_{iz}\ell_{i-1z})\lambda_{i,3} + J_{iz}, \tag{5.26c}$$

$$\rho_{i,4} = \partial \rho_i/\partial \ell_{i-1x} = (J_{ix}\ell_{i-1x} + J_{iy}\ell_{i-1y} + J_{iz}\ell_{i-1z})\lambda_{i,4} + J_{ix}\lambda_i, \tag{5.26d}$$

$$\rho_{i,5} = \partial \rho_i/\partial \ell_{i-1y} = (J_{ix}\ell_{i-1x} + J_{iy}\ell_{i-1y} + J_{iz}\ell_{i-1z})\lambda_{i,5} + J_{iy}\lambda_i, \tag{5.26e}$$

$$\rho_{i,6} = \partial \rho_i/\partial \ell_{i-1z} = (J_{ix}\ell_{i-1x} + J_{iy}\ell_{i-1y} + J_{iz}\ell_{i-1z})\lambda_{i,6} + J_{iz}\lambda_i \tag{5.26f}$$

$$[\partial \sigma_i/\partial x_v] = \begin{bmatrix}\sigma_{i,1} & \sigma_{i,2} & \sigma_{i,3} & \sigma_{i,4} & \sigma_{i,5} & \sigma_{i,6}\end{bmatrix}, \tag{5.27}$$

$$\sigma_{i,1} = \partial \sigma_i/\partial P_{i-1x} = (I_{ix}\ell_{i-1x} + I_{iy}\ell_{i-1y} + I_{iz}\ell_{i-1z})\lambda_{i,1} + I_{ix}, \tag{5.27a}$$

$$\sigma_{i,2} = \partial \sigma_i/\partial P_{i-1y} = (I_{ix}\ell_{i-1x} + I_{iy}\ell_{i-1y} + I_{iz}\ell_{i-1z})\lambda_{i,2} + I_{iy}, \tag{5.27b}$$

$$\sigma_{i,3} = \partial \sigma_i/\partial P_{i-1z} = (I_{ix}\ell_{i-1x} + I_{iy}\ell_{i-1y} + I_{iz}\ell_{i-1z})\lambda_{i,3} + I_{iz}, \tag{5.27c}$$

$$\sigma_{i,4} = \partial \sigma_i/\partial \ell_{i-1x} = (I_{ix}\ell_{i-1x} + I_{iy}\ell_{i-1y} + I_{iz}\ell_{i-1z})\lambda_{i,4} + I_{ix}\lambda_i, \tag{5.27d}$$

$$\sigma_{i,5} = \partial \sigma_i/\partial \ell_{i-1y} = (I_{ix}\ell_{i-1x} + I_{iy}\ell_{i-1y} + I_{iz}\ell_{i-1z})\lambda_{i,5} + I_{iy}\lambda_i, \tag{5.27e}$$

$$\sigma_{i,6} = \partial \sigma_i/\partial \ell_{i-1z} = (I_{ix}\ell_{i-1x} + I_{iy}\ell_{i-1y} + I_{iz}\ell_{i-1z})\lambda_{i,6} + I_{iz}\lambda_i. \tag{5.27f}$$

To compute Eq. (5.24), the term $[\partial \beta_i/\partial x_v]$ is also required. This term can be obtained by differentiating Eq. (2.19) to give

$$\left[\frac{\partial \beta_i}{\partial x_v}\right] = \frac{1}{(\sigma_i^2+\rho_i^2+\tau_i^2)\sqrt{(\sigma_i^2+\rho_i^2)}}\left\{(\sigma_i^2+\rho_i^2)\left[\frac{\partial \tau_i}{\partial x_v}\right] - \tau_i\left(\sigma_i\left[\frac{\partial \sigma_i}{\partial x_v}\right] + \rho_i\left[\frac{\partial \rho_i}{\partial x_v}\right]\right)\right\}, \tag{5.28}$$

with

$$[\partial \tau_i/\partial x_v] = \begin{bmatrix}\tau_{i,1} & \tau_{i,2} & \tau_{i,3} & \tau_{i,4} & \tau_{i,5} & \tau_{i,6}\end{bmatrix}, \tag{5.29}$$

$$\tau_{i,1} = \partial \tau_i/\partial P_{i-1x} = (K_{ix}\ell_{i-1x} + K_{iy}\ell_{i-1y} + K_{iz}\ell_{i-1z})\lambda_{i,1} + K_{ix}, \tag{5.29a}$$

$$\tau_{i,2} = \partial\tau_i/\partial P_{i-1y} = \left(K_{ix}\ell_{i-1x} + K_{iy}\ell_{i-1y} + K_{iz}\ell_{i-1z}\right)\lambda_{i,2} + K_{iy}, \quad (5.29b)$$

$$\tau_{i,3} = \partial\tau_i/\partial P_{i-1z} = \left(K_{ix}\ell_{i-1x} + K_{iy}\ell_{i-1y} + K_{iz}\ell_{i-1z}\right)\lambda_{i,3} + K_{iz}, \quad (5.29c)$$

$$\tau_{i,4} = \partial\tau_i/\partial \ell_{i-1x} = \left(K_{ix}\ell_{i-1x} + K_{iy}\ell_{i-1y} + K_{iz}\ell_{i-1z}\right)\lambda_{i,4} + K_{ix}\lambda_i, \quad (5.29d)$$

$$\tau_{i,5} = \partial\tau_i/\partial \ell_{i-1y} = \left(K_{ix}\ell_{i-1x} + K_{iy}\ell_{i-1y} + K_{iz}\ell_{i-1z}\right)\lambda_{i,5} + K_{iy}\lambda_i, \quad (5.29e)$$

$$\tau_{i,6} = \partial\tau_i/\partial \ell_{i-1z} = \left(K_{ix}\ell_{i-1x} + K_{iy}\ell_{i-1y} + K_{iz}\ell_{i-1z}\right)\lambda_{i,6} + K_{iz}\lambda_i. \quad (5.29f)$$

(c) When $\bar{\ell}_i$ is the unit directional vector of the reflected ray at a spherical boundary surface, $\partial\bar{\ell}_i/\partial\bar{R}_{i-1} = \left[\partial\bar{\ell}_i/\partial x_v\right]$, $x_v \in \bar{R}_{i-1}$, can be obtained by differentiating Eq. (2.25) to give

$$\frac{\partial\bar{\ell}_i}{\partial\bar{R}_{i-1}} = \begin{bmatrix} \partial\ell_{ix}/\partial x_v \\ \partial\ell_{iy}/\partial x_v \\ \partial\ell_{iz}/\partial x_v \end{bmatrix} = 2C\theta_i \begin{bmatrix} \partial n_{ix}/\partial x_v \\ \partial n_{iy}/\partial x_v \\ \partial n_{iz}/\partial x_v \end{bmatrix} + \begin{bmatrix} 0 & 0 & 0 & 1 & 0 & 0 \\ 0 & 0 & 0 & 0 & 1 & 0 \\ 0 & 0 & 0 & 0 & 0 & 1 \end{bmatrix} + \begin{bmatrix} 2n_{ix} \\ 2n_{iy} \\ 2n_{iz} \end{bmatrix} \left[\frac{\partial(C\theta_i)}{\partial x_v}\right], \quad (5.30)$$

where $\left[\partial n_{ix}/\partial x_v \quad \partial n_{iy}/\partial x_v \quad \partial n_{iz}/\partial x_v\right]^T$ and $\left[\partial(C\theta_i)/\partial x_v\right]$ are respectively given in Eqs. (5.24) and (5.23).

Combining Eqs. (5.18) and (5.22) for a refraction process, or Eqs. (5.18) and (5.30) for a reflection process, one obtains the Jacobian matrix $\partial\bar{R}_i/\partial\bar{R}_{i-1}$ of the refracted/or reflected ray $\bar{R}_i$ with respect to the incoming ray $\bar{R}_{i-1}$ at a spherical boundary surface as

$$\frac{\partial\bar{R}_i}{\partial\bar{R}_{i-1}} = \bar{M}_i = \begin{bmatrix} \dfrac{\partial\bar{P}_i}{\partial\bar{P}_{i-1}} & \dfrac{\partial\bar{P}_i}{\partial\bar{\ell}_{i-1}} \\ \dfrac{\partial\bar{\ell}_i}{\partial\bar{P}_{i-1}} & \dfrac{\partial\bar{\ell}_i}{\partial\bar{\ell}_{i-1}} \end{bmatrix}. \quad (5.31)$$

An optical system may contain spherical and flat boundary surfaces. To compute the change of the ray refracted/or reflected by the gth boundary surface in an optical system, i.e. $\Delta\bar{R}_g$, one has to label its boundary surfaces from i = 0 to i = n (e.g., Fig. 2.1 and Fig. 3.12). If there is not any changes of boundary variables in $\bar{X}_i$ (i = 1,2,…,n), $\Delta\bar{R}_g$ can be obtained by successively using the $\bar{M}_i$ matrices of Eqs. (5.7) and (5.31):

$$\Delta\bar{R}_g = \frac{\partial\bar{R}_g}{\partial\bar{X}_0}\Delta\bar{X}_0 = \bar{M}_g\bar{M}_{g-1}\ldots\bar{M}_2\bar{M}_1\bar{S}_0\Delta\bar{X}_0$$
$$= \begin{bmatrix} \dfrac{\partial\bar{P}_g}{\partial\bar{P}_0} & \dfrac{\partial\bar{P}_g}{\partial(\alpha_0,\beta_0)} \\ \dfrac{\partial\bar{l}_g}{\partial\bar{P}_0} & \dfrac{\partial\bar{R}_g}{\partial(\alpha_0,\beta_0)} \end{bmatrix} \begin{bmatrix} \Delta\bar{P}_0 \\ \Delta\alpha_0 \\ \Delta\beta_0 \end{bmatrix}$$
$$= \begin{bmatrix} \partial P_{gx}/\partial P_{0x} & \partial P_{gx}/\partial P_{0y} & \partial P_{gx}/\partial P_{0z} & \partial P_{gx}/\partial\alpha_0 & \partial P_{gx}/\partial\beta_0 \\ \partial P_{gy}/\partial P_{0x} & \partial P_{gy}/\partial P_{0y} & \partial P_{gy}/\partial P_{0z} & \partial P_{gy}/\partial\alpha_0 & \partial P_{gy}/\partial\beta_0 \\ \partial P_{gz}/\partial P_{0x} & \partial P_{gz}/\partial P_{0y} & \partial P_{gz}/\partial P_{0z} & \partial P_{gz}/\partial\alpha_0 & \partial P_{gz}/\partial\beta_0 \\ \partial\ell_{gx}/\partial P_{0x} & \partial\ell_{gx}/\partial P_{0y} & \partial\ell_{gx}/\partial P_{0z} & \partial\ell_{gx}/\partial\alpha_0 & \partial\ell_{gx}/\partial\beta_0 \\ \partial\ell_{gy}/\partial P_{0x} & \partial\ell_{gy}/\partial P_{0y} & \partial\ell_{gy}/\partial P_{0z} & \partial\ell_{gy}/\partial\alpha_0 & \partial\ell_{gy}/\partial\beta_0 \\ \partial\ell_{gz}/\partial P_{0x} & \partial\ell_{gz}/\partial P_{0y} & \partial\ell_{gz}/\partial P_{0z} & \partial\ell_{gz}/\partial\alpha_0 & \partial\ell_{gz}/\partial\beta_0 \end{bmatrix} \begin{bmatrix} \Delta P_{0x} \\ \Delta P_{0y} \\ \Delta P_{0z} \\ \Delta\alpha_0 \\ \Delta\beta_0 \end{bmatrix}. \tag{5.32}$$

Note that Eq. (5.10), which is $\bar{S}_0 = \partial\bar{R}_0/\partial\bar{X}_0$, is used in Eq. (5.32). When rays are traced through an optical system, it is generally desired to send a source ray $\bar{R}_0$ from a specific source point $\bar{P}_0$ through a specific point $\bar{P}_g$ on the aperture if the aperture is labeled as the gth boundary surface in a given optical system. The Newton–Raphson method can reach this task. Newton–Raphson method requires a suitable initial estimate (say $\bar{\ell}'_0 = \bar{\ell}'_0(\alpha'_0, \beta'_0)$) of the unit directional vector. Its rate and final accuracy of convergence can be improved by adding the initial estimate

**Table 5.1** The polar coordinates $\alpha_{0/\text{chief}}$ and $\beta_{0/\text{chief}}$ of the chief ray originating from different $\bar{P}_0$ of the system in Fig. 2.1

| $\bar{P}_0$ \ polar coordinates | $\alpha_{0/\text{chief}}$ | $\beta_{0/\text{chief}}$ |
|---|---|---|
| $[0\ \ -507\ \ 262.2187\ \ 1]^T$ | 0° | −26.3291° |
| $[0\ \ -507\ \ 250\ \ 1]^T$ | 0° | −25.2623° |
| $[0\ \ -507\ \ 200\ \ 1]^T$ | 0° | −20.6930° |
| $[0\ \ -507\ \ 170\ \ 1]^T$ | 0° | −17.8041° |
| $[0\ \ -507\ \ 150\ \ 1]^T$ | 0° | −15.8225° |
| $[0\ \ -507\ \ 100\ \ 1]^T$ | 0° | −10.7010° |
| $[0\ \ -507\ \ 50\ \ 1]^T$ | 0° | −5.3982° |
| $[0\ \ -507\ \ 0\ \ 1]^T$ | 0° | 0.0000° |

**Table 5.2** The upper- and lower limits of ray fans on meridional plane and sagittal cone for different $\bar{P}_0$ of the system in Fig. 2.1

| $\bar{P}_0$ \ limits | ray fan on meridional plane | | ray fan on sagittal cone | |
|---|---|---|---|---|
| | lower limit of $\beta_0$ | upper limit of $\beta_0$ | lower limit of $\alpha_0$ | upper limit of $\alpha_0$ |
| $[0\ \ -507\ \ 262.2187\ \ 1]^T$ | −28.1013° | −25.6858° | −1.8776° | 1.8776° |
| $[0\ \ -507\ \ 250\ \ 1]^T$ | −27.0501° | −24.5320° | −1.9557° | 1.9557° |
| $[0\ \ -507\ \ 200\ \ 1]^T$ | −22.5441° | −19.5887° | −1.9978° | 1.9978° |
| $[0\ \ -507\ \ 170\ \ 1]^T$ | −19.6904° | −16.4649° | −1.9906° | 1.9906° |
| $[0\ \ -507\ \ 150\ \ 1]^T$ | −17.7301° | −14.3248° | −1.9866° | 1.9866° |
| $[0\ \ -507\ \ 100\ \ 1]^T$ | −12.6511° | −8.8119° | −1.9786° | 1.9786° |
| $[0\ \ -507\ \ 50\ \ 1]^T$ | −7.3711° | −3.4512° | −1.9739° | 1.9739° |
| $[0\ \ -507\ \ 0\ \ 1]^T$ | −1.9700° | 1.9700° | −1.9700° | 1.9700° |

$\bar{\ell}'_0$ (or the updated estimated in last iteration) with the product of $\left(\partial\bar{P}_g/\partial(\alpha'_0,\beta'_0)\right)^{-1}$ and $-\Delta\bar{P}_g$, the error of incidence point. By applying Eq. (5.32) to the Newton–Raphson method, it is possible to determine the chief and marginal rays accurately and obtain the one-to-one relationship between $\bar{\ell}_0$ (or polar coordinates $\alpha_0$ and $\beta_0$) and the incidence point $\bar{P}_g$ at the aperture for a given source point $\bar{P}_0$. Table 5.1 shows the polar coordinates $\alpha_{0/\text{chief}}$ and $\beta_{0/\text{chief}}$ of the chief ray $\bar{R}_{0/\text{chief}} = \left[\bar{P}_0\ \ \bar{\ell}_{0/\text{chief}}\right]^T$, where $\bar{\ell}_{0/\text{chief}} = \left[C\beta_{0/\text{chief}}C(90^o+\alpha_{0/\text{chief}})\ \ C\beta_{0/\text{chief}}S(90^o+\alpha_{0/\text{chief}})\ S\beta_{0/\text{chief}}\ \ 0\right]^T$ for different source point $\bar{P}_0$ of the system in Fig. 2.1 by Newton–Raphson method. Table 5.2 lists the upper- and lower limits of ray fans on meridional plane and sagittal cone for different source point $\bar{P}_0$ of that system.

In Chap. 8 we need $\partial\bar{R}_i/\partial\bar{X}_0$, the Jacobian matrix of the ith ray with respect to the variable vector $\bar{X}_0$ of source ray $\bar{R}_0$, to investigate the wavefront shape of a given optical system. It can be obtained by setting g = i into Eq. (5.32) and defining $\Delta\bar{R}_i = \partial\bar{R}_i/\partial\bar{X}_0\Delta\bar{X}_0$, to give

$$\frac{\partial\bar{R}_i}{\partial\bar{X}_0} = \begin{bmatrix} \partial P_{ix}/\partial P_{0x} & \partial P_{ix}/\partial P_{0y} & \partial P_{ix}/\partial P_{0z} & \partial P_{ix}/\partial\alpha_0 & \partial P_{ix}/\partial\beta_0 \\ \partial P_{iy}/\partial P_{0x} & \partial P_{iy}/\partial P_{0y} & \partial P_{iy}/\partial P_{0z} & \partial P_{iy}/\partial\alpha_0 & \partial P_{iy}/\partial\beta_0 \\ \partial P_{iz}/\partial P_{0x} & \partial P_{iz}/\partial P_{0y} & \partial P_{iz}/\partial P_{0z} & \partial P_{iz}/\partial\alpha_0 & \partial P_{iz}/\partial\beta_0 \\ \partial\ell_{ix}/\partial P_{0x} & \partial\ell_{ix}/\partial P_{0y} & \partial\ell_{ix}/\partial P_{0z} & \partial\ell_{ix}/\partial\alpha_0 & \partial\ell_{ix}/\partial\beta_0 \\ \partial\ell_{iy}/\partial P_{0x} & \partial\ell_{iy}/\partial P_{0y} & \partial\ell_{iy}/\partial P_{0z} & \partial\ell_{iy}/\partial\alpha_0 & \partial\ell_{iy}/\partial\beta_0 \\ \partial\ell_{iz}/\partial P_{0x} & \partial\ell_{iz}/\partial P_{0y} & \partial\ell_{iz}/\partial P_{0z} & \partial\ell_{iz}/\partial\alpha_0 & \partial\ell_{iz}/\partial\beta_0 \end{bmatrix} \tag{5.33}$$
$$= \bar{M}_i\bar{M}_{i-1}\cdots\bar{M}_2\bar{M}_1\bar{S}_0.$$

*Example 5.5* The matrices $\overline{S}_0$, $\overline{M}_1$, and $\overline{M}_2$ for system in Fig. 2.9 are given by

$$\overline{S}_0 = \begin{bmatrix} 1 & 0 & 0 & 0 & 0 \\ 0 & 1 & 0 & 0 & 0 \\ 0 & 0 & 1 & 0 & 0 \\ 0 & 0 & 0 & -0.9962 & 0 \\ 0 & 0 & 0 & 0 & -0.0872 \\ 0 & 0 & 0 & 0 & 0.9962 \end{bmatrix}, \tag{5.34a}$$

$$\overline{M}_1 = \begin{bmatrix} 1 & 0 & 0 & 10.3895 & 0 & 0 \\ 0 & -0.0105 & 0.1202 & 0 & -2.2978 & 1.0573 \\ 0 & -0.0884 & 1.0105 & 0 & -1.1100 & 10.4820 \\ -0.0068 & 0 & 0 & 0.5964 & 0 & 0 \\ 0 & 0 & 0.0001 & 0 & 0.4422 & 0.0280 \\ 0 & 0.0006 & -0.0069 & 0 & 0.0361 & 0.5916 \end{bmatrix}, \tag{5.34b}$$

$$\overline{M}_2 = \begin{bmatrix} 1 & 0 & 0 & 9.4667 & 0 & 0 \\ 0 & 0.0011 & -0.0608 & 0 & 0.9072 & -0.5595 \\ 0 & -0.0182 & 0.9989 & 0 & -0.1556 & 9.4566 \\ -0.0050 & 0 & 0 & 1.4526 & 0 & 0 \\ 0 & 0 & 0 & 0 & 2.2498 & 0.0455 \\ 0 & 0.0001 & -0.0050 & 0 & 0.0467 & 1.4553 \end{bmatrix}. \tag{5.34c}$$

## 5.4 The Jacobian Matrix of Ray $\overline{R}_i$ with Respect to Boundary Variable Vector $\overline{X}_i$ at a Flat Boundary Surface

Sections 5.2 and 5.3 determined the Jacobian matrix, $\overline{M}_i = \partial\overline{R}_i/\partial\overline{R}_{i-1}$, of the refracted/or reflected ray $\overline{R}_i$ with respect to its incoming ray $\overline{R}_{i-1}$. However, the change of any boundary variable may also cause the change of the refracted/or reflected ray $\overline{R}_i$ (Fig. 5.5). To distinguish it from the matrix $\overline{M}_i = \partial\overline{R}_i/\partial\overline{R}_{i-1}$, we designate the Jacobian matrix $\partial\overline{R}_i/\partial\overline{X}_i$ of the refracted/or reflected ray $\overline{R}_i$ with respect to the boundary variable vector $\overline{X}_i$ as $\overline{S}_i = \partial\overline{R}_i/\partial\overline{X}_i$, where $\overline{X}_i$ is given by Eq. (2.28). Again, $\overline{S}_i$ will be divided into two parts, namely $\partial\overline{P}_i/\partial\overline{X}_i$ and $\partial\overline{\ell}_i/\partial\overline{X}_i$, as discussed in the following.

(a) $\partial\overline{P}_i/\partial\overline{X}_i = [\partial\overline{P}_i/\partial x_v]$, $x_v \in \overline{X}_i$, is obtained by differentiating Eq. (2.36) with respect to $\overline{X}_i$ to give

$$\frac{\partial \overline{P}_i}{\partial \overline{X}_i} = \left[\frac{\partial \overline{P}_i}{\partial x_v}\right] = \begin{bmatrix} \partial P_{ix}/\partial x_v \\ \partial P_{iy}/\partial x_v \\ \partial P_{iz}/\partial x_v \end{bmatrix} = \begin{bmatrix} \ell_{i-1x} \\ \ell_{i-1y} \\ \ell_{i-1z} \end{bmatrix} \left[\frac{\partial \lambda_i}{\partial x_v}\right], \tag{5.35}$$

where

$$\frac{\partial \lambda_i}{\partial \overline{X}_i} = \left[\frac{\partial \lambda_i}{\partial x_v}\right] = -\frac{1}{E_i}\left[\frac{\partial D_i}{\partial x_v}\right] + \frac{D_i}{E_i^2}\left[\frac{\partial E_i}{\partial x_v}\right] \tag{5.36}$$

with $D_i$ and $E_i$ are respectively given in Eqs. (2.39) and (2.40), and

$$\left[\frac{\partial D_i}{\partial x_v}\right] = \Big[-J_{ix} \quad -J_{iy} \quad -J_{iz} \quad (P_{i-1x} - t_{ix})\frac{\partial J_{ix}}{\partial \omega_{ix}} + (P_{i-1y} - t_{iy})\frac{\partial J_{iy}}{\partial \omega_{ix}} + (P_{i-1z} - t_{iz})\frac{\partial J_{iz}}{\partial \omega_{ix}}$$
$$(P_{i-1x} - t_{ix})\frac{\partial J_{ix}}{\partial \omega_{iy}} + (P_{i-1y} - t_{iy})\frac{\partial J_{iy}}{\partial \omega_{iy}} + (P_{i-1z} - t_{iz})\frac{\partial J_{iz}}{\partial \omega_{iy}}$$
$$(P_{i-1x} - t_{ix})\frac{\partial J_{ix}}{\partial \omega_{iz}} + (P_{i-1y} - t_{iy})\frac{\partial J_{iy}}{\partial \omega_{iz}} + (P_{i-1z} - t_{iz})\frac{\partial J_{iz}}{\partial \omega_{iz}} \quad 0 \quad 0 \quad 0\Big], \tag{5.37}$$

$$\left[\frac{\partial E_i}{\partial x_v}\right] = \Big[0 \quad 0 \quad 0 \quad \ell_{i-1x}\frac{\partial J_{ix}}{\partial \omega_{ix}} + \ell_{i-1y}\frac{\partial J_{iy}}{\partial \omega_{ix}} + \ell_{i-1z}\frac{\partial J_{iz}}{\partial \omega_{ix}}$$
$$\ell_{i-1x}\frac{\partial J_{ix}}{\partial \omega_{iy}} + \ell_{i-1y}\frac{\partial J_{iy}}{\partial \omega_{iy}} + \ell_{i-1z}\frac{\partial J_{iz}}{\partial \omega_{iy}} \quad \ell_{i-1x}\frac{\partial J_{ix}}{\partial \omega_{iz}}$$
$$+ \ell_{i-1y}\frac{\partial J_{iy}}{\partial \omega_{iz}} + \ell_{i-1z}\frac{\partial J_{iz}}{\partial \omega_{iz}} \quad 0 \quad 0 \quad 0\Big]. \tag{5.38}$$

The required terms ($\partial J_{ix}/\partial \omega_{ix}$, $\partial J_{iy}/\partial \omega_{ix}$, and so on) in Eqs. (5.37) and (5.38) are the components of $\partial {}^0\overline{A}_i/\partial(t_{ix}, t_{iy}, t_{iz}, \omega_{ix}, \omega_{iy}, \omega_{iz})$, which can be obtained by directly differentiating ${}^0\overline{A}_i$ in Eq. (2.8) with respect to pose variables, and are given in Appendix A.

(b) When $\overline{\ell}_i$ is the unit directional vector of the reflected ray at a flat boundary surface, $\partial \overline{\ell}_i/\partial \overline{X}_i = [\partial \overline{\ell}_i/\partial x_v]$, $x_v \in \overline{X}_i$, is derived directly by differentiating Eq. (2.44) to give

$$\frac{\partial \overline{\ell}_i}{\partial \overline{X}_i} = \left[\frac{\partial \overline{\ell}_i}{\partial x_v}\right] = \begin{bmatrix} \partial \ell_{ix}/\partial x_v \\ \partial \ell_{iy}/\partial x_v \\ \partial \ell_{iz}/\partial x_v \end{bmatrix} = 2E_i \begin{bmatrix} \partial J_{ix}/\partial x_v \\ \partial J_{iy}/\partial x_v \\ \partial J_{iz}/\partial x_v \end{bmatrix} + 2\begin{bmatrix} J_{ix} \\ J_{iy} \\ J_{iz} \end{bmatrix}\left[\frac{\partial E_i}{\partial x_v}\right], \tag{5.39}$$

where $E_i$ and $[\partial E_i/\partial x_v]$ are respectively given in Eq. (2.40) and (5.38), and

$$\begin{bmatrix} \partial J_{ix}/\partial x_v \\ \partial J_{iy}/\partial x_v \\ \partial J_{iz}/\partial x_v \end{bmatrix} = \begin{bmatrix} 0 & 0 & 0 & \partial J_{ix}/\partial \omega_{ix} & \partial J_{ix}/\partial \omega_{iy} & \partial J_{ix}/\partial \omega_{iz} & 0 & 0 & 0 \\ 0 & 0 & 0 & \partial J_{iy}/\partial \omega_{ix} & \partial J_{iy}/\partial \omega_{iy} & \partial J_{iy}/\partial \omega_{iz} & 0 & 0 & 0 \\ 0 & 0 & 0 & \partial J_{iz}/\partial \omega_{ix} & \partial J_{iz}/\partial \omega_{iy} & \partial J_{iz}/\partial \omega_{iz} & 0 & 0 & 0 \end{bmatrix}. \tag{5.40}$$

Again, the required components of Eq. (5.40) can be found in Appendix A.

(c) When $\bar{\ell}_i$ is the unit directional vector of the refracted ray at a flat boundary surface, $\partial\bar{\ell}_i/\partial\bar{X}_i = [\partial\bar{\ell}_i/\partial x_v]$, $x_v \in \bar{X}_i$, can be obtained directly by differentiating Eq. (2.45) to give

$$\frac{\partial\bar{\ell}_i}{\partial\bar{X}_i} = \left[\frac{\partial\bar{\ell}_i}{\partial x_v}\right] = \begin{bmatrix} \partial\ell_{ix}/\partial x_v \\ \partial\ell_{iy}/\partial x_v \\ \partial\ell_{iz}/\partial x_v \end{bmatrix} = \left(s_i\sqrt{1 - N_i^2 + (N_iE_i)^2} - N_iE_i\right)\begin{bmatrix} \partial J_{ix}/\partial x_v \\ \partial J_{iy}/\partial x_v \\ \partial J_{iz}/\partial x_v \end{bmatrix}$$
$$+ \begin{bmatrix} \ell_{i-1x} \\ \ell_{i-1y} \\ \ell_{i-1z} \end{bmatrix}\left[\frac{\partial N_i}{\partial x_v}\right] + \left(\frac{s_iN_i\left(E_i^2 - 1\right)}{\sqrt{1 - N_i^2 + (N_iE_i)^2}} - E_i\right)\begin{bmatrix} J_{ix} \\ J_{iy} \\ J_{iz} \end{bmatrix}\left[\frac{\partial N_i}{\partial x_v}\right]$$
$$+ \left(\frac{s_iN_i^2E_i}{\sqrt{1 - N_i^2 + (N_iE_i)^2}} - N_i\right)\begin{bmatrix} J_{ix} \\ J_{iy} \\ J_{iz} \end{bmatrix}\left[\frac{\partial E_i}{\partial x_v}\right], \tag{5.41}$$

where $E_i$, $[\partial E_i/\partial x_v]$, $\begin{bmatrix} \partial J_{ix}/\partial x_v & \partial J_{iy}/\partial x_v & \partial J_{iz}/\partial x_v \end{bmatrix}^T$ are respectively given in Eqs. (2.40), (5.38), (5.40), and

$$\left[\frac{\partial N_i}{\partial x_v}\right] = \begin{bmatrix} 0 & 0 & 0 & 0 & 0 & 0 & \frac{1}{\xi_i} & \frac{-N_i}{\xi_i} & 0 \end{bmatrix}. \tag{5.42}$$

The change in the reflected/refracted ray $\Delta\bar{R}_i$ at a flat boundary surface due to the change in the boundary variable vector $\Delta\bar{X}_i$ can be obtained from the product of matrix $\bar{S}_i$ and $\Delta\bar{X}_i$, that is

$$\Delta\bar{R}_i = \bar{S}_i\Delta\bar{X}_i. \tag{5.43}$$

By summing up Eqs. (5.8) and (5.43), the change in the refracted/or refracted ray (i.e., $\Delta\bar{R}_i$) due to changes in the incoming ray $\bar{R}_{i-1}$ and boundary variable vector $\bar{X}_i$ (i.e., $\Delta\bar{R}_{i-1}$ and $\Delta\bar{X}_i$) is obtained as (see Fig. 5.11)

$$\Delta\bar{R}_i = \begin{bmatrix} \partial\bar{P}_i/\partial\bar{P}_{i-1} & \partial\bar{P}_i/\partial\bar{\ell}_{i-1} \\ \partial\bar{\ell}_i/\partial\bar{P}_{i-1} & \partial\bar{\ell}_i/\partial\bar{\ell}_{i-1} \end{bmatrix}\begin{bmatrix} \Delta\bar{P}_{i-1} \\ \Delta\bar{\ell}_{i-1} \end{bmatrix} + \begin{bmatrix} \partial\bar{P}_i/\partial\bar{X}_i \\ \partial\bar{\ell}_i/\partial\bar{X}_i \end{bmatrix}\Delta\bar{X}_i = \bar{M}_i\Delta\bar{R}_{i-1} + \bar{S}_i\Delta\bar{X}_i. \tag{5.44}$$

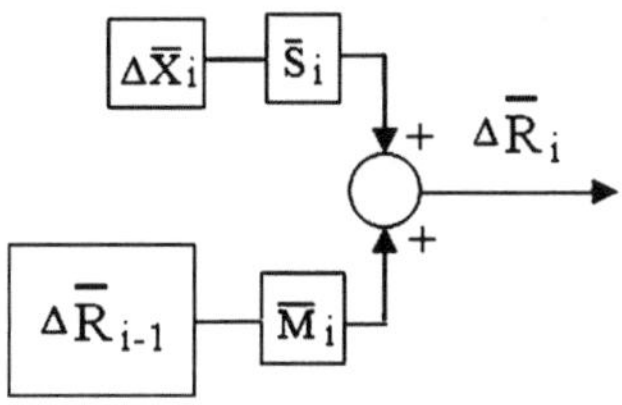


**Fig. 5.11** The change of the ray, $\Delta\bar{R}_i$, is the sum of $\bar{M}_i\Delta\bar{R}_{i-1}$ and $\bar{S}_i\Delta\bar{X}_i$

One can apply Eq. (5.44) successively and use $\Delta\overline{R}_0 = \overline{S}_0\Delta\overline{X}_0$ to compute the change of a ray, say $\Delta\overline{R}_g$, for a prism, if all its boundary surfaces are flat. The full details of this issue will be addressed in Sect. 5.6 for an optical system containing both flat and spherical boundary surfaces.

*Example 5.6* The expressions of $\Delta\overline{R}_1$ and $\Delta\overline{R}_2$ for system shown in Fig. 2.14 are respectively given by

$$\Delta\overline{R}_1 = \overline{M}_1\Delta\overline{R}_0+\overline{S}_1\Delta\overline{X}_1 = \overline{M}_1\left(\overline{S}_0\Delta\overline{X}_0\right)+\overline{S}_1\Delta\overline{X}_1 \tag{5.45a}$$

and

$$\begin{aligned}\Delta\overline{R}_2 &= \overline{M}_2\Delta\overline{R}_1+\overline{S}_2\Delta\overline{X}_2 = \overline{M}_2\left(\overline{M}_1\Delta\overline{R}_0+\overline{S}_1\Delta\overline{X}_1\right)+\overline{S}_2\Delta\overline{X}_2\\ &= \overline{M}_2\overline{M}_1\overline{S}_0\Delta\overline{X}_0+\overline{M}_2\overline{S}_1\Delta\overline{X}_1+\overline{S}_2\Delta\overline{X}_2.\end{aligned} \tag{5.45b}$$

Their flow charts are respectively given in Figs. 5.12 and 5.13.

*Example 5.7* The matrices $\overline{S}_i$(i = 1 and i = 2) for Fig. 2.14 are given by

$$\overline{S}_1 = \begin{bmatrix} 0 & 0 & 0 & 0 & 0 & 0 & 0 & 0 & 0\\ 0 & 1 & 0 & -5.8749 & 0 & 0 & 0 & 0 & 0\\ 0 & 0.0875 & 0 & -0.5140 & 0 & 0 & 0 & 0 & 0\\ 0 & 0 & 0 & 0 & 0 & -0.3342 & 0 & 0 & 0\\ 0 & 0 & 0 & -0.0195 & 0 & 0 & -0.0034 & 0.0023 & 0\\ 0 & 0 & 0 & 0.3342 & 0 & 0 & 0.0581 & -0.0387 & 0 \end{bmatrix}, \tag{5.46a}$$

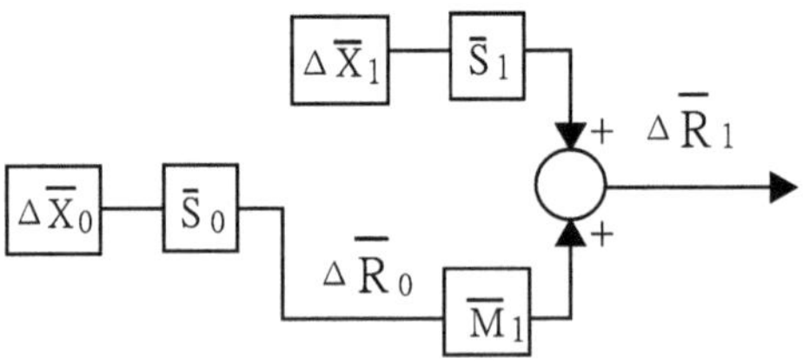


**Fig. 5.12** $\Delta\overline{R}_1$ is the sum of $\overline{M}_1\overline{S}_0\Delta\overline{X}_0$, the change of source ray, and $\overline{S}_1\Delta\overline{X}_1$, the change due to $\Delta\overline{X}_1$

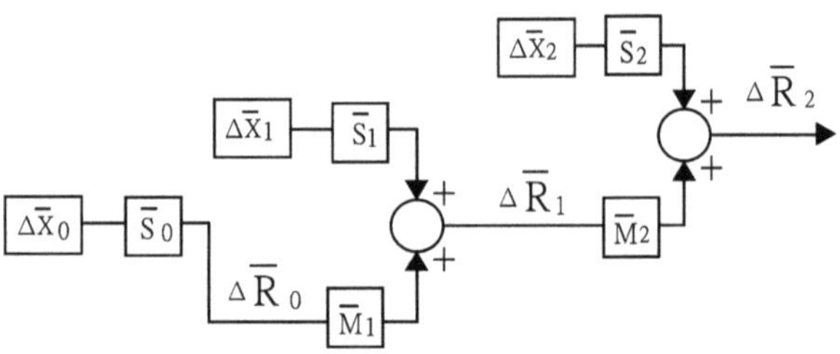


**Fig. 5.13** $\Delta\overline{R}_2$ is the sum of $\overline{M}_2\overline{M}_1\overline{S}_0\Delta\overline{X}_0$, $\overline{M}_2\overline{S}_1\Delta\overline{X}_1$, and $\overline{S}_2\Delta\overline{X}_2$

$$\overline{S}_2 = \begin{bmatrix} 0 & 0 & 0 & 0 & 0 & 0 & 0 & 0 & 0 \\ 0 & 1 & 0 & -6.4569 & 0 & 0 & 0 & 0 & 0 \\ 0 & 0.0582 & 0 & -0.3758 & 0 & 0 & 0 & 0 & 0 \\ 0 & 0 & 0 & 0 & 0 & 0.5013 & 0 & 0 & 0 \\ 0 & 0 & 0 & 0.0439 & 0 & 0 & -0.0051 & 0.0076 & 0 \\ 0 & 0 & 0 & -0.5013 & 0 & 0 & 0.0581 & -0.0872 & 0 \end{bmatrix}. \tag{5.46b}$$

## 5.5 The Jacobian Matrix of Ray $\overline{R}_i$ with Respect to Boundary Variable Vector $\overline{X}_i$ at a Spherical Boundary Surface

The objective of this section is to determine the Jacobian matrix $\overline{S}_i = \partial\overline{R}_i/\partial\overline{X}_i$ of a refracted/or reflected ray $\overline{R}_i$ with respect to the boundary variable vector $\overline{X}_i$ of a spherical boundary surface. Again, we will divide $\overline{S}_i$ into two parts: $\partial\overline{P}_i/\partial\overline{X}_i$ and $\partial\overline{\ell}_i/\partial\overline{X}_i$, as discussed in the following.

(a) $\partial\overline{P}_i/\partial\overline{X}_i = \left[\partial\overline{P}_i/\partial x_v\right]$, $x_v \in \overline{X}_i$, is obtained by differentiating $\overline{P}_i = \overline{P}_{i-1}+\overline{\ell}_{i-1}\lambda_i$ given in Eq. (2.10), i.e.,

$$\frac{\partial\overline{P}_i}{\partial\overline{X}_i} = \left[\frac{\partial\overline{P}_i}{\partial x_v}\right] = \begin{bmatrix} \partial P_{ix}/\partial x_v \\ \partial P_{iy}/\partial x_v \\ \partial P_{iz}/\partial x_v \end{bmatrix} = \begin{bmatrix} \ell_{i-1x} \\ \ell_{i-1y} \\ \ell_{i-1z} \end{bmatrix} \left[\frac{\partial\lambda_i}{\partial x_v}\right], \tag{5.47}$$

where

$$\frac{\partial\lambda_i}{\partial\overline{X}_i} = \left[\frac{\partial\lambda_i}{\partial x_v}\right] = \left(-1 \pm \frac{D_i}{\sqrt{\left(D_i^2 - E_i\right)}}\right)\left[\frac{\partial D_i}{\partial x_v}\right] \mp \frac{1}{2\sqrt{\left(D_i^2 - E_i\right)}}\left[\frac{\partial E_i}{\partial x_v}\right] \tag{5.48}$$

with $D_i$ and $E_i$ are respectively given in Eqs. (2.16) and (2.17), and

$$\left[\frac{\partial D_i}{\partial x_v}\right] = [-\ell_{i-1x} \quad -\ell_{i-1y} \quad -\ell_{i-1z} \quad 0 \quad 0 \quad 0 \quad 0 \quad 0 \quad 0], \tag{5.49}$$

$$\left[\frac{\partial E_i}{\partial x_v}\right] = \left[2(t_{ix} - P_{i-1x}) \quad 2\left(t_{iy} - P_{i-1y}\right) \quad 2(t_{iz} - P_{i-1z}) \quad 0 \quad 0 \quad 0 \quad 0 \quad 0 \quad -2R_i\right]. \tag{5.50}$$

(b) When $\overline{\ell}_i$ is the unit directional vector of the reflected ray at a spherical boundary surface, $\partial\overline{\ell}_i/\partial\overline{X}_i = \left[\partial\overline{\ell}_i/\partial x_v\right]$, $x_v \in \overline{X}_i$, can be directly derived by differentiating Eq. (2.25), i.e.,

$$\frac{\partial \bar{\ell}_i}{\partial \overline{X}_i}=\left[\frac{\partial \bar{\ell}_i}{\partial x_v}\right]=\begin{bmatrix}\partial \ell_{ix}/\partial x_v\\ \partial \ell_{iy}/\partial x_v\\ \partial \ell_{iz}/\partial x_v\end{bmatrix}=\begin{bmatrix}2n_{ix}\\ 2n_{iy}\\ 2n_{iz}\end{bmatrix}\left[\frac{\partial (C\theta_i)}{\partial x_v}\right]+2C\theta_i\begin{bmatrix}\partial n_{ix}/\partial x_v\\ \partial n_{iy}/\partial x_v\\ \partial n_{iz}/\partial x_v\end{bmatrix}. \tag{5.51}$$

The terms, $[\partial n_{ix}/\partial x_v]$, $[\partial n_{iy}/\partial x_v]$, and $[\partial n_{iz}/\partial x_v]$ of Eq. (5.51) are computed by differentiating Eq. (2.9) with respect to $x_v$ to give

$$\begin{aligned}\left[\frac{\partial n_{ix}}{\partial x_v}\right]=s_i\Big[0\ \ 0\ \ 0\ \ &\left(\frac{\partial I_{ix}}{\partial \omega_{ix}}C\beta_i C\alpha_i+\frac{\partial J_{ix}}{\partial \omega_{ix}}C\beta_i S\alpha_i+\frac{\partial K_{ix}}{\partial \omega_{ix}}S\beta_i\right)\ \left(\frac{\partial I_{ix}}{\partial \omega_{iy}}C\beta_i C\alpha_i+\frac{\partial J_{ix}}{\partial \omega_{iy}}C\beta_i S\alpha_i+\frac{\partial K_{ix}}{\partial \omega_{iy}}S\beta_i\right)\\ &\left(\frac{\partial I_{ix}}{\partial \omega_{iz}}C\beta_i C\alpha_i+\frac{\partial J_{ix}}{\partial \omega_{iz}}C\beta_i S\alpha_i+\frac{\partial K_{ix}}{\partial \omega_{iz}}S\beta_i\right)\ \ 0\ \ 0\ \ 0\Big]\\ &+s_i(-I_{ix}S\beta_i C\alpha_i-J_{ix}S\beta_i S\alpha_i+K_{ix}C\beta_i)\left[\frac{\partial \beta_i}{\partial x_v}\right]+s_i(-I_{ix}C\beta_i S\alpha_i+J_{ix}C\beta_i C\alpha_i)\left[\frac{\partial \alpha_i}{\partial x_v}\right],\end{aligned} \tag{5.52a}$$

$$\begin{aligned}\left[\frac{\partial n_{iy}}{\partial x_v}\right]=s_i\Big[0\ \ 0\ \ 0\ \ &\left(\frac{\partial I_{iy}}{\partial \omega_{ix}}C\beta_i C\alpha_i+\frac{\partial J_{iy}}{\partial \omega_{ix}}C\beta_i S\alpha_i+\frac{\partial K_{iy}}{\partial \omega_{ix}}S\beta_i\right)\ \left(\frac{\partial I_{iy}}{\partial \omega_{iy}}C\beta_i C\alpha_i+\frac{\partial J_{iy}}{\partial \omega_{iy}}C\beta_i S\alpha_i+\frac{\partial K_{iy}}{\partial \omega_{iy}}S\beta_i\right)\\ &\left(\frac{\partial I_{iy}}{\partial \omega_{iz}}C\beta_i C\alpha_i+\frac{\partial J_{iy}}{\partial \omega_{iz}}C\beta_i S\alpha_i+\frac{\partial K_{iy}}{\partial \omega_{iz}}S\beta_i\right)\ \ 0\ \ 0\ \ 0\Big]\\ &+s_i(-I_{iy}S\beta_i C\alpha_i-J_{iy}S\beta_i S\alpha_i+K_{iy}C\beta_i)\left[\frac{\partial \beta_i}{\partial x_v}\right]+s_i(-I_{iy}C\beta_i S\alpha_i+J_{iy}C\beta_i C\alpha_i)\left[\frac{\partial \alpha_i}{\partial x_v}\right],\end{aligned} \tag{5.52b}$$

$$\begin{aligned}\left[\frac{\partial n_{iz}}{\partial x_v}\right]=s_i\Big[0\ \ 0\ \ 0\ \ &\left(\frac{\partial I_{iz}}{\partial \omega_{ix}}C\beta_i C\alpha_i+\frac{\partial J_{iz}}{\partial \omega_{ix}}C\beta_i S\alpha_i+\frac{\partial K_{iz}}{\partial \omega_{ix}}S\beta_i\right)\ \left(\frac{\partial I_{iz}}{\partial \omega_{iy}}C\beta_i C\alpha_i+\frac{\partial J_{iz}}{\partial \omega_{iy}}C\beta_i S\alpha_i+\frac{\partial K_{iz}}{\partial \omega_{iy}}S\beta_i\right)\\ &\left(\frac{\partial I_{iz}}{\partial \omega_{iz}}C\beta_i C\alpha_i+\frac{\partial J_{iz}}{\partial \omega_{iz}}C\beta_i S\alpha_i+\frac{\partial K_{iz}}{\partial \omega_{iz}}S\beta_i\right)\ \ 0\ \ 0\ \ 0\Big]\\ &+s_i(-I_{iz}S\beta_i C\alpha_i-J_{iz}S\beta_i S\alpha_i+K_{iz}C\beta_i)\left[\frac{\partial \beta_i}{\partial x_v}\right]+s_i(-I_{iz}C\beta_i S\alpha_i+J_{iz}C\beta_i C\alpha_i)\left[\frac{\partial \alpha_i}{\partial x_v}\right].\end{aligned} \tag{5.52c}$$

The required terms $[\partial \alpha_i/\partial x_v]$ and $[\partial \beta_i/\partial x_v]$, in Eqs. (5.52a) to (5.52c) can be obtained by differentiating Eqs. (2.18) and (2.19), respectively, with respect to $\overline{X}_i$, that is,

$$\left[\frac{\partial \alpha_i}{\partial x_v}\right]=\frac{1}{\sigma_i^2+\rho_i^2}\left(\sigma_i\left[\frac{\partial \rho_i}{\partial x_v}\right]-\rho_i\left[\frac{\partial \sigma_i}{\partial x_v}\right]\right), \tag{5.53}$$

$$\left[\frac{\partial \beta_i}{\partial x_v}\right]=\frac{1}{(\sigma_i^2+\rho_i^2+\tau_i^2)\sqrt{(\sigma_i^2+\rho_i^2)}}\left\{(\sigma_i^2+\rho_i^2)\left[\frac{\partial \tau_i}{\partial x_v}\right]-\tau_i\left(\sigma_i\left[\frac{\partial \sigma_i}{\partial x_v}\right]+\rho_i\left[\frac{\partial \rho_i}{\partial x_v}\right]\right)\right\}, \tag{5.54}$$

where $[\partial \sigma_i/\partial x_v]$, $[\partial \rho_i/\partial x_v]$, and $[\partial \tau_i/\partial x_v]$ are obtained from Eqs. (2.12), (2.13), and (2.14), respectively, as follows:

$$\left[\frac{\partial\sigma_i}{\partial x_v}\right] = \Big[-I_{ix} \quad -I_{iy} \quad -I_{iz} \quad (P_{ix}-t_{ix})\frac{\partial I_{ix}}{\partial\omega_{ix}} + (P_{iy}-t_{iy})\frac{\partial I_{iy}}{\partial\omega_{ix}}$$
$$+(P_{iz}-t_{iz})\frac{\partial I_{iz}}{\partial\omega_{ix}} (P_{ix}-t_{ix})\frac{\partial I_{ix}}{\partial\omega_{iy}} + (P_{iy}-t_{iy})\frac{\partial I_{iy}}{\partial\omega_{iy}} + (P_{iz}-t_{iz})\frac{\partial I_{iz}}{\partial\omega_{iy}} \quad (P_{ix}-t_{ix})\frac{\partial I_{ix}}{\partial\omega_{iz}}$$
$$+ (P_{iy}-t_{iy})\frac{\partial I_{iy}}{\partial\omega_{iz}} (P_{iz}-t_{iz})\frac{\partial I_{iz}}{\partial\omega_{iz}} \quad 0 \quad 0 \quad 0\Big] + (I_{ix}\ell_{i-1x} + I_{iy}\ell_{i-1y} + I_{iz}\ell_{i-1z})\left[\frac{\partial\lambda_i}{\partial x_v}\right], \tag{5.55}$$

$$\left[\frac{\partial\rho_i}{\partial x_v}\right] = \Big[-J_{ix} \quad -J_{iy} \quad -J_{iz} \quad (P_{ix}-t_{ix})\frac{\partial J_{ix}}{\partial\omega_{ix}} + (P_{iy}-t_{iy})\frac{\partial J_{iy}}{\partial\omega_{ix}} + (P_{iz}-t_{iz})\frac{\partial J_{iz}}{\partial\omega_{ix}}(P_{ix}-t_{ix})\frac{\partial J_{ix}}{\partial\omega_{iy}}$$
$$+ (P_{iy}-t_{iy})\frac{\partial J_{iy}}{\partial\omega_{iy}} + (P_{iz}-t_{iz})\frac{\partial J_{iz}}{\partial\omega_{iy}} \quad (P_{ix}-t_{ix})\frac{\partial J_{ix}}{\partial\omega_{iz}} + (P_{iy}-t_{iy})\frac{\partial J_{iy}}{\partial\omega_{iz}}$$
$$+ (P_{iz}-t_{iz})\frac{\partial J_{iz}}{\partial\omega_{iz}} \quad 0 \quad 0 \quad 0] + (J_{ix}l_{i-x} + J_{iy}l_{i-1y} + J_{iz}l_{i-1z})\left[\frac{\partial\lambda_i}{\partial x_v}\right], \tag{5.56}$$

$$\left[\frac{\partial\tau_i}{\partial x_v}\right] = \Big[-K_{ix} \quad -K_{iy} \quad -K_{iz}(P_{ix}-t_{ix})\frac{\partial K_{ix}}{\partial\omega_{ix}} + (P_{iy}-t_{iy})\frac{\partial K_{iy}}{\partial\omega_{ix}} + (P_{iz}-t_{iz})\frac{\partial K_{iz}}{\partial\omega_{ix}}(P_{ix}-t_{ix})\frac{\partial K_{ix}}{\partial\omega_{ix}}$$
$$+ (P_{iy}-t_{iy})\frac{\partial K_{iy}}{\partial\omega_{iy}} + (P_{iz}-t_{iz})\frac{\partial K_{iz}}{\partial\omega_{iy}} \quad (P_{ix}-t_{ix})\frac{\partial K_{ix}}{\partial\omega_{iz}} + (P_{iy}-t_{iy})\frac{\partial K_{iy}}{\partial\omega_{iz}}$$
$$+ (P_{iz}-t_{iz})\frac{\partial K_{iz}}{\partial\omega_{iz}} \quad 0 \quad 0 \quad 0\Big] + (K_{ix}\ell_{i-1x} + K_{iy}\ell_{i-1y} + K_{iz}\ell_{i-1z})\left[\frac{\partial\lambda_i}{\partial x_v}\right]. \tag{5.57}$$

The term $[\partial\lambda_i/\partial x_v]$ of Eqs. (5.55), (5.56), and (5.57) is given in Eq. (5.48). Again, the required terms ($\partial I_{ix}/\partial\omega_{ix}$, $\partial I_{iy}/\partial\omega_{ix}$, and so on) can be found in the Appendix A.

(c) When $\bar{\ell}_i$ is the unit directional vector of the refracted ray at a spherical boundary surface, $\partial\bar{\ell}_i/\partial\bar{X}_i = [\partial\bar{\ell}_i/\partial x_v]$, $x_v \in \bar{X}_i$, can be obtained directly by differentiating Eq. (2.27) to give

$$\frac{\partial\bar{\ell}_i}{\partial\bar{X}_i} = \left[\frac{\partial\bar{\ell}_i}{\partial x_v}\right] = \begin{bmatrix}\partial\ell_{ix}/\partial x_v \\ \partial\ell_{iy}/\partial x_v \\ \partial\ell_{iz}/\partial x_v\end{bmatrix} = \left(-\sqrt{1-N_i^2+(N_iC\theta_i)^2} + N_iC\theta_i\right)\begin{bmatrix}\partial n_{ix}/\partial x_v \\ \partial n_{iy}/\partial x_v \\ \partial n_{iz}/\partial x_v\end{bmatrix} + \begin{bmatrix}\ell_{i-1x} \\ \ell_{i-1y} \\ \ell_{i-1z}\end{bmatrix}\left[\frac{\partial N_i}{\partial x_v}\right]$$
$$+ \left(\frac{-N_i^2C\theta_i}{\sqrt{1-N_i^2+(N_iC\theta_i)^2}} + N_i\right)\begin{bmatrix}n_{ix} \\ n_{iy} \\ n_{iz}\end{bmatrix}\left[\frac{\partial(C\theta_i)}{\partial x_v}\right] + \left(C\theta_i + \frac{(N_i - N_iC\theta_i^2)}{\sqrt{1-N_i^2+(N_iC\theta_i)^2}}\right)\begin{bmatrix}n_{ix} \\ n_{iy} \\ n_{iz}\end{bmatrix}\left[\frac{\partial N_i}{\partial x_v}\right]. \tag{5.58}$$

The required terms $[\partial n_{ix}/\partial x_v]$, $[\partial n_{iy}/\partial x_v]$, $[\partial n_{iz}/\partial x_v]$, and $[\partial N_i/\partial x_v]$, of Eq. (5.58) are respectively given in Eqs. (5.52a), (5.52b), (5.52c), and (5.42). The term $[\partial(C\theta_i)/\partial x_v]$ of Eq. (5.58) is computed from Eq. (2.20) to give

$$\left[\frac{\partial(C\theta_i)}{\partial x_v}\right] = -s_i\left(\ell_{i-1x}\left[\frac{\partial n_{ix}}{\partial x_v}\right] + \ell_{i-1y}\left[\frac{\partial n_{iy}}{\partial x_v}\right] + \ell_{i-1z}\left[\frac{\partial n_{iz}}{\partial x_v}\right]\right), \tag{5.59}$$

where $[\partial n_{ix}/\partial x_v]$, $[\partial n_{iy}/\partial x_v]$, and $[\partial n_{iz}/\partial x_v]$ of Eq. (5.59) are given in Eqs. (5.52a)–(5.52c).

One can combine Eqs. (5.47) and (5.51) for a reflection process or Eqs. (5.47) and (5.58) for a refraction process, to give $\bar{S}_i$ at a spherical boundary surface:

$$\bar{S}_i = \frac{\partial \bar{R}_i}{\partial \bar{X}_i} = \begin{bmatrix} \partial \bar{P}_i/\partial \bar{X}_i \\ \partial \bar{\ell}_i/\partial \bar{X}_i \end{bmatrix}. \tag{5.60}$$

Similar with the case of a flat boundary surface, one can employ Eqs. (5.31) and (5.60) to give the following equation for determining the change of the refracted/or refracted ray, $\Delta\bar{R}_i$, due to $\Delta\bar{R}_{i-1}$ and $\Delta\bar{X}_i$ at a spherical boundary surface:

$$\begin{aligned}\Delta\bar{R}_i &= \begin{bmatrix} \partial \bar{P}_i/\partial \bar{P}_{i-1} & \partial \bar{P}_i/\partial \bar{\ell}_{i-1} \\ \partial \bar{\ell}_i/\partial \bar{P}_{i-1} & \partial \bar{\ell}_i/\partial \bar{\ell}_{i-1} \end{bmatrix} \begin{bmatrix} \Delta\bar{P}_{i-1} \\ \Delta\bar{\ell}_{i-1} \end{bmatrix} + \begin{bmatrix} \partial \bar{P}_i/\partial \bar{X}_i \\ \partial \bar{\ell}_i/\partial \bar{X}_i \end{bmatrix} \Delta\bar{X}_i \\ &= \bar{M}_i \Delta\bar{R}_{i-1} + \bar{S}_i \Delta\bar{X}_i. \end{aligned} \tag{5.61}$$

Now, one can apply Eq. (5.44) and (5.61) successively and $\Delta\bar{R}_0 = \bar{S}_0 \Delta\bar{X}_0$ to compute the change of a ray for an optical system containing both flat and spherical boundary surfaces. The full details of this issue will be addressed in the following section.

*Example 5.8* The matrices $\bar{S}_i$ (i = 1 and i = 2) for Fig. 2.9 are given by

$$\bar{S}_1 = \begin{bmatrix} 0 & 0 & 0 & 0 & 0 & 0 & 0 & 0 & 0 \\ 0 & 1.0105 & -0.1202 & 0 & 0 & 0 & 0 & 0 & -1.0176 \\ 0 & 0.0884 & -0.0105 & 0 & 0 & 0 & 0 & 0 & -0.0890 \\ 0.0068 & 0 & 0 & 0 & 0 & 0 & 0 & 0 & 0 \\ 0 & 0 & -0.0001 & 0 & 0 & 0 & -0.0025 & 0.0017 & 0 \\ 0 & -0.0006 & 0.0069 & 0 & 0 & 0 & 0.1374 & -0.0916 & 0.0014 \end{bmatrix}, \tag{5.62a}$$

$$\bar{S}_2 = \begin{bmatrix} 0 & 0 & 0 & 0 & 0 & 0 & 0 & 0 & 0 \\ 0 & 0.9989 & 0.0608 & 0 & 0 & 0 & 0 & 0 & 1.0007 \\ 0 & 0.0182 & 0.0011 & 0 & 0 & 0 & 0 & 0 & 0.0182 \\ 0.0050 & 0 & 0 & 0 & 0 & 0 & 0 & 0 & 0 \\ 0 & 0 & 0 & 0 & 0 & 0 & -0.0001 & 0.0002 & 0 \\ 0 & -0.0001 & 0.0050 & 0 & 0 & 0 & -0.0427 & 0.0641 & 0.0002 \end{bmatrix}. \tag{5.62b}$$

## 5.6 The Jacobian Matrix of Ray $\overline{R}_i$ with Respect to System Variable Vector $\overline{X}_{sys}$

This section addresses the determination of the Jacobian matrix of a ray $\overline{R}_g$ at the gth ($0 \le g \le n$) boundary surface in an optical system with respect to the system variable vector $\overline{X}_{sys}$. Recall that Eqs. (5.44) and (5.61) express the change of the refracted/or reflected ray $\Delta\overline{R}_i$ in terms of $\Delta\overline{R}_{i-1}$ and $\Delta\overline{X}_i$, where $\overline{X}_i$ is the variable vector of a boundary surface. By using these two equations successively with $i = g$ to $i = 0$, one thus obtains the change of the ray $\overline{R}_g$ at the $g^{th}$ boundary surface, given as (see Fig. 5.14):

$$\begin{aligned}\Delta\overline{R}_g &= \overline{M}_g\Delta\overline{R}_{g-1} + \overline{S}_g\Delta\overline{X}_g = \overline{M}_g\left(\overline{M}_{g-1}\Delta\overline{R}_{g-2} + \overline{S}_{g-1}\Delta\overline{X}_{g-1}\right) + \overline{S}_g\Delta\overline{X}_g \\ &= \overline{M}_g\overline{M}_{g-1}\Delta\overline{R}_{g-2} + \overline{M}_g\overline{S}_{g-1}\Delta\overline{X}_{g-1} + \overline{S}_g\Delta\overline{X}_g \\ &= \ldots \\ &= \overline{M}_g\overline{M}_{g-1}\overline{M}_{g-2}\ldots\ldots\overline{M}_2\overline{M}_1\overline{S}_0\Delta\overline{X}_0 + \overline{M}_g\overline{M}_{g-1}\overline{M}_{g-2}\ldots\ldots\overline{M}_3\overline{M}_2\overline{S}_1\Delta\overline{X}_1 + \ldots \\ &\quad + \overline{M}_g\overline{M}_{g-1}\overline{S}_{g-2}\Delta\overline{X}_{g-2} + \overline{M}_g\overline{S}_{g-1}\Delta\overline{X}_{g-1} + \overline{S}_g\Delta\overline{X}_g \\ &= \sum_{u=0}^{u=g}\overline{M}_g\overline{M}_{g-1}\ldots\ldots\overline{M}_{u+2}\overline{M}_{u+1}\overline{S}_u\Delta\overline{X}_u.\end{aligned} \tag{5.63}$$

One can present Eq. (5.63) in term of the change of system variable vector, $\Delta\overline{X}_{sys}$, by the following chain rule:

$$\Delta\overline{X}_i = \frac{\partial\overline{X}_i}{\partial\overline{X}_{sys}}\Delta\overline{X}_{sys}. \tag{5.64}$$

$\partial\overline{X}_i/\partial\overline{X}_{sys}$ ($i = 0$ to $i = g$), which will be discussed in details in next section, is the Jacobain matrix between boundary variable vector $\overline{X}_i$ and system variable vector $\overline{X}_{sys}$. The Jacobian matrix $\partial\overline{R}_g/\partial\overline{X}_{sys}$ can therefore obtained from Eq. (5.63) if we define $\Delta\overline{R}_g = \partial\overline{R}_g/\partial\overline{X}_{sys}\Delta\overline{X}_{sys}$, to give

$$\frac{\partial\overline{R}_g}{\partial\overline{X}_{sys}} = \sum_{u=0}^{u=g}\overline{M}_g\overline{M}_{g-1}\ldots\ldots\overline{M}_{u+2}\overline{M}_{u+1}\overline{S}_u\frac{\partial\overline{X}_u}{\partial\overline{X}_{sys}}. \tag{5.65}$$

*Example 5.9* The expression of $\Delta\overline{R}_2$ for system shown in Fig. 2.14 is given by (see Fig. 5.15):

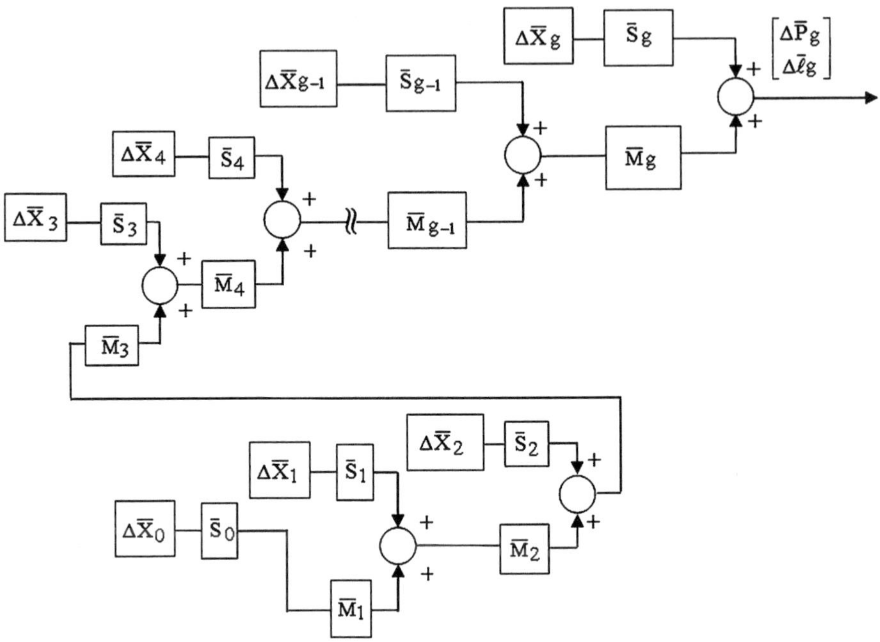


**Fig. 5.14** The determination of $\Delta\overline{R}_g$ for an optical system

$$\Delta\overline{R}_2 = \overline{M}_2\overline{M}_1\overline{S}_0\Delta\overline{X}_0 + \overline{M}_2\overline{S}_1\Delta\overline{X}_1 + \overline{S}_2\Delta\overline{X}_2$$

$$= \overline{M}_2\overline{M}_1\overline{S}_0\frac{\partial\overline{X}_0}{\partial\overline{X}_{sys}}\Delta\overline{X}_{sys} + \overline{M}_2\overline{S}_1\frac{\partial\overline{X}_1}{\partial\overline{X}_{sys}}\Delta\overline{X}_{sys} + \overline{S}_2\frac{\partial\overline{X}_2}{\partial\overline{X}_{sys}}\Delta\overline{X}_{sys}$$

$$= \begin{bmatrix} 1 & 0 & 0 & -16.65 & 0 & 0 & 0 & 0 & 0 & 0 & -3.35 & 0 & 0 & 0 \\ 0 & 0 & 0 & 0 & 0 & 0 & 1 & 0 & -6.48 & 0 & 0 & 0 & 0 & 1 \\ 0 & -0.09 & 1 & 0 & 19.75 & 0 & 0.09 & 0 & 2.81 & 0 & 0 & 0.58 & -0.39 & 0.06 \\ 0 & 0 & 0 & -1.00 & 0 & 0 & 0 & 0 & 0 & 0 & 0 & 0 & 0 & 0 \\ 0 & 0 & 0 & 0 & -0.09 & 0 & 0 & 0 & 0 & 0 & 0 & 0 & 0 & 0 \\ 0 & 0 & 0 & 0 & 1.00 & 0 & 0 & 0 & 0 & 0 & 0 & 0 & 0 & 0 \end{bmatrix}\Delta\overline{X}_{sys}, \quad (5.66a)$$

where the system variable vector $\overline{X}_{sys}$ is given in Eq. (3.28). In Eq. (5.66a), the leading coefficients represent the relative effect of each variable change in determining the overall value of $\Delta\overline{R}_2$, respectively. Thus, if the sensitivity of a particular variable is equal to zero, small changes in its value have no effect on $\Delta\overline{R}_2$. The Jacobian matrix $\partial\overline{R}_2/\partial\overline{X}_{sys}$ can be obtained from Eq. (5.66a) to give

$$\frac{\partial\overline{R}_2}{\partial\overline{X}_{sys}} = \overline{M}_2\overline{M}_1\overline{S}_0\frac{\partial\overline{X}_0}{\partial\overline{X}_{sys}} + \overline{M}_2\overline{S}_1\frac{\partial\overline{X}_1}{\partial\overline{X}_{sys}} + \overline{S}_2\frac{\partial\overline{X}_2}{\partial\overline{X}_{sys}}. \quad (5.66b)$$

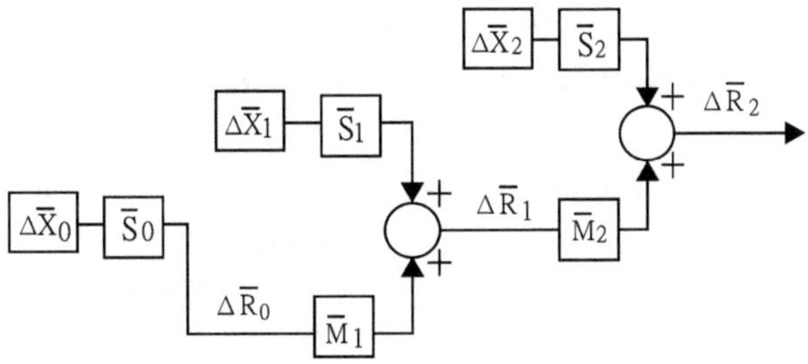


**Fig. 5.15** $\Delta\overline{R}_2$ of Fig. 2.14 is the sum of $\overline{M}_2\overline{M}_1\overline{S}_0\Delta\overline{X}_0$, $\overline{M}_2\overline{S}_1\Delta\overline{X}_1$, and $\overline{S}_2\Delta\overline{X}_2$

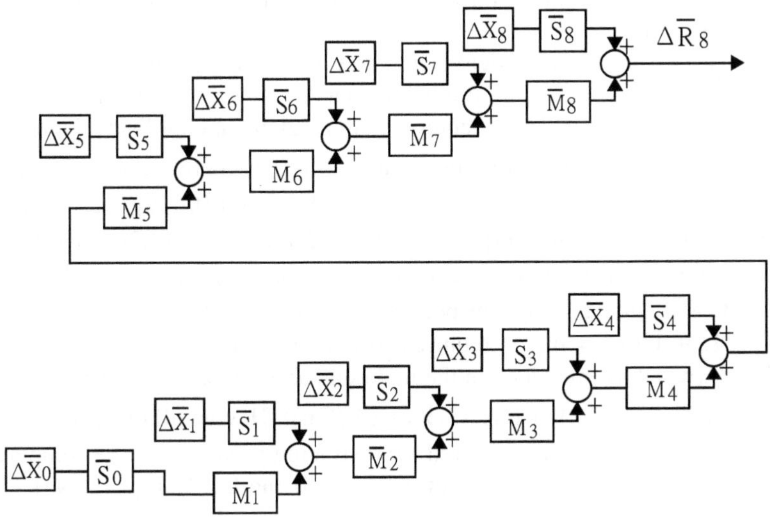


**Fig. 5.16** The flow chart of the Jacobian matrix $\partial\overline{R}_8/\partial\overline{X}_{sys}$ of the system in Fig. 2.1

*Example 5.10* The Jacobian matrix $\partial\overline{R}_8/\partial\overline{X}_{sys}$ of the system in Fig. 2.1 is given by (see Fig. 5.16):

$$\begin{aligned}\frac{\partial\overline{R}_8}{\partial\overline{X}_{sys}} &= \overline{M}_8\overline{M}_7\overline{M}_6\overline{M}_5\overline{M}_4\overline{M}_3\overline{M}_2\overline{M}_1\overline{S}_0\frac{\partial\overline{X}_0}{\partial\overline{X}_{sys}} + \overline{M}_8\overline{M}_7\overline{M}_6\overline{M}_5\overline{M}_4\overline{M}_3\overline{M}_2\overline{S}_1\frac{\partial\overline{X}_1}{\partial\overline{X}_{sys}} \\ &+ \overline{M}_8\overline{M}_7\overline{M}_6\overline{M}_5\overline{M}_4\overline{M}_3\overline{S}_2\frac{\partial\overline{X}_2}{\partial\overline{X}_{sys}} + \overline{M}_8\overline{M}_7\overline{M}_6\overline{M}_5\overline{M}_4\overline{S}_3\frac{\partial\overline{X}_3}{\partial\overline{X}_{sys}} + \overline{M}_8\overline{M}_7\overline{M}_6\overline{M}_5\overline{S}_4\frac{\partial\overline{X}_4}{\partial\overline{X}_{sys}} \\ &+ \overline{M}_8\overline{M}_7\overline{M}_6\overline{S}_5\frac{\partial\overline{X}_5}{\partial\overline{X}_{sys}} + \overline{M}_8\overline{M}_7\overline{S}_6\frac{\partial\overline{X}_6}{\partial\overline{X}_{sys}} + \overline{M}_8\overline{S}_7\frac{\partial\overline{X}_7}{\partial\overline{X}_{sys}} + \overline{S}_8\frac{\partial\overline{X}_8}{\partial\overline{X}_{sys}}.\end{aligned} \tag{5.67}$$

## 5.7 The Jacobian Matrix Between Boundary Variable Vector $\overline{X}_i$ and System Variable Vector $\overline{X}_{sys}$

Many merit functions have been proposed based on the boundary variable vectors $\overline{X}_i$ (i = 0 to i = n for system with n boundary surfaces). Consequently, transferring the computed merit function into the system variable vector $\overline{X}_{sys}$ is still highly challenging. Accordingly, this section proposes a numerical method for determining the Jacobian matrix between $\overline{X}_i$ and $\overline{X}_{sys}$. The proposed methodology can be implemented in computer codes, and provides a basis for the future development of numerical techniques for computing the derivatives of the merit function of optical systems.

As stated in Chap. 3, an optical element (e.g., element j shown in Fig. 3.1) is a block of optical material with multiple boundary surfaces. Defining the pose of this jth optical element in 3-D space requires both an element coordinate frame $(xyz)_{ej}$ and the pose matrix ${}^0\overline{A}_{ej}$. Although many different methods are available for specifying ${}^0\overline{A}_{ej}$, this book computes ${}^0\overline{A}_{ej}$ as ${}^0\overline{A}_{ej} = \text{tran}(t_{ejx}, t_{ejy}, t_{ejz})\ \text{rot}(\overline{z}, \omega_{ejz})\ \text{rot}(\overline{y}, \omega_{ejy})\text{rot}(\overline{x}, \omega_{ejx})$ (see Eqs. (3.22a)–(3.22f) or Eqs. (3.29a)–(3.29d)). As shown in Fig. 3.1, element j contains $L_j$ boundary surfaces labeled from $i = m_j - L_j + 1$ to $i = m_j$, respectively. Consequently, pose matrix ${}^{ej}\overline{A}_i$ $(i = m_j - L_j + 1$ to $i = m_j)$ is required to define the position and orientation of each $(xyz)_i$ (which is imbedded in the ith boundary surface) with respect to $(xyz)_{ej}$ (see Eqs. (3.23a)–(3.23k) or Eqs. (3.30a)–(3.30m)). Having all these matrices ${}^0\overline{A}_{ej}$ and ${}^{ej}\overline{A}_i$, the pose matrix ${}^0\overline{A}_i$ required for raytracing purposes can be computed via the following matrix multiplication (see Eqs. (2.8) and (2.34), Eqs. (3.24a)–(3.24k) or Eqs. (3.31a)–(3.31m):

$$
{}^0A_i = {}^0A_{ej}{}^{ej}A_i = \begin{bmatrix} I_{ix} & J_{ix} & K_{ix} & t_{ix} \\ I_{iy} & J_{iy} & K_{iy} & t_{iy} \\ I_{iz} & J_{iz} & K_{iz} & t_{iz} \\ 0 & 0 & 0 & 1 \end{bmatrix}. \tag{5.68}
$$

As shown in Fig. 2.1, the boundary coordinate frame $(xyz)_i$ (i = 1 to i = n) and element coordinate frame $(xyz)_{ej}$ (j = 1 to j = k) for an axis-symmetrical optical system are in the same directions. This feature makes that the Jacobian matrix $\partial\overline{X}_i/\partial\overline{X}_{sys}$ of such systems can be calculated straightforwardly. However, the determination of $\partial\overline{X}_i/\partial\overline{X}_{sys}$ for optical systems with prisms (e.g. Fig. 3.12) is comparatively difficult. In the following discussions of our approach in computing $\partial\overline{X}_i/\partial\overline{X}_{sys}$, we take the system of Fig. 3.12 for illustration purposes. The system has just k = 4 optical elements and n = 13 boundary surfaces. However, the modeling method described in this section is applicable to all optical systems, irrespective of the number of elements (k) or boundary surfaces (n).

Let the optical elements in Fig. 3.12 be labeled sequentially from j = 0 to j = k = 4 and the boundary surfaces be labeled from i = 0 to i = n=13 (see Fig. 3.12 to Fig. 3.16). The system variable vector $\overline{X}_{sys}$ of Eq. (3.33) consists of four groups of variables, namely (1) $\overline{X}_0 = [P_{0x} \quad P_{0y} \quad P_{0z} \quad \alpha_0 \quad \beta_0]^T$: the five independent variables of the source ray $\overline{R}_0$; (2) $\overline{X}_\xi$: the refractive indices (e.g., $\overline{X}_\xi = [\xi_{e1} \quad \xi_{air} \quad \xi_{e2} \quad \xi_{glue} \quad \xi_{e3}]^T$ for system in Fig. 3.12); (3) $\overline{X}_\kappa$: the curvatures $\kappa_i$ of the spherical boundary surfaces (e.g., $\overline{X}_\kappa = [\kappa_1 \quad \kappa_2]^T$ for system in Fig. 3.12); and (4) $\overline{X}_{rest}$: all the remaining system variables (i.e.,$\overline{X}_{rest} = \overline{X}_{sys} - \overline{X}_{e0} - \overline{X}_\xi - \overline{X}_R$). The Jacobian matrix between the boundary variable vector $\overline{X}_i$ (Eq. (2.28)) and the system variable vector $\overline{X}_{sys}$ (Eq. (3.33)) can be partitioned into 16 sub-matrices, i.e.,

$$\frac{\partial \overline{X}_i}{\partial \overline{X}_{sys}} = \begin{bmatrix} \partial x_u \\ \partial x_v \end{bmatrix} = [J_i(u,v)] = \begin{bmatrix} \frac{\partial(t_{ix},t_{iy},t_{iz})}{\partial \overline{X}_0} & \frac{\partial(t_{ix},t_{iy},t_{iz})}{\partial \overline{X}_\xi} & \frac{\partial(t_{ix},t_{iy},t_{iz})}{\partial \overline{X}_\kappa} & \frac{\partial(t_{ix},t_{iy},t_{iz})}{\partial \overline{X}_{rest}} \\ \frac{\partial(\omega_{ix},\omega_{iy},\omega_{iz})}{\partial \overline{X}_0} & \frac{\partial(\omega_{ix},\omega_{iy},\omega_{iz})}{\partial \overline{X}_\xi} & \frac{\partial(\omega_{ix},\omega_{iy},\omega_{iz})}{\partial \overline{X}_\kappa} & \frac{\partial(\omega_{ix},\omega_{iy},\omega_{iz})}{\partial \overline{X}_{rest}} \\ \frac{\partial(\xi_{i-1},\xi_i)}{\partial \overline{X}_0} & \frac{\partial(\xi_{i-1},\xi_i)}{\partial \overline{X}_\xi} & \frac{\partial(\xi_{i-1},\xi_i)}{\partial \overline{X}_\kappa} & \frac{\partial(\xi_{i-1},\xi_i)}{\partial \overline{X}_{rest}} \\ \frac{\partial R_i}{\partial \overline{X}_0} & \frac{\partial R_i}{\partial \overline{X}_\xi} & \frac{\partial R_i}{\partial \overline{X}_\kappa} & \frac{\partial R_i}{\partial \overline{X}_{rest}} \end{bmatrix}_{9\times 41}, \tag{5.69}$$

where $J_i(u,v) = \partial x_u/\partial x_v$, $x_u \in \overline{X}_i$ *and* $x_v \in \overline{X}_{sys}$, (v = 1 to v = 41; If i = 0, then u = 1 to u = 5; if $i \in \{1,2,..,13\}$, then u = 1 to u = 9 for the system shown in Fig. 3.12) specifies the rate of change of the uth boundary variable in $\overline{X}_i$ given a small infinitesimal change in the vth variable in $\overline{X}_{sys}$. In this book, the Jacobian matrix $\partial \overline{X}_i/\partial \overline{X}_{sys}$ is determined using the following procedure:

(1) Initialize the current components of $[J_i(u,v)] = \partial \overline{X}_i/\partial \overline{X}_{sys}$ (i = 0 to i = n) to zero.

(2) The components of $\partial \overline{X}_0/\partial \overline{X}_{sys} = [J_0(u,v)]$ ($u \in \{1,2,\ldots,5\}$,v = 1 to v = 41) with non-zero values are

$$J_0(1,1) = J_0(2,2) = J_0(3,3) = J_0(4,4) = J_0(5,5) = 1. \tag{5.70}$$

(3) The variables in $\overline{X}_0$ are independent of the variables in $\overline{X}_i$ (i = 1 to i = 13). Therefore, the following results are obtained for the first column of Eq. (5.69):

$$\frac{\partial(t_{ix},t_{iy},t_{iz})}{\partial \overline{X}_0} = \overline{0}_{3\times 5}, \tag{5.71}$$

$$\frac{\partial(\omega_{ix},\omega_{iy},\omega_{iz})}{\partial \overline{X}_0} = \overline{0}_{3\times 5}, \tag{5.72}$$

$$\frac{\partial(\xi_{i-1}, \xi_i)}{\partial \overline{X}_0} = \overline{0}_{2\times 5}, \tag{5.73}$$

$$\frac{\partial R_i}{\partial \overline{X}_0} = \overline{0}_{1\times 5}. \tag{5.74}$$

(4) $\xi_{i-1}$ and $\xi_i$ (i = 1 to i = 13) are independent variables, i.e., they are not functions of $\overline{X}_{rest}$. Thus, the following result is obtained:

$$\frac{\partial(\xi_{i-1}, \xi_i)}{\partial \overline{X}_{rest}} = \overline{0}_{2\times 29}. \tag{5.75}$$

Furthermore, no variable in $\overline{X}_\xi$ is used when computing the six boundary pose variables, and therefore the following equations are obtained:

$$\frac{\partial(t_{ix}, t_{iy}, t_{iz})}{\partial \overline{X}_\xi} = \overline{0}_{3\times 5}, \tag{5.76}$$

$$\frac{\partial(\omega_{ix}, \omega_{iy}, \omega_{iz})}{\partial \overline{X}_\xi} = \overline{0}_{3\times 5}. \tag{5.77}$$

(5) A dimensional analysis reveals that the angular pose parameters $\omega_{ix}, \omega_{iy},$ and$\omega_{iz}$ (i = 1 to i = 13) are independent of the curvatures $\kappa_i$ of the boundary surfaces. Therefore, it follows that

$$\frac{\partial(\omega_{ix}, \omega_{iy}, \omega_{iz})}{\partial \overline{X}_\kappa} = \overline{0}_{3\times 2}. \tag{5.78}$$

(6) The radii of the system elements, i.e., $R_i$ (i = 1 to i = 13), are independent of $\overline{X}_\xi$. Consequently, it follows that

$$\frac{\partial R_i}{\partial \overline{X}_\xi} = \overline{0}_{1\times 5}, \tag{5.79}$$

Furthermore, the refractive indices of the system elements, i.e., $\xi_{i-1}$ and $\xi_i$ (i = 1 to i = 13), are independent of the curvatures $\kappa_i$ of the boundary surfaces. In other words,

$$\frac{\partial(\xi_{i-1}, \xi_i)}{\partial \overline{X}_\kappa} = \overline{0}_{2\times 2}. \tag{5.80}$$

(7) Similarly, $R_i$ (i = 1 to i = 13) are independent of the variables in $\overline{X}_{rest}$. Thus,

$$\frac{\partial R_i}{\partial \overline{X}_{rest}} = \overline{0}_{1\times 29}. \tag{5.81}$$

(8) Determine $\partial(\xi_{i-1}, \xi_i)/\partial \overline{X}_\xi$ (i = 1 to i = 13).

If $\xi$, a variable in$\overline{X}_\xi$, is the vth component of $\overline{X}_{sys}$, then $\partial(\xi_{i-1}, \xi_i)/\partial\overline{X}_\xi$ can be determined from the following functions:

$$J_i(7, v) = \frac{\partial\xi_{i-1}}{\partial\xi} = \begin{cases} 1 & \xi_{i-1} = \xi \\ 0 & \xi_{i-1} \neq \xi \end{cases}, \tag{5.82}$$

$$J_i(8, v) = \frac{\partial\xi_i}{\partial\xi} = \begin{cases} 1 & \xi_i = \xi \\ 0 & \xi_i \neq \xi \end{cases}. \tag{5.83}$$

(9) Determine $\partial R_i/\partial\overline{X}_\kappa (i = 1 \text{ to } i = 13)$.

If $\kappa$, a variable in $\overline{X}_\kappa$, is the vth component of $\overline{X}_{sys}$, then $\partial\overline{R}_i/\partial\overline{X}_\kappa$ can be determined from the function

$$J_i(9, v) = \frac{\partial R_i}{\partial\kappa} = \begin{cases} -1/\kappa^2 & R_i = 1/\kappa \\ 0 & R_i \neq 1/\kappa \end{cases}. \tag{5.84}$$

(10) Determine $\partial(t_{ix}, t_{iy}, t_{iz})/\partial\overline{X}_\kappa$, $\partial(t_{ix}, t_{iy}, t_{iz})/\partial\overline{X}_{rest}$ and $\partial(\omega_{ix}, \omega_{iy}, \omega_{iz})/\partial\overline{X}_{rest}$.

To establish these terms, it is first necessary to obtain $\partial({}^0\overline{A}_i)/\partial\overline{X}_{sys} = \partial({}^0\overline{A}_i)/\partial x_v$ by differentiating Eq. (5.68) with respect to the vth component of $\overline{X}_{sys}$. This gives

$$\frac{\partial({}^0\overline{A}_i)}{\partial\overline{X}_{sys}} = \frac{\partial({}^0\overline{A}_i)}{\partial x_v} = \frac{\partial({}^0\overline{A}_{ej})}{\partial x_v}\,{}^{ej}\overline{A}_i + {}^0\overline{A}_{ej}\,\frac{\partial({}^{ej}\overline{A}_i)}{\partial x_v}$$
$$= \begin{bmatrix} \partial I_{ix}/\partial x_v & \partial J_{ix}/\partial x_v & \partial K_{ix}/\partial x_v & \partial t_{ix}/\partial x_v \\ \partial I_{iy}/\partial x_v & \partial J_{iy}/\partial x_v & \partial K_{iy}/\partial x_v & \partial t_{iy}/\partial x_v \\ \partial I_{iz}/\partial x_v & \partial J_{iz}/\partial x_v & \partial K_{iz}/\partial x_v & \partial t_{iz}/\partial x_v \\ 0 & 0 & 0 & 0 \end{bmatrix}, \tag{5.85}$$

where $\partial({}^0\overline{A}_{ej})/\partial x_v$ and $\partial({}^{ej}\overline{A}_i)/\partial x_v$ are the first derivative matrices of ${}^0\overline{A}_{ej}$ and ${}^{ej}\overline{A}_i$, respectively, with respect to the vth component of $\overline{X}_{sys}$. Computing Eq. (5.85) is a significant challenge when $\overline{X}_{sys}$ contains many components. In the present study, $\partial({}^0\overline{A}_{ej})/\partial x_v$ and $\partial({}^{ej}\overline{A}_i)/\partial x_v$ are obtained using the method described in Appendix B, and $d({}^0\overline{A}_i)/d\overline{X}_{sys}$ is then determined from Eq. (5.85). In computing these sub-matrix (i.e., $\partial(t_{ix}, t_{iy}, t_{iz})/\partial\overline{X}_\kappa$, $\partial(t_{ix}, t_{iy}, t_{iz})/\partial\overline{X}_{rest}$, and $\partial(\omega_{ix}, \omega_{iy}, \omega_{iz})/\partial\overline{X}_{rest}$ of Eq. (5.69)), $\partial\omega_{ix}/\partial\overline{X}_{sys}$, $\partial\omega_{iy}/\partial\overline{X}_{sys}$, $\partial\omega_{iz}/\partial\overline{X}_{sys}$, $\partial t_{ix}/\partial\overline{X}_{sys}$, $\partial t_{iy}/\partial\overline{X}_{sys}$, and $\partial t_{iz}/\partial\overline{X}_{sys}$ can be obtained by differentiating Eqs. (1.34 )–(1.39) respectively, to give:

$$\frac{\partial\omega_{iz}}{\partial x_v} = \frac{G_i}{F_i}, \tag{5.86}$$

where $F_i$ and $G_i$ are respectively given in Eqs. (D.1a) and (D.1b) in the Appendix D.

$$\frac{\partial \omega_{iy}}{\partial x_v} = \frac{G_i + H_i}{F_i}, \tag{5.87}$$

in which $F_i$, $G_i$, and $H_i$ are respectively given in Eqs. (D.2a), (D.2b), and (D.2c) in the Appendix D.

$$\frac{\partial \omega_{ix}}{\partial x_v} = \frac{G_i H_i - U_i Q_i}{F_i}, \tag{5.88}$$

where $F_i$, $G_i$, $H_i$, $U_i$, and $Q_i$ are respectively given in Eqs. (D.3a), (D.3b), (D.3c), (D.3d), and (D.3e) in the Appendix D.

$$\frac{\partial t_{ix}}{\partial x_v} = \frac{\partial t_{ix}}{\partial x_v}, \tag{5.89}$$

$$\frac{\partial t_{iy}}{\partial x_v} = \frac{\partial t_{iy}}{\partial x_v}, \tag{5.90}$$

$$\frac{\partial t_{iz}}{\partial x_v} = \frac{\partial t_{iz}}{\partial x_v}. \tag{5.91}$$

Equations (5.89), (5.90), and (5.91) just show that $\partial t_{ix}/\partial \bar{X}_{sys}$, $\partial t_{iy}/\partial \bar{X}_{sys}$, and $\partial t_{iz}/\partial \bar{X}_{sys}$ can be obtained directly by differentiating the (1,4)th, (2,4)th, and (3,4)th components of ${}^0\bar{A}_i$ computed by Eq. (5.68).

*Example 5.11* The source ray is labeled as the 0th boundary surface and $\bar{X}_0$ in Eq. (2.3) is considered as its boundary variable vector. If we take system in Fig. 2.1 as example, $\partial \bar{X}_0/\partial \bar{X}_{sys}$ is given by

$$\frac{\partial \bar{X}_0}{\partial \bar{X}_{sys}} = \begin{bmatrix} 1 & 0 & 0 & 0 & 0 & \bar{0}_{1x54} \\ 0 & 1 & 0 & 0 & 0 & \bar{0}_{1x54} \\ 0 & 0 & 1 & 0 & 0 & \bar{0}_{1x54} \\ 0 & 0 & 0 & 1 & 0 & \bar{0}_{1x54} \\ 0 & 0 & 0 & 0 & 1 & \bar{0}_{1x54} \end{bmatrix}. \tag{5.92}$$

*Example 5.12* The non-zero values of $J_i(7, v)$ and $J_i(8, v)$ (i = 1 to i = 11 and v = 1 to v = 59) for the system in Fig. 2.1 are

$$\begin{aligned} J_1(7,12) &= J_2(7,13) = J_3(7,12) = J_4(7,23) = J_5(7,12) = J_6(7,12) = J_7(7,12) \\ &= J_8(7,40) = J_9(7,12) = J_{10}(7,50) = J_{11}(7,12) = J_{12}(7,12) = J_1(8,13) \\ &= J_2(8,12) = J_3(8,23) = J_4(8,12) = J_5(8,12) = J_6(8,12) = J_7(8,40) \\ &= J_8(8,12) = J_9(8,50) = J_{10}(8,12) = J_{11}(8,12) = J_{12}(8,12) = 1. \end{aligned} \tag{5.93}$$

*Example 5.13* The non-zero values of $J_i(7, v)$ and $J_i(8, v)$ (i = 1 to i = 13 and v = 1 to v = 41) for the system in Fig. 3.12 are as follows:

$$\begin{aligned}J_1(7,12) =& J_2(7,13) = J_3(7,12) = J_4(7,22) = J_5(7,22) = J_6(7,22) = J_7(7,21)\\ =& J_8(7,31) = J_9(7,31) = J_{10}(7,31) = J_{11}(7,31) = J_{12}(7,31) = J_{13}(7,12) = 1,\end{aligned}$$

$$\begin{aligned}J_1(8,13) =& J_2(8,12) = J_3(8,22) = J_4(8,22) = J_5(8,22) = J_6(8,21) = J_7(8,31)\\ =& J_8(8,31) = J_9(8,31) = J_{10}(8,31) = J_{11}(8,31) = J_{12}(8,12) = J_{13}(8,12) = 1.\end{aligned} \tag{5.94}$$

*Example 5.14* The non-zero values of $J_i(9, v) = \partial R_i/\partial \kappa_i$ (i = 1 to i = 12 and v = 1 to v = 59) for the system of Fig. 2.1 are:

$$\begin{aligned}J_1(9,15) =& -1/\kappa_1^2,\ J_2(9,16) = -1/\kappa_2^2,\ J_3(9,25) = -1/\kappa_3^2,\ J_4(9,26)\\ =& -1/\kappa_4^2,\ J_7(9,42) = -1/\kappa_7^2,\\ J_8(9,43) =& -1/\kappa_8^2,\ J_9(9,52) = -1/\kappa_9^2,\ \text{and}\ J_{10}(9,53) = -1/\kappa_{10}^2.\end{aligned} \tag{5.95}$$

*Example 5.15* The non-zero values of $J_i(9, v) = \partial R_i/\partial \kappa$ (i = 1 to i = 13 and v = 1 to v = 41) for the system shown in Fig. 3.12 are

$$J_1(9,15) = -1/\kappa_1^2\ \text{and}\ J_2(9,16) = -1/\kappa_2^2. \tag{5.96}$$

*Example 5.16* The non-zero values of $J_{11}(u, v)$ (u = 1 to u = 9 and v = 1 to v = 59) for system shown in Fig. 2.1 are

$$\begin{aligned}J_{11}(1,54) &= J_{11}(2,7) = J_{11}(2,14) = J_{11}(2,18) = J_{11}(2,24) = J_{11}(2,28)\\ &= J_{11}(2,33) = J_{11}(2,35) = J_{11}(2,41)\\ &= J_{11}(2,45) = J_{11}(2,51) = J_{11}(2,55) = J_{11}(3,56) = J_{11}(4,57)\\ &= J_{11}(5,58) = J_{11}(6,59) = J_{11}(7,12) = J_{11}(8,12) = 1,\end{aligned}$$

$$J_{11}(1,58) = -J_{11}(2,57) = -8, J_{11}(2,59) = -J_{11}(3,58) = 5. \tag{5.97}$$

*Example 5.17* The non-zero values of $J_i(7, v)$ and $J_i(8, v)$ (i = 1 to i = 13 and v = 1 to v = 41) for the system shown in Fig. 3.12 are

$$\begin{aligned} J_1(7,12) =& J_2(7,13) = J_3(7,12) = J_4(7,22) = J_5(7,22) = J_6(7,22) = J_7(7,21) \\ =& J_8(7,31) = J_9(7,31) = J_{10}(7,31) = J_{11}(7,31) = J_{12}(7,31) = J_{13}(7,12) = 1, \end{aligned}$$

$$\begin{aligned} J_1(8,13) =& J_2(8,12) = J_3(8,22) = J_4(8,22) = J_5(8,22) = J_6(8,21) = J_7(8,31) \\ =& J_8(8,31) = J_9(8,31) = J_{10}(8,31) = J_{11}(8,31) = J_{12}(8,12) = J_{13}(8,12) = 1. \end{aligned} \tag{5.98}$$

*Example 5.18* The non-zero values of $J_i(9,v) = \partial R_i/\partial \overline{X}_{sys}$ (i = 1 to i = 13 and v = 1 to v = 41) for the system of Fig. 3.12 are

$$J_1(9,15) = -1/\kappa_1^2 \text{ and } J_2(9,16) = -1/\kappa_2^2. \tag{5.99}$$

*Example 5.19* The non-zero values of $J_{10}(u,v) = \partial \overline{X}_{10}/\partial \overline{X}_{sys}$(u = 1 to u = 9 and v = 1 to v = 41) for the system shown in Fig. 3.12 are

$$\begin{aligned} J_{10}(2,7) =& J_{10}(2,14) = J_{10}(2,17) = J_{10}(2,23) = J_{10}(3,33) = J_{10}(4,29) = -J_{10}(3,25) \\ =& -J_{10}(5,27) = J_{10}(5,35) = J_{10}(6,30) = J_{10}(6,34) = J_{10}(7,31) = J_{10}(8,31) = 1, \end{aligned}$$

$$J_{10}(1,24) = 0.5, J_{10}(2,32) = \sqrt{2},\ J_{10}(1,29) = -J_{10}(2,27) = J_{10}(2,35) = 10\sqrt{2}. \tag{5.100}$$

*Example 5.20* The non-zero values of $J_i(u,v)$ (i = 1 to i = 2, u = 1 to u = 9, and v = 1 to v = 14) for system shown in Fig. 2.9 are

$$\begin{aligned} J_1(1,6) &= J_1(2,7) = J_1(2,15) = J_1(3,8) = J_1(4,9) = J_1(5,10) \\ &= J_1(6,11) = J_1(7,12) = J_1(8,13) = J_1(9,15) \\ &= J_2(1,6) = J_2(2,7) = J_2(2,14) = -J_2(2,16) = J_2(3,8) \\ &= J_2(4,9) = J_2(5,10) = J_2(6,11) = J_2(7,13) \\ &= J_2(8,12) = J_2(9,16) = 1, J_2(1,11) = -J_2(3,9) = 140. \end{aligned} \tag{5.101}$$

*Example 5.21* The non-zero values of $J_i(u,v)$ (i = 1 to i = 2, u = 1 to u = 9, and v = 1 to v = 14) for system shown in Fig. 2.14 are

$$\begin{aligned} J_1(1,6) &= J_1(2,7) = J_1(3,8) = J_1(4,9) = J_1(5,10) = J_1(6,11) \\ &= J_1(7,12) = J_1(8,13) = J_2(1,6) \\ &= J_2(2,7) = J_2(2,14) = J_2(3,8) = J_2(4,9) = J_2(5,10) \\ &= J_2(6,11) = J_2(7,13) = J_2(8,12) = 1, \\ J_2(1,11) &= -J_2(3,9) = -10. \end{aligned} \tag{5.102}$$

*Example 5.22* When the source ray is $\overline{R}_0 = [0 \quad 0 \quad 0 \quad 0 \quad 1 \quad 0]^T$, then the change of the exit ray of element j = 2 in the system in Fig. 3.12 is given by:

$$\Delta\overline{R}_6 = \begin{bmatrix} 0.0244 \\ 0 \\ 0 \\ -0.0216 \\ 0 \\ 0 \end{bmatrix} \Delta P_{0x} + \begin{bmatrix} 0 \\ -0.0173 \\ 0.0173 \\ 0 \\ 0.0153 \\ -0.0153 \end{bmatrix} \Delta P_{0z} + \begin{bmatrix} -45.4274 \\ 0 \\ 0 \\ -0.7542 \\ 0 \\ 0 \end{bmatrix} \Delta\alpha_0 + \begin{bmatrix} 0 \\ -32.122 \\ 32.122 \\ 0 \\ -0.5333 \\ 0.5333 \end{bmatrix} \Delta\beta_0 + \begin{bmatrix} 0.9756 \\ 0 \\ 0 \\ 0.0216 \\ 0 \\ 0 \end{bmatrix} \Delta t_{e1x}$$
$$+ \begin{bmatrix} 0 \\ 1 \\ 0 \\ 0 \\ 0 \\ 0 \end{bmatrix} (\Delta v_1 + \Delta q_{e1} + \Delta v_2) + \begin{bmatrix} 0 \\ -0.6898 \\ 0.6898 \\ 0 \\ -0.0153 \\ 0.0153 \end{bmatrix} \Delta t_{e1z} + \begin{bmatrix} 0 \\ 24.2821 \\ -24.2821 \\ 0 \\ 0.563 \\ -0.563 \end{bmatrix} \Delta\omega_{e1x} + \begin{bmatrix} 34.3401 \\ 0 \\ 0 \\ 0.7962 \\ 0 \\ 0 \end{bmatrix} \Delta\omega_{e1z}$$
$$+ \begin{bmatrix} 0 \\ 7.8210 \\ -0.7499 \\ 0 \\ 0 \\ 0 \end{bmatrix} \Delta\omega_{e2x} + \begin{bmatrix} -3.5355 \\ 3.5355 \\ 1.4645 \\ 0.7071 \\ 0 \\ 0 \end{bmatrix} \Delta\omega_{e2y} + \begin{bmatrix} 4.5961 \\ -5 \\ 0 \\ 0.2929 \\ 0 \\ 0 \end{bmatrix} \Delta\omega_{e2z}$$
$$+ \begin{bmatrix} 0 \\ -0.7071 \\ -0.2929 \\ 0 \\ 0 \\ 0 \end{bmatrix} \Delta h_{e2} + \begin{bmatrix} 0 \\ 1.2071 \\ -0.2071 \\ 0 \\ 0 \\ 0 \end{bmatrix} \Delta m_{e2} + \begin{bmatrix} 0 \\ 32.6777 \\ -15.6066 \\ 0 \\ 2.0506 \\ -2.0506 \end{bmatrix} \Delta\psi_{e2} + \begin{bmatrix} 0 \\ 10 \\ -10 \\ 0 \\ 1.8385 \\ -1.8385 \end{bmatrix} \Delta\sigma_{e2}. \tag{5.103}$$

*Example 5.23* The change of the ray $\overline{R}_2$for the system in Fig. 2.9 is

$$\Delta\bar{R}_2 = \begin{bmatrix} 0.9360 & 0 & 0 & -15.9746 & 0 & 0.0640 & 0 & 0 & 0 & 0 \\ 0 & 0.005 & -0.057 & 0 & -0.977 & 0 & 0.995 & 0.057 & -8.515 & 0 \\ 0 & -0.082 & 0.942 & 0 & 16.050 & 0 & 0.082 & 0.059 & -0.155 & 0 \\ -0.015 & 0 & 0 & -0.811 & 0 & 0.015 & 0 & 0 & 0 & 0 \\ 0 & 0 & 0 & 0 & 0.003 & 0 & 0 & 0 & -0.002 & 0 \\ 0 & 0.001 & -0.015 & 0 & 0.800 & 0 & -0.001 & 0.015 & -0.703 & 0 \end{bmatrix.$$
$$\left.\begin{matrix} 0 & 0 & 0 & 0 & 0 & 0 \\ 0 & -0.079 & 0.053 & 1.000 & 0 & 0.002 \\ 0 & 1.300 & -0.866 & 0.018 & 0.007 & 0 \\ 0.701 & 0 & 0 & 0 & 0 & 0 \\ 0 & 0 & 0 & 0 & 0 & 0 \\ 0 & 0.264 & -0.176 & 0 & 0.001 & 0 \end{matrix}\right] \Delta\bar{X}_{sys}, \tag{5.104}$$

where the system variable vector $\bar{X}_{sys}$ is given in Eq. (3.27).

## 5.8 The Hessian Matrix Between Boundary Variable Vector $\bar{X}_i$ and System Variable Vector $\bar{X}_{sys}$

When optimizing optical systems and analyzing the resulting image, it is necessary to calculate the higher order derivatives of various optical quantities (e.g., the exit ray, the optical path length) such that their effects on a merit function of the system can be evaluated and the most influential design variables identified [24]. As shown in Eq. (5.1), $\bar{R}_i$ is mathematically a recursive function with the given function $\bar{R}_0 = \bar{R}_0(\bar{X}_0)$. Consequently, the Hessian matrix of $\bar{R}_i$ with respect to the system variable vector $\bar{X}_{sys}$ can be determined by:

$$\begin{aligned} \frac{\partial^2\bar{R}_i}{\partial\bar{X}_{sys}^2} &= \left[\frac{\partial^2\bar{R}_i}{\partial x_w \partial x_v}\right] \\ &= \left[\left(\frac{\partial\bar{R}_{i-1}}{\partial x_w}\right)^T \quad \left(\frac{\partial\bar{X}_i}{\partial x_w}\right)^T\right] \begin{bmatrix} \frac{\partial^2\bar{R}_i}{\partial\bar{R}_{i-1}\partial\bar{R}_{i-1}} & \frac{\partial^2\bar{R}_i}{\partial\bar{R}_{i-1}\partial\bar{X}_i} \\ \frac{\partial^2\bar{R}_i}{\partial\bar{X}_i\partial\bar{R}_{i-1}} & \frac{\partial^2\bar{R}_i}{\partial\bar{X}_i\partial\bar{X}_i} \end{bmatrix} \begin{bmatrix} \frac{\partial\bar{R}_{i-1}}{\partial x_v} \\ \frac{\partial\bar{X}_i}{\partial x_v} \end{bmatrix} \\ &\quad + \left[\frac{\partial\bar{R}_i}{\partial\bar{R}_{i-1}} \quad \frac{\partial\bar{R}_i}{\partial\bar{X}_i}\right] \begin{bmatrix} \frac{\partial^2\bar{R}_{i-1}}{\partial x_w \partial x_v} \\ \frac{\partial^2\bar{X}_i}{\partial x_w \partial x_v} \end{bmatrix}, \end{aligned} \tag{5.105}$$

where $x_v \in \bar{X}_{sys}$ and $x_w \in \bar{X}_{sys}$. It is noted from Eq. (5.105) that it is impossible to determine $\partial^2\bar{R}_i/\partial\bar{X}_{sys}^2$ without first having $[\partial^2\bar{X}_i/\partial x_w\partial x_v] = \partial^2\bar{X}_i/\partial\bar{X}_{sys}^2$. Therefore, in the following we will present the methodology for determining $\partial^2\bar{X}_i/\partial\bar{X}_{sys}^2$ such that, in the future, the Hessian matrix of a chosen merit function can be computed.

Let $\partial^2\overline{X}_i/\partial\overline{X}_{sys}^2 = [\partial x_u/\partial x_w\partial x_v] = [H_i(u, v, w)]$, $x_u \in \overline{X}_i$ *and* $x_v \in \overline{X}_{sys}$, $x_w \in \overline{X}_{sys}$ (v = 1 to v = 41 and w = 1 to w = 41. If i = 0, then u = 1 to u = 5; if $i \in \{1, 2, .., 13\}$, u = 1 to u = 9 for the system in Fig. 3.12) denote the second-order differentiation of the uth boundary variable in $\overline{X}_i$ with respect to the vth and wth variables in $\overline{X}_{sys}$. In this book, $H_i(u, v, w)$is determined via the following procedure:

(1) Initialize the current components of $[H_i(u, v, w)]$ to zero.

(2) Determine $H_i(u, v, w)$by differentiating $J_i(u, v)$ with respect to the wth component of vector $\overline{X}_{sys}$. From the carefully examination of Eqs. (5.70) to (5.84), the following non-zero components of $H_i(u, v, w)$are obtained:

$$H_i(9, v, v) = \frac{\partial^2 R_i}{\partial\kappa^2} = \begin{cases} 2/\kappa^3 & R_i = 1/\kappa \\ 0 & R_i \neq 1/\kappa\,. \end{cases} \tag{5.106}$$

*Example 5.24* The non-zero values of $H_i(9, v, v) = \partial^2 R_i/\partial\kappa^2$ (i = 1 to i = 13 and v = 1 to v = 41) for the illustrative system in Fig. 3.12 are

$$H_1(9, 15, 15) = 2/\kappa_1^3 \text{ and } H_2(9, 16, 16) = 2/\kappa_2^3. \tag{5.107}$$

(3) Determine $\partial^2(t_{ix}, t_{iy}, t_{iz})/\partial\overline{X}_\kappa^2$, $\partial^2(t_{ix}, t_{iy}, t_{iz})/\partial\overline{X}_{rest}^2$, and $\partial^2(\omega_{ix}, \omega_{iy}, \omega_{iz})/\partial\overline{X}_{rest}^2$.

To establish these terms, it is first necessary to obtain $\partial^2({}^0\overline{A}_i)/\partial\overline{X}_{sys}^2 = \partial^2({}^0\overline{A}_i)/\partial x_v\partial x_w$, i.e., the differentiation of ${}^i\overline{A}_0$ with respect to variables $x_v$ and $x_w$ in $\overline{X}_{sys}$, by differentiating Eq. (5.85). This gives

$$\begin{aligned}\frac{\partial^2({}^0\overline{A}_i)}{\partial\overline{X}_{sys}^2} &= \left[\frac{\partial^2({}^0\overline{A}_i)}{\partial x_w\partial x_v}\right] = \left[\frac{\partial^2({}^0\overline{A}_{ej})}{\partial x_w\partial x_v}\right]{}^{ej}\overline{A}_i + {}^0\overline{A}_{ej}\left[\frac{\partial^2({}^{ej}\overline{A}_i)}{\partial x_w\partial x_v}\right] + \left[\frac{\partial{}^0\overline{A}_{ej}}{\partial x_v}\right]\left[\frac{\partial{}^{ej}\overline{A}_i}{\partial x_w}\right] + \left[\frac{\partial{}^0\overline{A}_{ej}}{\partial x_w}\right]\left[\frac{\partial{}^{ej}\overline{A}_i}{\partial x_v}\right]\\ &= \begin{bmatrix} \partial I_{ix}^2/\partial x_w\partial x_v & \partial J_{ix}^2/\partial x_w\partial x_v & \partial K_{ix}^2/\partial x_w\partial x_v & \partial t_{ix}^2/\partial x_w\partial x_v \\ \partial I_{iy}^2/\partial x_w\partial x_v & \partial J_{iy}^2/\partial x_w\partial x_v & \partial K_{iy}^2/\partial x_w\partial x_v & \partial t_{iy}^2/\partial x_w\partial x_v \\ \partial I_{iz}^2/\partial x_w\partial x_v & \partial J_{iz}^2/\partial x_w\partial x_v & \partial K_{iz}^2/\partial x_w\partial x_v & \partial t_{iz}^2/\partial x_w\partial x_v \\ 0 & 0 & 0 & 0 \end{bmatrix},\end{aligned} \tag{5.108}$$

where $\partial^2({}^0\overline{A}_{ej})/\partial x_w\partial x_v$ and $\partial^2({}^{ej}\overline{A}_i)/\partial x_w\partial x_v$ (v = 1 to v = 41 and w = 1 to w = 41) are the second-order derivative matrices of ${}^0\overline{A}_{ej}$ and ${}^{ej}\overline{A}_i$, respectively, with respect to the vth and wth components (i.e., $x_v$ and $x_w$) of $\overline{X}_{sys}$. Note that in implementing Eq. (5.108) using computer code, $\partial^2({}^0\overline{A}_{ej})/\partial x_v\partial x_w$ and

$\partial^2 \left({}^{ej}\overline{A}_i\right) / \partial x_v \partial x_w$ (v = 1 to v = 41 and w = 1 to w = 41) are determined using the method described in Appendix C.

By directly differentiating Eq. (5.86), one obtains

$$\frac{\partial^2 \omega_{iz}}{\partial x_w \partial x_v} = \frac{F_i \frac{\partial G_i}{\partial x_w} - G_i \frac{\partial F_i}{\partial x_w}}{F_i^2}, \tag{5.109}$$

where $F_i$,[ $G_i$,$\partial F_i / \partial x_w$, and $\partial G_i / \partial x_w$ are respectively given in Eqs. (D.1a), (D.1b), (D.4a), and (D.4b) in the Appendix D.

Furthermore, from Eq. (5.87), one obtains

$$\frac{\partial^2 \omega_{iy}}{\partial x_w \partial x_v} = \frac{F_i \left(\frac{\partial G_i}{\partial x_w} + \frac{\partial H_i}{\partial x_w}\right) - (G_i + H_i) \frac{\partial F_i}{\partial x_w}}{F_i^2}, \tag{5.110}$$

where $F_i$,$G_i$,$H_i$,$\partial F_i / \partial x_w$,$\partial G_i / \partial x_w$, and $\partial H_i / \partial x_w$ are respectively given in Eqs. (D.2a), (D.2b), (D.2c), (D.5a), (D.5b), and (D.5c) in the Appendix D.

Finally, from Eq. (5.88), one has

$$\frac{\partial^2 \omega_{ix}}{\partial x_w \partial x_v} = \frac{F_i \left[H_i \frac{\partial G_i}{\partial x_w} + G_i \frac{\partial H_i}{\partial x_w} - \left(Q_i \frac{\partial U_i}{\partial x_w} + U_i \frac{\partial Q_i}{\partial x_w}\right)\right] - (G_i H_i - U_i Q_i) \frac{\partial F_i}{\partial x_w}}{F_i^2}, \tag{5.111}$$

where $F_i$,G]i ,$H_i$,$U_i$,$Q_i$,$\partial F_i / \partial x_w$,$\partial G_i / \partial x_w$,$\partial H_i / \partial x_w$,$\partial U_i / \partial x_w$, and $\partial Q_i / \partial x_w$ are respectively given in Eqs. (D.3a), (D.3b), (D.3c), (D.3d), (D.3e), (D.6a), (D.6b), (D.6c), (D.6d), and (D.6e) in the Appendix D.

$$\frac{\partial^2 t_{ix}}{\partial \overline{X}_{sys}^2} = \frac{\partial^2 t_{ix}}{\partial x_w \partial x_v}, \tag{5.112}$$

$$\frac{\partial^2 t_{iy}}{\partial \overline{X}_{sys}^2} = \frac{\partial^2 t_{iy}}{\partial x_w \partial x_v}, \tag{5.113}$$

$$\frac{\partial^2 t_{iz}}{\partial \overline{X}_{sys}^2} = \frac{\partial^2 t_{iz}}{\partial x_w \partial x_v}. \tag{5.114}$$

Equations (5.112), (5.113), and (5.114) just show that $\partial^2 t_{ix} \Big/ \partial \overline{X}_{sys}^2$, $\partial^2 t_{iy} \Big/ \partial \overline{X}_{sys}^2$ and $\partial^2 t_{iz} \Big/ \partial \overline{X}_{sys}^2$ are the (1,4)th, (2,4)th and (3,4)th components of $\partial^2 \left({}^0\overline{A}_i\right) \Big/ \partial \overline{X}_{sys}^2$ computed by Eq. (5.108).

*Example 5.25* The non-zero values of $H_{10}(u, v, w)$ for the system in Fig. 3.11 are:

$$\begin{aligned} H_{10}(1,29,33) &= H_{10}(1,33,29) = -H_{10}(2,27,33) = -H_{10}(2,33,27) \\ &= H_{10}(2,33,35) = H_{10}(2,35,33) \\ &= -H_{10}(4,27,34) = -H_{10}(4,34,27) = H_{10}(4,34,35) \\ &= H_{10}(4,35,34) = -H_{10}(5,29,34) \\ &= -H_{10}(5,34,29) = -H_{10}(6,27,29) = -H_{10}(6,29,27) \\ &= H_{10}(6,29,35) = H_{10}(6,35,29) = 1, \end{aligned}$$

$$\begin{aligned} H_{10}(1,27,30) &= H_{10}(1,30,27) = -H_{10}(1,30,35) = -H_{10}(1,35,30) \\ &= H_{10}(2,29,30) = H_{10}(2,30,29) \\ &= -H_{10}(3,27,27) = H_{10}(3,27,35) = H_{10}(3,35,27) \\ &= -H_{10}(3,29,29) = -H_{10}(3,35,35) = 10\sqrt{2}. \end{aligned} \quad (5.115)$$

# Appendix A

The differentiation of ${}^0\overline{A}_i$ shown in Eq. (2.8) with respect to pose variables are given by:

$$\frac{\partial({}^0\overline{A}_i)}{\partial t_{ix}} = \begin{bmatrix} \partial I_{ix}/\partial t_{ix} & \partial J_{ix}/\partial t_{ix} & \partial K_{ix}/\partial t_{ix} & \partial t_{ix}/\partial t_{ix} \\ \partial I_{iy}/\partial t_{ix} & \partial J_{iy}/\partial t_{ix} & \partial K_{iy}/\partial t_{ix} & \partial t_{iy}/\partial t_{ix} \\ \partial I_{iz}/\partial t_{ix} & \partial J_{iz}/\partial t_{ix} & \partial K_{iz}/\partial t_{ix} & \partial t_{iz}/\partial t_{ix} \\ 0 & 0 & 0 & 0 \end{bmatrix} = \begin{bmatrix} 0 & 0 & 0 & 1 \\ 0 & 0 & 0 & 0 \\ 0 & 0 & 0 & 0 \\ 0 & 0 & 0 & 0 \end{bmatrix}, \quad (A.1)$$

$$\frac{\partial({}^0\overline{A}_i)}{\partial t_{iy}} = \begin{bmatrix} \partial I_{ix}/\partial t_{iy} & \partial J_{ix}/\partial t_{iy} & \partial K_{ix}/\partial t_{iy} & \partial t_{ix}/\partial t_{iy} \\ \partial I_{iy}/\partial t_{iy} & \partial J_{iy}/\partial t_{iy} & \partial K_{iy}/\partial t_{iy} & \partial t_{iy}/\partial t_{iy} \\ \partial I_{iz}/\partial t_{iy} & \partial J_{iz}/\partial t_{iy} & \partial K_{iz}/\partial t_{iy} & \partial t_{iz}/\partial t_{iy} \\ 0 & 0 & 0 & 0 \end{bmatrix} = \begin{bmatrix} 0 & 0 & 0 & 0 \\ 0 & 0 & 0 & 1 \\ 0 & 0 & 0 & 0 \\ 0 & 0 & 0 & 0 \end{bmatrix}, \quad (A.2)$$

$$\frac{\partial({}^0\overline{A}_i)}{\partial t_{iz}} = \begin{bmatrix} \partial I_{ix}/\partial t_{iz} & \partial J_{ix}/\partial t_{iz} & \partial K_{ix}/\partial t_{iz} & \partial t_{ix}/\partial t_{iz} \\ \partial I_{iy}/\partial t_{iz} & \partial J_{iy}/\partial t_{iz} & \partial K_{iy}/\partial t_{iz} & \partial t_{iy}/\partial t_{iz} \\ \partial I_{iz}/\partial t_{iz} & \partial J_{iz}/\partial t_{iz} & \partial K_{iz}/\partial t_{iz} & \partial t_{iz}/\partial t_{iz} \\ 0 & 0 & 0 & 0 \end{bmatrix} = \begin{bmatrix} 0 & 0 & 0 & 0 \\ 0 & 0 & 0 & 0 \\ 0 & 0 & 0 & 1 \\ 0 & 0 & 0 & 0 \end{bmatrix}, \quad (A.3)$$

$$\frac{\partial({}^{0}\overline{A}_{i})}{\partial\omega_{ix}}=\begin{bmatrix}\partial I_{ix}/\partial\omega_{ix} & \partial J_{ix}/\partial\omega_{ix} & \partial K_{ix}/\partial\omega_{ix} & \partial t_{ix}/\partial\omega_{ix}\\ \partial I_{iy}/\partial\omega_{ix} & \partial J_{iy}/\partial\omega_{ix} & \partial K_{iy}/\partial\omega_{ix} & \partial t_{iy}/\partial\omega_{ix}\\ \partial I_{iz}/\partial\omega_{ix} & \partial J_{iz}/\partial\omega_{ix} & \partial K_{iz}/\partial\omega_{ix} & \partial t_{iz}/\partial\omega_{ix}\\ 0 & 0 & 0 & 0\end{bmatrix}$$
$$=\begin{bmatrix}0 & C\omega_{iz}S\omega_{iy}C\omega_{ix}+S\omega_{iz}S\omega_{ix} & -C\omega_{iz}S\omega_{iy}S\omega_{ix}+S\omega_{iz}C\omega_{ix} & 0\\ 0 & S\omega_{iz}S\omega_{iy}C\omega_{ix}-C\omega_{iz}S\omega_{ix} & -S\omega_{iz}S\omega_{iy}S\omega_{ix}-C\omega_{iz}C\omega_{ix} & 0\\ 0 & C\omega_{iy}C\omega_{ix} & -C\omega_{iy}S\omega_{ix} & 0\\ 0 & 0 & 0 & 0\end{bmatrix}, \tag{A.4}$$

$$\frac{\partial({}^{0}\overline{A}_{i})}{\partial\omega_{iy}}=\begin{bmatrix}\partial I_{ix}/\partial\omega_{iy} & \partial J_{ix}/\partial\omega_{iy} & \partial K_{ix}/\partial\omega_{iy} & \partial t_{ix}/\partial\omega_{iy}\\ \partial I_{iy}/\partial\omega_{iy} & \partial J_{iy}/\partial\omega_{iy} & \partial K_{iy}/\partial\omega_{iy} & \partial t_{iy}/\partial\omega_{iy}\\ \partial I_{iz}/\partial\omega_{iy} & \partial J_{iz}/\partial\omega_{iy} & \partial K_{iz}/\partial\omega_{iy} & \partial t_{iz}/\partial\omega_{iy}\\ 0 & 0 & 0 & 0\end{bmatrix}$$
$$=\begin{bmatrix}-C\omega_{iz}S\omega_{iy} & C\omega_{iz}C\omega_{iy}S\omega_{ix} & C\omega_{iz}C\omega_{iy}C\omega_{ix} & 0\\ -S\omega_{iz}S\omega_{iy} & S\omega_{iz}C\omega_{iy}S\omega_{ix} & S\omega_{iz}C\omega_{iy}C\omega_{ix} & 0\\ -C\omega_{iy} & -S\omega_{iy}S\omega_{ix} & -S\omega_{iy}C\omega_{ix} & 0\\ 0 & 0 & 0 & 0\end{bmatrix}, \tag{A.5}$$

$$\frac{\partial({}^{0}\overline{A}_{i})}{\partial\omega_{iz}}=\begin{bmatrix}\partial I_{ix}/\partial\omega_{iz} & \partial J_{ix}/\partial\omega_{iz} & \partial K_{ix}/\partial\omega_{iz} & \partial t_{ix}/\partial\omega_{iz}\\ \partial I_{iy}/\partial\omega_{iz} & \partial J_{iy}/\partial\omega_{iz} & \partial K_{iy}/\partial\omega_{iz} & \partial t_{iy}/\partial\omega_{iz}\\ \partial I_{iz}/\partial\omega_{iz} & \partial J_{iz}/\partial\omega_{iz} & \partial K_{iz}/\partial\omega_{iz} & \partial t_{iz}/\partial\omega_{iz}\\ 0 & 0 & 0 & 0\end{bmatrix}$$
$$=\begin{bmatrix}-S\omega_{iz}C\omega_{iy} & -S\omega_{iz}S\omega_{iy}S\omega_{ix}-C\omega_{iz}C\omega_{ix} & -S\omega_{iz}S\omega_{iy}C\omega_{ix}+C\omega_{iz}S\omega_{ix} & 0\\ C\omega_{iz}C\omega_{iy} & C\omega_{iz}S\omega_{iy}S\omega_{ix}-S\omega_{iz}C\omega_{ix} & C\omega_{iz}S\omega_{iy}C\omega_{ix}+S\omega_{iz}S\omega_{ix} & 0\\ 0 & 0 & 0 & 0\\ 0 & 0 & 0 & 0\end{bmatrix}. \tag{A.6}$$

# Appendix B

There are various means of specifying ${}^{0}\overline{A}_{ej}$ [e.g., Eqs. (3.29a)–(3.29d)] and ${}^{ej}\overline{A}_{i}$ [e.g., Eqs. (3.30a)–(3.30m)] based on different sequences of the rotation and translation operators in Sect. 1.5. However, if ${}^{0}\overline{A}_{ej}$ is to be computed using computer code, a decision must be taken as to which form to apply. In the present study, the following formulation is applied:

$$
{}^{h}\overline{\mathrm{A}}_{\mathrm{g}} = tran(\mathrm{t_x}, \mathrm{t_y}, \mathrm{t_z})\, rot(\overline{\mathrm{z}}, \omega_{\mathrm{z}})\, rot(\overline{\mathrm{y}}, \omega_{\mathrm{y}})\, rot(\overline{\mathrm{x}}, \omega_{\mathrm{x}})
$$

$$
= \begin{bmatrix} I_{\mathrm{x}} & J_{\mathrm{x}} & K_{\mathrm{x}} & \mathrm{t_x} \\ I_{\mathrm{y}} & J_{\mathrm{y}} & K_{\mathrm{y}} & \mathrm{t_y} \\ I_{\mathrm{z}} & J_{\mathrm{z}} & K_{\mathrm{z}} & \mathrm{t_z} \\ 0 & 0 & 0 & 1 \end{bmatrix} = \begin{bmatrix} \mathrm{C}\omega_{\mathrm{y}}\mathrm{C}\omega_{\mathrm{z}} & \mathrm{S}\omega_{\mathrm{x}}\mathrm{S}\omega_{\mathrm{y}}\mathrm{C}\omega_{\mathrm{z}} - \mathrm{C}\omega_{\mathrm{x}}\mathrm{S}\omega_{\mathrm{z}} & \mathrm{C}\omega_{\mathrm{x}}\mathrm{S}\omega_{\mathrm{y}}\mathrm{C}\omega_{\mathrm{z}} - \mathrm{S}\omega_{\mathrm{x}}\mathrm{S}\omega_{\mathrm{z}} & \mathrm{t_x} \\ \mathrm{C}\omega_{\mathrm{y}}\mathrm{S}\omega_{\mathrm{z}} & \mathrm{C}\omega_{\mathrm{x}}\mathrm{C}\omega_{\mathrm{z}} + \mathrm{S}\omega_{\mathrm{x}}\mathrm{S}\omega_{\mathrm{y}}\mathrm{S}\omega_{\mathrm{z}} & -\mathrm{S}\omega_{\mathrm{x}}\mathrm{C}\omega_{\mathrm{z}} + \mathrm{C}\omega_{\mathrm{x}}\mathrm{S}\omega_{\mathrm{y}}\mathrm{S}\omega_{\mathrm{z}} & \mathrm{t_y} \\ -\mathrm{S}\omega_{\mathrm{y}} & \mathrm{S}\omega_{\mathrm{x}}\mathrm{C}\omega_{\mathrm{y}} & \mathrm{C}\omega_{\mathrm{x}}\mathrm{C}\omega_{\mathrm{y}} & \mathrm{t_z} \\ 0 & 0 & 0 & 1 \end{bmatrix}, \tag{B.1}
$$

where ${}^{h}\overline{\mathrm{A}}_{\mathrm{g}}$ can be ${}^{0}\overline{\mathrm{A}}_{\mathrm{ej}}$ or ${}^{ej}\overline{\mathrm{A}}_{\mathrm{i}}$.

In Eqs. (3.29a)–(3.29d) and Eqs. (3.30a)–(3.30m), the arguments in matrices ${}^{0}\overline{\mathrm{A}}_{\mathrm{ej}}$ and ${}^{ej}\overline{\mathrm{A}}_{\mathrm{i}}$ can be written as $\omega_{\mathrm{x}} = \mathrm{a}_0 + \sum_{\mathrm{v}} = 1^{41}\mathrm{a_v x_v}$, $\omega_{\mathrm{y}} = \mathrm{b}_0 + \sum_{\mathrm{v}} = 1^{41}\mathrm{b_v x_v}$and $\omega_{\mathrm{z}} = \mathrm{c}_0 + \sum_{\mathrm{v}} = 1^{41}c_{\mathrm{v}}\mathrm{x_v}$, respectively, where $\mathrm{x_v}$is the vth component of $\overline{\mathrm{X}}_{\mathrm{sys}}$, and $\mathrm{a_v}$, $\mathrm{b_v}$ and $\mathrm{c_v}$ ({\text{v}} = 1 to {\text{v}} = 41) are known constants. Therefore, $\partial({}^{h}\overline{\mathrm{A}}_{\mathrm{g}})/\partial \mathrm{x_v}$ (where ${}^{\mathrm{h}}\overline{\mathrm{A}}_{\mathrm{g}}$ can be ${}^{0}\overline{\mathrm{A}}_{\mathrm{ej}}$ or ${}^{ej}\overline{\mathrm{A}}_{\mathrm{i}}$) can be determined directly by differentiating Eq. (B.1) with respect to $\mathrm{x_v}$ to give

$$
\frac{\partial({}^{h}\overline{\mathrm{A}}_{\mathrm{g}})}{\partial \mathrm{x_v}} = \begin{bmatrix} \partial \mathrm{I_x}/\partial \mathrm{x_v} & \partial \mathrm{J_x}/\partial \mathrm{x_v} & \partial \mathrm{K_x}/\partial \mathrm{x_v} & \partial \mathrm{t_x}/\partial \mathrm{x_v} \\ \partial \mathrm{I_y}/\partial \mathrm{x_v} & \partial \mathrm{J_y}/\partial \mathrm{x_v} & \partial \mathrm{K_y}/\partial \mathrm{x_v} & \partial \mathrm{t_y}/\partial \mathrm{x_v} \\ \partial \mathrm{I_z}/\partial \mathrm{x_v} & \partial \mathrm{J_z}/\partial \mathrm{x_v} & \partial \mathrm{K_z}/\partial \mathrm{x_v} & \partial \mathrm{t_z}/\partial \mathrm{x_v} \\ 0 & 0 & 0 & 0 \end{bmatrix}, \tag{B.2}
$$

where

$$
\frac{\partial \mathrm{I_x}}{\partial \mathrm{x_v}} = -\mathrm{b_v}\mathrm{S}\omega_{\mathrm{y}}\mathrm{C}\omega_{\mathrm{z}} - \mathrm{c_v}\mathrm{C}\omega_{\mathrm{y}}\mathrm{S}\omega_{\mathrm{z}}, \tag{B.3}
$$

$$
\frac{\partial \mathrm{I_y}}{\partial \mathrm{x_v}} = -\mathrm{b_v}\mathrm{S}\omega_{\mathrm{y}}\mathrm{S}\omega_{\mathrm{z}} + \mathrm{c_v}\mathrm{C}\omega_{\mathrm{y}}\mathrm{C}\omega_{\mathrm{z}}, \tag{B.4}
$$

$$
\frac{\partial \mathrm{I_y}}{\partial \mathrm{x_v}} = -\mathrm{b_v}\mathrm{S}\omega_{\mathrm{y}}, \tag{B.5}
$$

$$
\frac{\partial \mathrm{J_x}}{\partial \mathrm{x_v}} = \mathrm{a_v}\mathrm{C}\omega_{\mathrm{x}}\mathrm{S}\omega_{\mathrm{y}}\mathrm{C}\omega_{\mathrm{z}} + \mathrm{b_v}\mathrm{S}\omega_{\mathrm{x}}\mathrm{C}\omega_{\mathrm{y}}\mathrm{C}\omega_{\mathrm{z}} - \mathrm{c_v}\mathrm{S}\omega_{\mathrm{x}}\mathrm{S}\omega_{\mathrm{y}}\mathrm{S}\omega_{\mathrm{z}} + \mathrm{a_v}\mathrm{S}\omega_{\mathrm{x}}\mathrm{S}\omega_{\mathrm{z}} - \mathrm{c_v}\mathrm{C}\omega_{\mathrm{x}}\mathrm{C}\omega_{\mathrm{z}}, \tag{B.6}
$$

$$
\frac{\partial \mathrm{J_y}}{\partial \mathrm{x_v}} = \mathrm{a_v}\mathrm{C}\omega_{\mathrm{x}}\mathrm{S}\omega_{\mathrm{y}}\mathrm{S}\omega_{\mathrm{z}} + \mathrm{b_v}\mathrm{S}\omega_{\mathrm{x}}\mathrm{C}\omega_{\mathrm{y}}\mathrm{S}\omega_{\mathrm{z}} + \mathrm{c_v}\mathrm{S}\omega_{\mathrm{x}}\mathrm{S}\omega_{\mathrm{y}}\mathrm{C}\omega_{\mathrm{z}} - \mathrm{a_v}\mathrm{S}\omega_{\mathrm{x}}\mathrm{C}\omega_{\mathrm{z}} - \mathrm{c_v}\mathrm{C}\omega_{\mathrm{x}}\mathrm{S}\omega_{\mathrm{z}}, \tag{B.7}
$$

$$
\frac{\partial \mathrm{J_y}}{\partial \mathrm{x_v}} = \mathrm{a_v}\mathrm{C}\omega_{\mathrm{x}}\mathrm{C}\omega_{\mathrm{y}} - \mathrm{b_v}\mathrm{S}\omega_{\mathrm{x}}\mathrm{S}\omega_{\mathrm{y}}, \tag{B.8}
$$

$$
\frac{\partial \mathrm{K_x}}{\partial \mathrm{x_v}} = -\mathrm{a_v}\mathrm{S}\omega_{\mathrm{x}}\mathrm{S}\omega_{\mathrm{y}}\mathrm{C}\omega_{\mathrm{z}} + \mathrm{b_v}\mathrm{C}\omega_{\mathrm{x}}\mathrm{C}\omega_{\mathrm{y}}\mathrm{C}\omega_{\mathrm{z}} - \mathrm{c_v}\mathrm{C}\omega_{\mathrm{x}}\mathrm{S}\omega_{\mathrm{y}}\mathrm{S}\omega_{\mathrm{z}} - \mathrm{a_v}\mathrm{C}\omega_{\mathrm{x}}\mathrm{S}\omega_{\mathrm{z}} - \mathrm{c_v}\mathrm{S}\omega_{\mathrm{x}}\mathrm{C}\omega_{\mathrm{z}}, \tag{B.9}
$$

$$\frac{\partial K_y}{\partial x_v} = -a_v S\omega_x S\omega_y S\omega_z + b_v C\omega_x C\omega_y S\omega_z + c_v C\omega_x S\omega_y C\omega_z - a_v C\omega_x C\omega_z + c_v S\omega_x S\omega_z, \tag{B.10}$$

$$\frac{\partial K_z}{\partial x_v} = -a_v S\omega_x C\omega_y - b_v C\omega_x S\omega_y, \tag{B.11}$$

$$\frac{\partial t_x}{\partial x_v} = \frac{\partial t_x}{\partial x_v}, \tag{B.12}$$

$$\frac{\partial t_y}{\partial x_v} = \frac{\partial t_y}{\partial x_v}, \tag{B.13}$$

$$\frac{\partial t_z}{\partial x_v} = \frac{\partial t_z}{\partial x_v}. \tag{B.14}$$

Note that Eqs. (B.12), (B.13) and (B.14) simply indicate that $\partial t_x/\partial x_v$, $\partial t_y/\partial x_v$and $\partial t_z/\partial x_v$ can be obtained by directly differentiating their corresponding expressions (e.g., $\partial t_x/\partial x_{24} = 0.5$ for ${}^0\overline{A}_{e2}$ from Eq. (3.29b) and $\partial t_y/\partial x_{14} = 1/\kappa_1^2$ for ${}^{e1}\overline{A}_2$ from Eq. (3.30b) for system in Fig. 3.12).

Having obtained ${}^0\overline{A}_{ej}$, ${}^{ej}\overline{A}_i$, $\partial({}^0\overline{A}_{ej})/\partial x_v$ and $\partial({}^{ej}\overline{A}_i)/\partial x_v$, one can compute $\partial({}^0\overline{A}_i)/\partial x_v$ from Eq. (5.85) and determine $\partial t_{ix}/\partial\overline{X}_{sys}$, $\partial t_{iy}/\partial\overline{X}_{sys}$, $\partial t_{iz}/\partial\overline{X}_{sys}$, $\partial\omega_{ix}/\partial\overline{X}_{sys}$, $\partial\omega_{iy}/\partial\overline{X}_{sys}$, and $\partial\omega_{iz}/\partial\overline{X}_{sys}$ accordingly.

# Appendix C

Differentiating $\partial({}^h\overline{A}_g)/\partial x_v$ in Eq. (B.2) (where ${}^h\overline{A}_g$ can be ${}^0\overline{A}_{ej}$ or ${}^{ej}\overline{A}_i$) with respect to $x_w$ in $\overline{X}_{sys}$ yields the following:

$$\frac{\partial^2({}^h\overline{A}_g)}{\partial x_w \partial x_v} = \begin{bmatrix} \partial^2 I_x/\partial x_w\partial x_v & \partial^2 J_x/\partial x_w\partial x_v & \partial^2 K_x/\partial x_w\partial x_v & \partial^2 t_x/\partial x_w\partial x_v \\ \partial^2 I_y/\partial x_w\partial x_v & \partial^2 J_y/\partial x_w\partial x_v & \partial^2 K_y/\partial x_w\partial x_v & \partial^2 t_y/\partial x_w\partial x_v \\ \partial^2 I_z/\partial x_w\partial x_v & \partial^2 J_z/\partial x_w\partial x_v & \partial^2 K_z/\partial x_w\partial x_v & \partial^2 t_z/\partial x_w\partial x_v \\ 0 & 0 & 0 & 0 \end{bmatrix}, \tag{C.1}$$

where

$$\frac{\partial^2 I_x}{\partial x_w \partial x_v} = -b_v^2 C\omega_y C\omega_z + 2b_v c_v S\omega_y S\omega_z - c_v^2 C\omega_y C\omega_z, \tag{C.2}$$

$$\frac{\partial^2 I_y}{\partial x_w \partial x_v} = -b_v^2 C\omega_y S\omega_z - 2b_v c_v S\omega_y C\omega_z - c_v^2 C\omega_y S\omega_z, \tag{C.3}$$

$$\frac{\partial^2 I_y}{\partial x_w \partial x_v} = -b_v^2 C\omega_y, \tag{C.4}$$

$$\begin{aligned}\frac{\partial^2 J_x}{\partial x_w \partial x_v} = &- a_v^2 S\omega_x S\omega_y C\omega_z + 2a_v b_v C\omega_x C\omega_y C\omega_z - 2a_v c_v C\omega_x S\omega_y S\omega_z - b_v^2 S\omega_x S\omega_y C\omega_z \\ &- 2b_v c_v S\omega_x C\omega_y S\omega_z - c_v^2 S\omega_x S\omega_y C\omega_z + a_v^2 C\omega_x S\omega_z + 2a_v c_v S\omega_x C\omega_z + c_v^2 C\omega_x S\omega_z,\end{aligned} \tag{C.5}$$

$$\begin{aligned}\frac{\partial^2 J_y}{\partial x_w \partial x_v} = &- a_v^2 S\omega_x S\omega_y S\omega_z + 2a_v b_v C\omega_x C\omega_y S\omega_z + 2a_v c_v C\omega_x S\omega_y C\omega_z - b_v^2 S\omega_x S\omega_y S\omega_z \\ &+ 2b_v c_v S\omega_x C\omega_y C\omega_z - c_v^2 S\omega_x S\omega_y S\omega_z - a_v^2 C\omega_x C\omega_z + 2a_v c_v S\omega_x S\omega_z - c_v^2 C\omega_x C\omega_z,\end{aligned} \tag{C.6}$$

$$\frac{\partial^2 J_y}{\partial x_w \partial x_v} = -a_v^2 S\omega_x C\omega_y - 2a_v b_v C\omega_x S\omega_y - b_v^2 S\omega_x C\omega_y, \tag{C.7}$$

$$\begin{aligned}\frac{\partial^2 K_x}{\partial x_w \partial x_v} = &- a_v^2 C\omega_x S\omega_y C\omega_z - 2a_v b_v S\omega_x C\omega_y C\omega_z + 2a_v c_v S\omega_x S\omega_y S\omega_z - b_v^2 C\omega_x S\omega_y C\omega_z \\ &- 2b_v c_v C\omega_x C\omega_y S\omega_z - c_v^2 C\omega_x S\omega_y C\omega_z + a_v^2 S\omega_x S\omega_z - 2a_v c_v C\omega_x C\omega_z + c_v^2 S\omega_x S\omega_z,\end{aligned} \tag{C.8}$$

$$\begin{aligned}\frac{\partial^2 K_y}{\partial x_w \partial x_v} = &- a_v^2 C\omega_x S\omega_y S\omega_z - 2a_v b_v S\omega_x C\omega_y S\omega_z - 2a_v c_v S\omega_x S\omega_y C\omega_z - b_v^2 C\omega_x S\omega_y S\omega_z \\ &+ 2b_v c_v C\omega_x C\omega_y C\omega_z - c_v^2 C\omega_x S\omega_y S\omega_z + a_v^2 S\omega_x C\omega_z + 2a_v c_v C\omega_x S\omega_z + c_v^2 S\omega_x C\omega_z,\end{aligned} \tag{C.9}$$

$$\frac{\partial^2 K_z}{\partial x_w \partial x_v} = -a_v a_v C\omega_x C\omega_y + 2a_v b_v S\omega_x S\omega_y - b_v^2 C\omega_x C\omega_y, \tag{C.10}$$

$$\frac{\partial^2 t_x}{\partial x_w \partial x_v} = \frac{\partial^2 t_x}{\partial x_w \partial x_v}, \tag{C.11}$$

$$\frac{\partial^2 t_y}{\partial x_w \partial x_v} = \frac{\partial^2 t_y}{\partial x_w \partial x_v}, \tag{C.12}$$

$$\frac{\partial^2 t_z}{\partial x_w \partial x_v} = \frac{\partial^2 t_z}{\partial x_w \partial x_v}. \tag{C.13}$$

Note that Eqs. (C.11), (C.12) and (C.13) simply indicate that $\partial^2 t_x / \partial x_w \partial x_v$, $\partial^2 t_y / \partial x_w \partial x_v$and $\partial^2 t_z / \partial x_w \partial x_v$ can be obtained by directly differentiating their corresponding expressions (e.g., $\partial^2 t_y / \partial x_{14}^2 = -2/\kappa_1^3$ for ${}^{e1}\overline{A}_2$ from Eq. (3.30b) for system in Fig. 3.12). Having obtained ${}^{0}\overline{A}_{ej}$, ${}^{ej}\overline{A}_{i}$, $\partial({}^{0}\overline{A}_{ej})/\partial x_v$, $\partial({}^{ej}\overline{A}_{i})/\partial x_v$, $\partial^2({}^{0}\overline{A}_{ej})/\partial x_w\partial x_v$and $\partial^2({}^{ej}\overline{A}_{i})/\partial x_w\partial x_v$, one can compute $\partial^2({}^{0}\overline{A}_{i})/\partial x_w\partial x_v$ from Eq. (5.108) and then determine $\partial^2 t_{ix}/\partial\overline{X}_{sys}^2$, $\partial^2 t_{iy}/\partial\overline{X}_{sys}^2$, $\partial^2 t_{iz}/\partial\overline{X}_{sys}^2$, $\partial^2\omega_{ix}/\partial\overline{X}_{sys}^2$, $\partial^2\omega_{iy}/\partial\overline{X}_{sys}^2$, and $\partial^2\omega_{iz}/\partial\overline{X}_{sys}^2$ accordingly.

## Appendix D

$$\frac{\partial\omega_{iz}}{\partial x_v} = \frac{G_i}{F_i}, \tag{D.1}$$

where

$$F_i = I_{ix}^2 + I_{iy}^2, \tag{D.1a}$$

$$G_i = I_{ix}\frac{\partial I_{iy}}{\partial x_v} - I_{iy}\frac{\partial I_{ix}}{\partial x_v}. \tag{D.1b}$$

$$\frac{\partial\omega_{iy}}{\partial x_v} = \frac{G_i + H_i}{F_i}, \tag{D.2}$$

in which

$$F_i = I_{iz}^2 + \left(I_{ix}C\omega_{iz} + I_{iy}S\omega_{iz}\right)^2, \tag{D.2a}$$

$$G_i = I_{iz}\left[(-I_{ix}S\omega_{iz} + I_{iy}C\omega_{iz})\frac{\partial\omega_{iz}}{\partial x_v} + (\frac{\partial I_{ix}}{\partial x_v}C\omega_{iz} + \frac{\partial I_{iy}}{\partial x_v}S\omega_{iz})\right], \tag{D.2b}$$

$$H_i = -\left(I_{ix}C\omega_{iz} + I_{iy}S\omega_{iz}\right)\frac{\partial I_{iz}}{\partial x_v}. \tag{D.2c}$$

$$\frac{\partial\omega_{ix}}{\partial x_v} = \frac{G_iH_i - U_iQ_i}{F_i}, \tag{D.3}$$

where

$$F_i = \left(K_{ix}S\omega_{iz} - K_{iy}C\omega_{iz}\right)^2 + \left(-J_{ix}S\omega_{iz} + J_{iy}C\omega_{iz}\right)^2, \tag{D.3a}$$

$$G_i = -J_{ix}S\omega_{iz} + J_{iy}C\omega_{iz}, \tag{D.3b}$$

$$H_i = \left(K_{ix}C\omega_{iz} + K_{iy}S\omega_{iz}\right)\frac{\partial\omega_{iz}}{\partial x_v} + \left(S\omega_{iz}\frac{\partial K_{ix}}{\partial x_v} - C\omega_{iz}\frac{\partial K_{iy}}{\partial x_v}\right), \tag{D.3c}$$

$$U_i = K_{ix}S\omega_{iz} - K_{iy}C\omega_{iz}, \tag{D.3d}$$

$$Q_{\mathrm{i}} = \left(-J_{\mathrm{ix}}\mathrm{C}\omega_{\mathrm{iz}} - J_{\mathrm{iy}}\mathrm{S}\omega_{\mathrm{iz}}\right)\frac{\partial \omega_{\mathrm{iz}}}{\partial \mathrm{x_v}} + \left(-\mathrm{S}\omega_{\mathrm{iz}}\frac{\partial \mathrm{J_{ix}}}{\partial \mathrm{x_v}} + \mathrm{C}\omega_{\mathrm{iz}}\frac{\partial \mathrm{J_{iy}}}{\partial \mathrm{x_v}}\right). \quad \text{(D.3e)}$$

$$\frac{\partial^2 \omega_{\mathrm{iz}}}{\partial \mathrm{x_w}\partial \mathrm{x_v}} = \frac{\mathrm{F_i}\frac{\partial \mathrm{G_i}}{\partial \mathrm{x_w}} - \mathrm{G_i}\frac{\partial \mathrm{F_i}}{\partial \mathrm{x_w}}}{\mathrm{F_i^2}}, \quad \text{(D.4)}$$

where $\mathrm{F_i}$,$\mathrm{G_i}$ are respectively given in Eqs. (D.1a), (D.1b) in this appendix, and

$$\frac{\partial \mathrm{F_i}}{\partial \mathrm{x_w}} = 2\left(\mathrm{I_{ix}}\frac{\partial \mathrm{I_{ix}}}{\partial \mathrm{x_w}} + \mathrm{I_{iy}}\frac{\partial \mathrm{I_{iy}}}{\partial \mathrm{x_w}}\right), \quad \text{(D.4a)}$$

$$\frac{\partial \mathrm{G_i}}{\partial \mathrm{x_w}} = \mathrm{I_{ix}}\frac{\partial^2 \mathrm{I_{iy}}}{\partial \mathrm{x_w}\partial \mathrm{x_v}} - \mathrm{I_{iy}}\frac{\partial^2 \mathrm{I_{ix}}}{\partial \mathrm{x_w}\partial \mathrm{x_v}} + \frac{\partial \mathrm{I_{ix}}}{\partial \mathrm{x_w}}\frac{\partial \mathrm{I_{iy}}}{\partial \mathrm{x_v}} - \frac{\partial \mathrm{I_{iy}}}{\partial \mathrm{x_w}}\frac{\partial \mathrm{I_{ix}}}{\partial \mathrm{x_v}}, \quad \text{(D.4b)}$$

$$\frac{\partial^2 \omega_{\mathrm{iy}}}{\partial \mathrm{x_w}\partial \mathrm{x_v}} = \frac{\mathrm{F_i}\left(\frac{\partial \mathrm{G_i}}{\partial \mathrm{x_w}} + \frac{\partial \mathrm{H_i}}{\partial \mathrm{x_w}}\right) - (\mathrm{G_i} + \mathrm{H_i})\frac{\partial \mathrm{F_i}}{\partial \mathrm{x_w}}}{\mathrm{F_i^2}}, \quad \text{(D.5)}$$

where $\mathrm{F_i}$, $\mathrm{G_i}$, $\mathrm{H_i}$ are respectively given in Eqs. (D.2a), (D.2b), (D.2c) in this appendix, and

$$\frac{\partial \mathrm{F_i}}{\partial \mathrm{x_w}} = 2I_{\mathrm{iz}}\frac{\partial \mathrm{I_{iz}}}{\partial \mathrm{x_w}} + 2\left(\mathrm{I_{ix}}\mathrm{C}\omega_{\mathrm{iz}} + \mathrm{I_{iy}}\mathrm{S}\omega_{\mathrm{iz}}\right)\left[\left(-\mathrm{I_{ix}}\mathrm{S}\omega_{\mathrm{iz}} + \mathrm{I_{iy}}\mathrm{C}\omega_{\mathrm{iz}}\right)\frac{\partial \omega_{\mathrm{iz}}}{\partial \mathrm{x_w}} + \left(\mathrm{C}\omega_{\mathrm{iz}}\frac{\partial \mathrm{I_{ix}}}{\partial \mathrm{x_w}} + \mathrm{S}\omega_{\mathrm{iz}}\frac{\partial \mathrm{I_{iy}}}{\partial \mathrm{x_w}}\right)\right], \quad \text{(D.5a)}$$

$$\begin{aligned}\frac{\partial \mathrm{G_i}}{\partial \mathrm{x_w}} = {} & \frac{\partial \mathrm{I_{iz}}}{\partial \mathrm{x_w}}\left[\left(-\mathrm{I_{ix}}\mathrm{S}\omega_{\mathrm{iz}} + \mathrm{I_{iy}}\mathrm{C}\omega_{\mathrm{iz}}\right)\frac{\partial \omega_{\mathrm{iz}}}{\partial \mathrm{x_v}} + \left(\mathrm{C}\omega_{\mathrm{iz}}\frac{\partial \mathrm{I_{ix}}}{\partial \mathrm{x_v}} + \mathrm{S}\omega_{\mathrm{iz}}\frac{\partial \mathrm{I_{iy}}}{\partial \mathrm{x_v}}\right)\right] \\ & + \mathrm{I_{iz}}\left[\left(-\mathrm{I_{ix}}\mathrm{C}\omega_{\mathrm{iz}} - \mathrm{I_{iy}}\mathrm{S}\omega_{\mathrm{iz}}\right)\frac{\partial \omega_{\mathrm{iz}}}{\partial \mathrm{x_w}}\frac{\partial \omega_{\mathrm{iz}}}{\partial \mathrm{x_v}} + \left(-\mathrm{S}\omega_{\mathrm{iz}}\frac{\partial \mathrm{I_{ix}}}{\partial \mathrm{x_w}} + \mathrm{C}\omega_{\mathrm{iz}}\frac{\partial \mathrm{I_{iy}}}{\partial \mathrm{x_w}}\right)\frac{\partial \omega_{\mathrm{iz}}}{\partial \mathrm{x_v}}\right. \\ & + \left(-\mathrm{I_{ix}}\mathrm{S}\omega_{\mathrm{iz}} + \mathrm{I_{iy}}\mathrm{C}\omega_{\mathrm{iz}}\right)\frac{\partial^2 \omega_{\mathrm{iz}}}{\partial \mathrm{x_w}\partial \mathrm{x_v}} + \left(-\mathrm{S}\omega_{\mathrm{iz}}\frac{\partial \mathrm{I_{ix}}}{\partial \mathrm{x_v}} + \mathrm{C}\omega_{\mathrm{iz}}\frac{\partial \mathrm{I_{iy}}}{\partial \mathrm{x_v}}\right)\frac{\partial \omega_{\mathrm{iz}}}{\partial \mathrm{x_w}} \\ & \left. + \left(\mathrm{C}\omega_{\mathrm{iz}}\frac{\partial^2 \mathrm{I_{ix}}}{\partial \mathrm{x_w}\partial \mathrm{x_v}} + \mathrm{S}\omega_{\mathrm{iz}}\frac{\partial^2 \mathrm{I_{iy}}}{\partial \mathrm{x_w}\partial \mathrm{x_v}}\right)\right],\end{aligned} \quad \text{(D.5b)}$$

$$\begin{aligned}\frac{\partial \mathrm{H_i}}{\partial \mathrm{x_w}} = {} & -\left(\mathrm{I_{ix}}\mathrm{C}\omega_{\mathrm{iz}} + \mathrm{I_{iy}}\mathrm{S}\omega_{\mathrm{iz}}\right)\frac{\partial^2 \mathrm{I_{iz}}}{\partial \mathrm{x_w}\partial \mathrm{x_v}} - \left(-\mathrm{I_{ix}}\mathrm{S}\omega_{\mathrm{iz}} + \mathrm{I_{iy}}\mathrm{C}\omega_{\mathrm{iz}}\right)\frac{\partial \omega_{\mathrm{iz}}}{\partial \mathrm{x_w}}\frac{\partial \mathrm{I_{iz}}}{\partial \mathrm{x_v}} \\ & - \left(\mathrm{C}\omega_{\mathrm{iz}}\frac{\partial \mathrm{I_{ix}}}{\partial \mathrm{x_w}} + \mathrm{S}\omega_{\mathrm{iz}}\frac{\partial \mathrm{I_{iy}}}{\partial \mathrm{x_w}}\right)\frac{\partial \mathrm{I_{iz}}}{\partial \mathrm{x_v}}.\end{aligned} \quad \text{(D.5c)}$$

$$\frac{\partial^2 \omega_{\mathrm{ix}}}{\partial \mathrm{x_w}\partial \mathrm{x_v}} = \frac{\mathrm{F_i}\left[\mathrm{H_i}\frac{\partial \mathrm{G_i}}{\partial \mathrm{x_w}} + \mathrm{G_i}\frac{\partial \mathrm{H_i}}{\partial \mathrm{x_w}} - \left(Q_{\mathrm{i}}\frac{\partial U_{\mathrm{i}}}{\partial \mathrm{x_w}} + \mathrm{U_i}\frac{\partial Q_{\mathrm{i}}}{\partial \mathrm{x_w}}\right)\right] - (\mathrm{G_i}\mathrm{H_i} - U_{\mathrm{i}}Q_{\mathrm{i}})\frac{\partial \mathrm{F_i}}{\partial \mathrm{x_w}}}{\mathrm{F_i^2}}, \quad \text{(D.6)}$$

where $F_i$, $G_i$, $H_i$, $U_i$, $Q_i$ are respectively given in Eqs. (D.3a), (D.3b), (D.3c), (D.3d), (D.3e) in this appendix, and

$$\begin{aligned}\frac{\partial F_i}{\partial x_w} =& 2(K_{ix}S\omega_{iz} - K_{iy}C\omega_{iz})\left[(K_{ix}C\omega_{iz} + K_{iy}S\omega_{iz})\frac{\partial \omega_{iz}}{\partial x_w} + \left(S\omega_{iz}\frac{\partial K_{ix}}{\partial x_w} - C\omega_{iz}\frac{\partial K_{iy}}{\partial x_w}\right)\right] \\ &+ 2(-J_{ix}S\omega_{iz} + J_{iy}C\omega_{iz})\left[(-J_{ix}C\omega_{iz} - J_{iy}S\omega_{iz})\frac{\partial \omega_{iz}}{\partial x_w} + \left(-S\omega_{iz}\frac{\partial J_{ix}}{\partial x_w} + C\omega_{iz}\frac{\partial J_{iy}}{\partial x_w}\right)\right],\end{aligned} \tag{D.6a}$$

$$\frac{\partial G_i}{\partial x_w} = (-J_{ix}C\omega_{iz} - J_{iy}S\omega_{iz})\frac{\partial \omega_{iz}}{\partial x_w} - S\omega_{iz}\frac{\partial J_{ix}}{\partial x_w} + C\omega_{iz}\frac{\partial J_{iy}}{\partial x_w}, \tag{D.6b}$$

$$\begin{aligned}\frac{\partial H_i}{\partial x_w} =& \left(C\omega_{iz}\frac{\partial K_{ix}}{\partial x_w} + S\omega_{iz}\frac{\partial K_{iy}}{\partial x_w}\right)\frac{\partial \omega_{iz}}{\partial x_v} + (-K_{ix}S\omega_{iz} + K_{iy}C\omega_{iz})\frac{\partial \omega_{iz}}{\partial x_w}\frac{\partial \omega_{iz}}{\partial x_v} + (K_{ix}C\omega_{iz} + K_{iy}S\omega_{iz})\frac{\partial^2 \omega_{iz}}{\partial x_w \partial x_v} \\ &+ \left(S\omega_{iz}\frac{\partial^2 K_{ix}}{\partial x_w \partial x_v} - C\omega_{iz}\frac{\partial^2 K_{iy}}{\partial x_w \partial x_v}\right) + \left(C\omega_{iz}\frac{\partial K_{ix}}{\partial x_v} + S\omega_{iz}\frac{\partial K_{iy}}{\partial x_v}\right)\frac{\partial \omega_{iz}}{\partial x_w},\end{aligned} \tag{D.6c}$$

$$\frac{\partial U_i}{\partial x_w} = (K_{ix}C\omega_{iz} + K_{iy}S\omega_{iz})\frac{\partial \omega_{iz}}{\partial x_w} + S\omega_{iz}\frac{\partial K_{ix}}{\partial x_w} - C\omega_{iz}\frac{\partial K_{iy}}{\partial x_w}, \tag{D.6d}$$

$$\begin{aligned}\frac{\partial Q_i}{\partial x_w} =& (-J_{ix}C\omega_{iz} - J_{iy}S\omega_{iz})\frac{\partial^2 \omega_{iz}}{\partial x_w \partial x_v} + (J_{ix}S\omega_{iz} - J_{iy}C\omega_{iz})\frac{\partial \omega_{iz}}{\partial x_w}\frac{\partial \omega_{iz}}{\partial x_v} \\ &- \left(C\omega_{iz}\frac{\partial J_{ix}}{\partial x_v} + S\omega_{iz}\frac{\partial J_{iy}}{\partial x_v}\right)\frac{\partial \omega_{iz}}{\partial x_w} + \left(-S\omega_{iz}\frac{\partial^2 J_{ix}}{\partial x_w \partial x_v} + C\omega_{iz}\frac{\partial^2 J_{iy}}{\partial x_w \partial x_v}\right).\end{aligned} \tag{D.6e}$$

## References


1. C.G. Wynne, P. Wormell, Lens design by computer. Appl. Opt. **2**, 1223–1238 (1963)
2. D.P. Fede, Automatic optical design, Appl. Opt. 2, 1209–1226 (1963)
3. M. Rimmer, Analysis of perturbed lens systems. Appl. Opt. **9**, 533–537 (1970)
4. H.H. Hopkins, H.J. Tiziani, A theoretical and experimental study of lens centering errors and their influence on optical image quality. Brit. J. Appl. Phys. **17**, 33–54 (1966)
5. T.B. Andersen, Optical aberration functions: chromatic aberrations and derivatives with respect to refractive indices for symmetrical systems. Appl. Opt. **21**, 4040–4044 (1982)
6. S.K. Gupta, R. Hradaynath, Angular tolerance on Dove prisms. Appl. Opt. **22**, 3146–3147 (1983)
7. J.F. Lee, C.Y. Leung, Method of calculating the alignment tolerance of a Porro prism resonator. Appl. Opt. **28**, 3691–3697 (1989)
8. B.D. Stone, Perturbations of optical systems. J. Opt. Soc. Am. A **14**, 2837–2849 (1997)
9. D.S. Grey, The inclusion of tolerance sensitivities in the merit function for lens optimization. SPIE **147**, 63–65 (1978)

10. E.G. Herrera, M. Strojnik, Interferometric tolerance determination for a Dove prism using exact ray trace. Opt. Commun. **281**, 897–905 (2008)
11. W. Mao, Adjustment of reflecting prisms. Opt. Eng. **34**, 79–82 (1995)
12. K.N. Chandler, On the effect of small errors in angles of corner-cube reflectors. J. Opt. Soc. Am. **50**, 203–206 (1960)
13. N. Lin, Orientation conjugation of reflecting prism rotation and second-order approximation of image rotation. Opt. Eng. **33**, 2400–2407 (1994)
14. E. Gutierrez, M. Strojnik, G. Paez, Tolerance determination for a Dove prism using exact ray trace. Proc. SPIE **6307**, 63070K (2006)
15. B.D. Stone, Determination of initial ray configurations for asymmetric systems. J. Opt. Soc. Am. A **14**, 3415–3429 (1997)
16. T. B. Andersen, Optical aberration functions: derivatives with respect to axial distances for symmetrical systems. Appl. Opt. 21, 1817–1823 (1982)
17. T.B. Andersen, Optical aberration functions: derivatives with respect to surface parameters for symmetrical systems. Appl. Opt. **24**, 1122–1129 (1985)
18. D.P. Feder, Calculation of an optical merit function and its derivatives with respect to the system parameters. J. Opt. Soc. Am. **47**, 913–925 (1957)
19. D.P. Feder, Differentiation of ray-tracing equations with respect to constructional parameters of rotationally symmetric systems. J. Opt. Soc. Am. **58**, 1494–1505 (1968)
20. O. Stavroudis, A simpler derivation of the formulas for generalized ray tracing. J. Opt. Soc. Am. 66, 1330–1333 (1976)
21. J. Kross, Differential ray tracing formulae for optical calculations: principles and applications. SPIE Opt. Des. Method Large Opt. 1013, 10–18 (1988)
22. W. Oertmann, Differential ray tracing formulae; applications especially to aspheric optical systems, SPIE Opt. Des. Method Large Opt., 1013, 20–26 (1988)
23. P.D. Lin, C.Y. Tsai, Determination of first-order derivatives of skew-ray at aspherical surface. J. Opt. Soc. Am. A **29**, 1141–1153 (2012)
24. C. Olson, R.N. Youngworth, Alignment analysis of optical systems using derivative information. Proc of SPIE **7068**, 1–10 (2008)

# Chapter 6
# Point Spread Function and Modulation Transfer Function

As stated in Sect. 3.5, the distribution of the ray density of the spot diagram formed in the image plane is called Point Spread Function (PSF). PSF plays an important role in the image formation theory, since it describes the impulse response of an optical system to a source point. However, literatures contain very few techniques for deriving the PSF of an optical system (e.g., [1]). Accordingly, Sects. 6.1 and 6.2 present a new method based on an irradiance model for computing the PSF of an optical system by considering the energy conservation along a single ray. It is shown that the irradiance method obtains a reliable and accurate PSF on the image plane. In addition, compared to the ray-counting method presented in Sect. 3.5, in which the quality of the PSF solution depends on the number of rays traced and the grid size used to mesh the image plane, the irradiance-based method requires just one ray tracing operation. Section 6.3 investigates the spot diagram by the irradiance method.

As shown in Sect. 3.6, the MTF is a measure of an optical system's ability to transfer the contrast from a specimen to the image plane at a specific spatial resolution. It can be computed either numerically by geometrical optics or measured experimentally by imaging a knife edge or a bar-target pattern of varying spatial frequency [2–11]. The literature deals with the MTF using ray-counting method [12] only for object brightness distribution functions (OBDF) oriented along the meridional or sagittal directions (see Fig. 3.20). Sections 6.4 and 6.5 extend the computation of MTF for an off-axis source point when the OBDF is oriented along any arbitrarily defined direction based on the irradiance model and ray-counting method. The former finds that the MTF is not a monotonic increasing or decreasing function when the direction of the OBDF is changing. Consequently, the extreme MTF values may occur when the OBDF is aligned with any direction between the meridional and sagittal directions. Four theorems are provided for the MTF and phase shift variations which take place when the OBDF is translated or rotated. It is found that the MTF and the phase shift are symmetrical or anti-symmetrical about certain directions. Thus, to observe all possible changes in the MTF and phase shift, it is sufficient to rotate the OBDF through a range of just 90°. The presented method based on the irradiance method for MTF computation does not count the number of ray hits on a grid, making the method immune to the

PD Lin, *New Computation Methods for Geometrical Optics*,
Springer Series in Optical Sciences 178, DOI: 10.1007/978-981-4451-79-6_6,

effects of grid size and thus improving traditional accuracy. Overall, our results confirm that the irradiance method provides an ideal solution for calculating the PSF and MTF of an optical system.

## 6.1 Jacobian Matrix Between In-plane Coordinates $(\mathbf{x_n}, \mathbf{z_n})$ of the Image Plane and the Polar Coordinates $(\boldsymbol{\alpha_0}, \boldsymbol{\beta_0})$ of a Source Ray

A source ray $\bar{R}_0$ originating at point source $\bar{P}_0 = [P_{0x} \quad P_{0y} \quad P_{0z} \quad 1]^T$ and traveling along the unit directional vector $\bar{\ell}_0 = [C\beta_0 C(90° + \alpha_0) \; C\beta_0 S(90° + \alpha_0) \; S\beta_0 \; 0]^T$ is shown in Fig. 2.2. Also shown by Eq. (2.31) (by setting i = n and $y_n = 0$) is that the in-plane coordinates of the image plane is ${}^n\bar{r}_n = [x_n \quad 0 \quad z_n \quad 1]^T$ when they are referred with respect to the boundary coordinate frame $(xyz)_n$. In determining the PSF of an optical system using the irradiance method, the Jacobian matrix $\partial(x_n, z_n)/\partial(\alpha_0, \beta_0)$ between in-plane coordinates $(x_n, z_n)$ and $(\alpha_0, \beta_0)$ is needed, in order to compute the infinitesimal area $d\pi_n = dx_n dz_n$ in an image plane by:

$$d\pi_n = dx_n dz_n = \left|\det\begin{bmatrix} \partial x_n/\partial\alpha_0 & \partial x_n/\partial\beta_0 \\ \partial z_n/\partial\alpha_0 & \partial z_n/\partial\beta_0 \end{bmatrix}\right| d\alpha_0 d\beta_0 = \left|\det\left(\frac{\partial(x_n, z_n)}{\partial(\alpha_0, \beta_0)}\right)\right| d\alpha_0 d\beta_0, \tag{6.1}$$

where | | is the absolute value of a scalar quantity. The Jacobian matrix $\partial(x_n, z_n)/\partial(\alpha_0, \beta_0)$ describes the extent to which a point $(x_n, z_n)$ is stretched in different directions at the image plane in the neighborhood of $(\alpha_0, \beta_0)$ (see Fig. 6.1). Literatures contain very few methods for deriving the PSF of optical systems due in large part to lack of the Jacobian matrices of Eq. (6.1). This difficulty now can be overcome. One can have the incidence point $\bar{P}_n$, which is expressed with respect to the world coordinate frame $(xyz)_0$, from Eq. (2.46) by setting g = n. Its expression with respect to the coordinate frame $(xyz)_n$ of the image plane can be obtained by transforming $\bar{P}_n$ to $(xyz)_n$ as follows:

$$\begin{aligned} {}^n\bar{P}_n &= \begin{bmatrix} {}^nP_{nx} \\ {}^nP_{ny} \\ {}^nP_{nz} \\ 1 \end{bmatrix} = {}^n\bar{A}_0\bar{P}_n = ({}^0\bar{A}_n)^{-1}\bar{P}_n \\ &= \begin{bmatrix} I_{nx} & I_{ny} & I_{nz} & -(I_{nx}t_{nx} + I_{ny}t_{ny} + I_{nz}t_{nz}) \\ J_{nx} & J_{ny} & J_{nz} & -(J_{nx}t_{nx} + J_{ny}t_{ny} + J_{nz}t_{nz}) \\ K_{nx} & K_{ny} & K_{nz} & -(K_{nx}t_{nx} + K_{ny}t_{ny} + K_{nz}t_{nz}) \\ 0 & 0 & 0 & 1 \end{bmatrix} \begin{bmatrix} P_{nx} \\ P_{ny} \\ P_{nz} \\ 1 \end{bmatrix} = \begin{bmatrix} x_n \\ 0 \\ z_n \\ 1 \end{bmatrix}. \end{aligned} \tag{6.2}$$

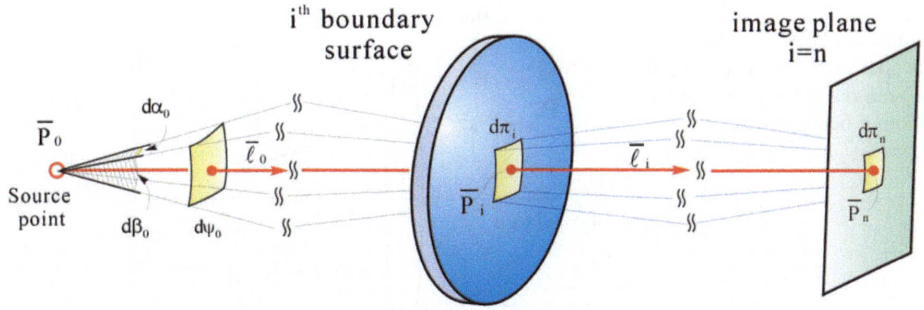


**Fig. 6.1** Source point $\bar{P}_0$ radiates light rays to an infinitesimal area $d\pi_n$ *centered* at $\bar{P}_n$ on the image plane

Now the required Jacobian matrix $\partial(x_n, z_n)/\partial(\alpha_0, \beta_0)$ in Eq. (6.1) can be obtained by taking differentiation of Eq. (6.2) with respect to $\alpha_0$ and $\beta_0$, to give

$$\begin{bmatrix} \partial x_n/\partial\alpha_0 & \partial x_n/\partial\beta_0 \\ \partial z_n/\partial\alpha_0 & \partial z_n/\partial\beta_0 \end{bmatrix} = \begin{bmatrix} I_{nx} & I_{ny} & I_{nz} & -(I_{nx}t_{nx} + I_{ny}t_{ny} + I_{nz}t_{nz}) \\ K_{nx} & K_{ny} & K_{nz} & -(K_{nx}t_{nx} + K_{ny}t_{ny} + K_{nz}t_{nz}) \end{bmatrix} \begin{bmatrix} \partial P_{nx}/\partial\alpha_0 & \partial P_{nx}/\partial\beta_0 \\ \partial P_{ny}/\partial\alpha_0 & \partial P_{ny}/\partial\beta_0 \\ \partial P_{nz}/\partial\alpha_0 & \partial P_{nz}/\partial\beta_0 \\ 0 & 0 \end{bmatrix}$$
$$= \begin{bmatrix} I_{nx} & I_{ny} & I_{nz} \\ K_{nx} & K_{ny} & K_{nz} \end{bmatrix} \begin{bmatrix} \partial P_{nx}/\partial\alpha_0 & \partial P_{nx}/\partial\beta_0 \\ \partial P_{ny}/\partial\alpha_0 & \partial P_{ny}/\partial\beta_0 \\ \partial P_{nz}/\partial\alpha_0 & \partial P_{nz}/\partial\beta_0 \end{bmatrix}. \tag{6.3}$$

where the second matrix of right-hand side of Eq. (6.3) can be obtained from Eq. (5.33) by setting $i = n$.

## 6.2 The Point Spread Function Based on Irradiance Method

The PSF in an optical system describes the irradiance distribution on the image plane associated with a source point $\bar{P}_0$. The derivation of the PSF is based on the assumption that $\bar{P}_0$ radiates uniformly with a constant intensity $I_0(\alpha_0, \beta_0) = I_0$ (in watts/steradian) in all directions (see Fig. 6.1). The energy flux $dF_0$ emitted from $\bar{P}_0$ into a solid angle $d\psi_0 = C\beta_0 d\alpha_0 d\beta_0$ along the ray tube $\bar{R}_0 = [P_{0x}\ P_{0y}\ P_{0z}\ C\beta_0 C(90^\circ + \alpha_0)\ C\beta_0 S(90^\circ + \alpha_0)\ S\beta_0]^T$ is given by

$$dF_0 = I_0 d\psi_0 = I_0 C\beta_0 d\alpha_0 d\beta_0. \tag{6.4}$$

Therefore, the total flux $F_0$ emitted from $\bar{P}_0$ and transmitted into the optical system is given by

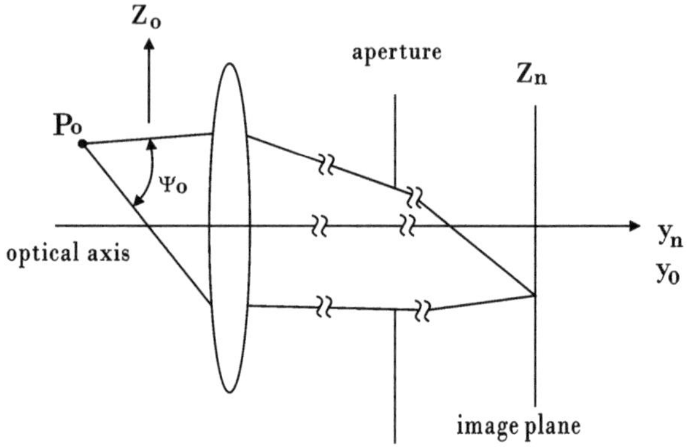


**Fig. 6.2** $\psi_0$ is the solid angle subtended by a ray cone with its apex at $\bar{P}_0$. All rays within $\psi_0$ can reach the image plane without being stopped by any stops

$$F_0 = I_0 \iint C\beta_0 d\alpha_0 d\beta_0 = I_0 \psi_0, \tag{6.5}$$

where $\psi_0$ (see Fig. 6.2) is the solid angle subtended by a ray cone with its apex at $\bar{P}_0$. All rays within $\psi_0$ can reach the image plane $\bar{r}_n$ without being blocked by any stops. If $\bar{P}_0$ is an on-axis point, the solid angle $\psi_0$ can be calculated directly with $\psi_0 = 2\pi(1 - C\sphericalangle)$, where $2\sphericalangle$ is the apex angle between the marginal rays. However, if $\bar{P}_0$ is an off-axis point, a numerical integration is required to compute $\psi_0$.

In order to normalize the PSF, an assumption is made that the total flux is equal to one (i.e., $F_0 = 1$), thus the following equation is obtained from Eq. (6.5):

$$I_0 = 1/\psi_0. \tag{6.6}$$

Defining $B_n(x_n, z_n)$ as the irradiance on an infinitesimal area $d\pi_n$ centered at the incidence point ${}^n\bar{P}_n$ on the image plane, the energy flux $dF_n$ received by $d\pi_n$ is given by $dF_n = B_n(x_n, z_n)d\pi_n$, where $d\pi_n$ is defined in Eq. (6.1). Assuming no transmission losses (i.e., $dF_0 = dF_n$), the following equation is obtained by applying the principle of energy flux conservation along this ray:

$$dF_0 = I_0 C\beta_0 d\alpha_0 d\beta_0 = dF_n = B_n(x_n, z_n)d\pi_n = B_n(x_n, z_n)\ dx_n dz_n. \tag{6.7}$$

Integrating both sides of Eq. (6.7) yields the following:

$$\iint B_n(x_n, z_n)d\pi_n = 1, \tag{6.8}$$

in other words, the total energy flux received at the image plane is normalized to unity.

Substituting Eqs. (6.1) and (6.6) into Eq. (6.7), the following expression is obtained for the irradiance (or PSF) $B_n(x_n, z_n)$on the image plane:

$$PSF = B_n(x_n, z_n) = \frac{C\beta_0}{\psi_0\left|\det\left(\frac{\partial(X_n, Z_n)}{\partial(\alpha_0, \beta_0)}\right)\right|} = \frac{C\beta_0}{\psi_0\left|\frac{\partial X_n}{\partial \alpha_0}\frac{\partial Z_n}{\partial \beta_0} - \frac{\partial X_n}{\partial \beta_0}\frac{\partial Z_n}{\partial \alpha_0}\right|}. \tag{6.9}$$

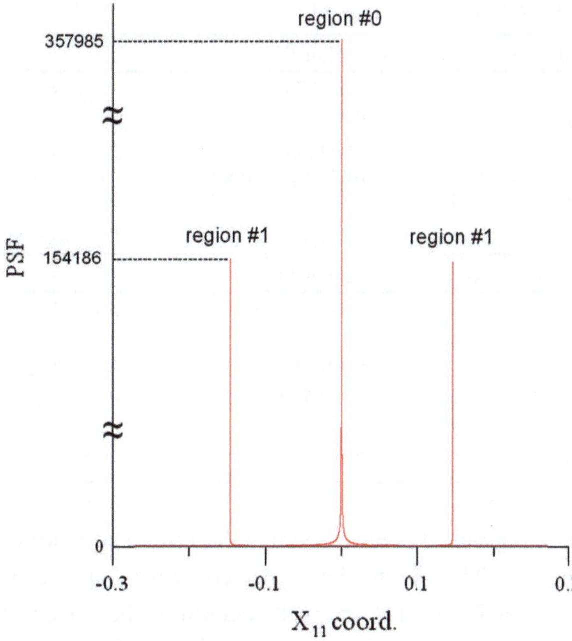


**Fig. 6.3** Distribution of PSF on an image plane as computed by irradiance method

Since Eq. (6.9) is a closed form containing no mathematical integration, only a single traced ray $\bar{R}_0$ suffices for the PSF at the incidence point $\bar{P}_n$ on the image plane for a given source ray $\bar{R}_0$. Thus, the proposed PSF method is far more computationally efficient than the ray-counting method addressed in Sect. 3.5 in which sufficient rays must be traced. Furthermore, Eq. (6.9) implies that PSF $= \infty$ when the Jacobian determinant is zero, yielding the position of caustic surface.

The line-spread function of an optical system, i.e. the impulse response of the optical system to an infinite number of source points along a line, can be obtained by integrating $B_n(x_n, z_n)$ along sections parallel to the direction of that line. One has to note that the derivations presented above are also valid for systems with non-axially symmetrical axes.

*Example 6.1* The solid angle is $\psi_0 = 0.005856$ steradian for the on-axis source point $\overline{\mathbf{P}}_0 = [0 \quad -507 \quad 0 \quad 1]^T$ of the system in Fig. 2.1.

*Example 6.2* Figure 6.3 presents the PSF of $\overline{P}_0 = [0 \quad -507 \quad 0 \quad 1]^T$ on the image plane as computed by the irradiance method. Note that the PSF is an axis-symmetrical function since Fig. 2.1 is an axis-symmetrical system and the source point $\overline{P}_0$ lies on the optical axis. Therefore, Fig. 6.3 only presents cross-sectional views rather than the entire PSF distribution. As shown in Fig. 6.3, the distribution contains three regions of high PSF, namely region #0 and two regions #1.

*Example 6.3* In order to show the detailed distribution of the PSF in the vicinity of the regions #0 and #1 of Fig. 6.3, Table 6.1 lists the PSF values at points in these

**Table 6.1** The PSF values of points in regions #0 and #1

| PSF values of points in region #0 | | | |
|---|---|---|---|
| $\alpha_0$ | $\pm 1.96762408781°$ | $\pm 1.9677240878°$ | $\pm 1.96792408780°$ |
| $\beta_0$ | 0° | 0° | 0° |
| $X_{11}$ | $-3.9343624 \times 10^{-5}$ | $-7.40454 \times 10^{-7}$ | $7.6483162 \times 10^{-5}$ |
| $\left\lvert \det\left(\frac{\partial(X_n,Z_n)}{\partial(\alpha_0,\beta_0)}\right)\right\rvert$ | 0.025342698582 | 0.000477001107 | 0.049280204562 |
| PSF | 6738.01 | 357985.04 | 3465.07 |
| **PSF values of points in region #1** | | | |
| $\alpha_0$ | $\pm 1.13897583371°$ | $\pm 1.113847583372°$ | $\pm 1.13857583372°$ |
| $\beta_0$ | 0° | 0° | 0° |
| $X_{11}$ | 0.146358838756 | 0.146358863068 | 0.146358865039 |
| $\left\lvert \det\left(\frac{\partial(X_n,Z_n)}{\partial(\alpha_0,\beta_0)}\right)\right\rvert$ | 0.056549447153 | 0.015524786362 | 0.001107490161 |
| PSF | 3019.64 | 10999.15 | 154185.99 85 |

three regions. It is observed that the PSF experiences a rapid rate of change in both regions. This is to be expected since all the rays originating from $\bar{P}_0 = [0 \quad -507 \quad 0 \quad 1]^T$ are well focused at the image plane, which is located at the cusp of caustic surface. Furthermore, the Jacobian determinant $|\partial(x_{11}, \partial z_{11})/\partial(\alpha_0, \beta_0)|$ is very small at these points, thus a ray incident upon an infinitesimal area of the image plane (i = 11) yields a high irradiance.

*Example 6.4* For comparison purposes, in Fig. 6.4 the PSF of Fig. 6.3 is normalized by setting its highest peak equal to unity and is then overlaid on that obtained by the ray-counting method by tracing 192539 rays and using a small grid size of 1/300 × 1/300 mm. Note that the ray-counting method is unable to show this "area effect" since it is based solely on the counting of rays. It is observed that the results obtained from the two methods are similar only when the following two conditions are satisfied: (1) the high peak PSF values obtained from the irradiance method in regions #0 and #1 are deleted; and (2) the ray-counting method is implemented using a large number of rays and a very small grid size. In other words, the accuracy of the PSF obtained from the ray-counting method depends significantly on the number of traced rays and the size of the grids used to mesh the image plane. On the contrary, these problems are avoided in the irradiance method since the PSF is computed using a single ray without the use of a grid system.

Overall, the present results suggest that it is difficult, if not impossible, to obtain an accurate estimation of the PSF distribution using the ray-counting method if the PSF distribution contains regions of rapid change (e.g. as shown in Table 6.1). In addition, the ray-counting method requires the tracing of many rays in order to estimate the PSF over the entire image plane, whereas the irradiance method requires just one tracing operation to determine the PSF for a single point on the image plane. In other words, the irradiance method is more computationally efficient than the conventional ray-counting method. However, it should be noted

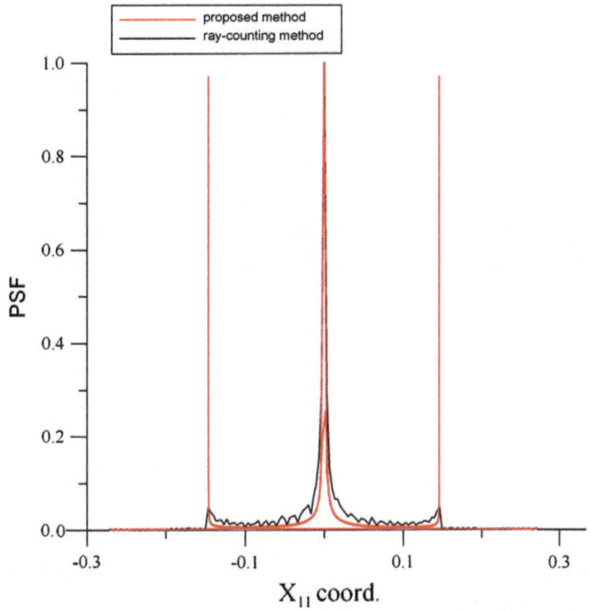


**Fig. 6.4** Distribution of PSF on an image plane computed by ray-counting method (192539 rays, grid size: 1/300 × 1/300 mm) and irradiance method, respectively

that the irradiance method still requires multiple ray tracing operations to construct the entire PSF figure since each figure comprises multiple source rays.

## 6.3 The Spot Diagram Based on Irradiance Method

We have presented the evaluation of the spot diagram by ray-counting method in Sect. 3.4. This section will address determination of the spot diagram on the image plane by irradiance method. To calculate the root mean square (rms) radius, it is first necessary to determine the centroid $\begin{bmatrix} x_{n/centroid} & y_{n/centroid} & z_{n/centroid} & 1 \end{bmatrix}^T$ of the image,

$$\begin{bmatrix} x_{n/centroid} \\ y_{n/centroid} \\ z_{n/centroid} \\ 1 \end{bmatrix} = \begin{bmatrix} \iint x_n B_n(x_n, z_n) d\pi_n / \iint B_n(x_n, z_n) d\pi_n \\ 0 \\ \iint z_n B_n(x_n, z_n) d\pi_n / \iint B_n(x_n, z_n) d\pi_n \\ 1 \end{bmatrix} = \begin{bmatrix} \iint x_n B_n(x_n, z_n) d\pi_n \\ 0 \\ \iint z_n B_n(x_n, z_n) d\pi_n \\ 1 \end{bmatrix}. \tag{6.10}$$

It is noted that Eq. (6.8) is incorporated within Eq. (6.10). Substituting Eqs. (6.1) and (6.9) into Eq. (6.10), the centroid of the image formed on the image plane is obtained as

$$\begin{bmatrix} x_{n/centroid} \\ y_{n/centroid} \\ z_{n/centroid} \\ 1 \end{bmatrix} = \frac{1}{\psi_0} \begin{bmatrix} \int x_n C\beta_0 d\alpha_0 d\beta_0 \\ 0 \\ \int z_n C\beta_0 d\alpha_0 d\beta_0 \\ 1 \end{bmatrix}. \tag{6.11}$$

The terms, $x_n$ and $z_n$, for a given source ray $\bar{R}_0$ are obtained from ray tracing, making the numerical computation of Eq. (6.11) is feasible. The rms radius of the spot size on the image plane provides a useful indication of the degree of blurring of the image in an optical system. It can be calculated from

$$\begin{aligned} rms^2 &= \iint \left[ \left(x_n - x_{n/centroid}\right)^2 + \left(z_n - z_{n/centroid}\right)^2 \right] B_n(x_n, z_n) d\pi_n \\ &= \iint \left(x_n^2 + z_n^2\right) B_n(x_n, z_n) d\pi_n - \left(x_{n/centroid}^2 + z_{n/centroid}^2\right) \\ &= \frac{1}{\psi_0} \iint \left(x_n^2 + z_n^2\right) C\beta_0 d\alpha_0 d\beta_0 - \left(x_{n/centroid}^2 + z_{n/centroid}^2\right). \end{aligned} \tag{6.12}$$

Conventionally, the image centroid and rms radius of the spot of a source point $\bar{P}_0$ are estimated using Eqs. (3.34) and (3.35). Although the accuracy of PSF from ray-counting method significantly depends on the number of rays and size of grids, our numerical results showed that ray-counting method can also offer accurate results for the centroid and rms radius of a spot diagram.

## 6.4 Theory of MTF for Any Arbitrary Direction of OBDF

In Sect. 3.6, the MTF calculation process for the on-axis source point of an axis-symmetrical system was addressed. The imaged spot of an on-axis source point in an axis-symmetrical optical system is symmetrical about the optical axis. Therefore, the MTF value of an on-axis source point remains unchanged as the OBDF is rotated. However, for an off-axis source point, the MTF values not only are lower than those of the on-axis source points, but also differ in the sagittal and meridional directions due to astigmatism and coma aberrations. Therefore, in seeking the extreme values of the MTF, it is required to calculate the MTF of off-axis source points for any arbitrary propagation direction of the OBDF.

Practical optical systems inevitably contain various aberrations. Thus, any source point $\bar{P}_0$ forms a spot on the image plane rather than a single image point since the rays passing through the aperture are focused over a region of the image plane. Importantly, the degree of blurring of an off-axis source point is not equal in all directions due to the effects of astigmatism and coma aberration. Figure 6.5 illustrates the same optical system as shown in Fig. 3.21, but with an off-axis source point $\bar{P}_0 = \begin{bmatrix} P_{0x} & P_{0y} & P_{0z} & 1 \end{bmatrix}^T$ rather than an on-axis source point. As shown, the source point forms a spot on the image plane around the imaged point $\begin{bmatrix} x_{n/chief} & 0 & z_{n/chief} & 1 \end{bmatrix}^T$ of the chief ray. In addition to coordinate frames $(xyz)_0$

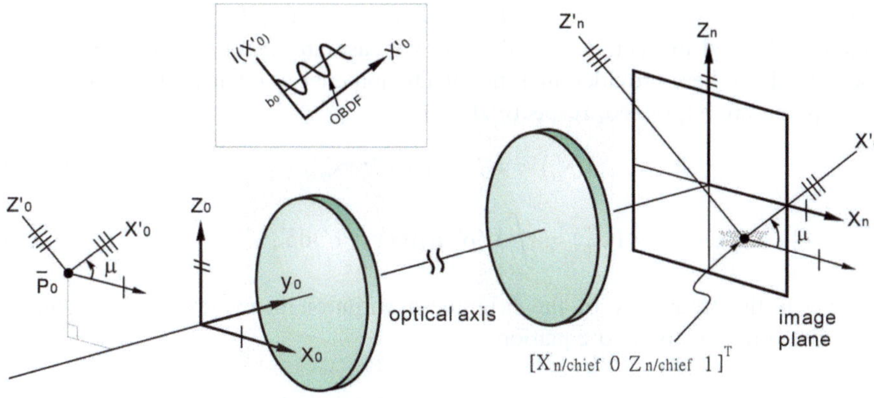


**Fig. 6.5** The spot diagram of an off-axis source point $\bar{P}_0$ on the image plane in an axis-symmetrical optical system. Note that the imaged spot is not symmetrical due to coma and astigmatism aberrations

and $(xyz)_n$, Fig. 6.5 also shows two parallel coordinate frames, $(x'y'z')_n$ and $(x'y'z')_0$. The origin of $(x'y'z')_0$ is at the source point $\bar{P}_0$, and the $x_0'$ axis is pointing to the propagation direction of the OBDF. If the source point $\bar{P}_0$ consists of alternating light and dark bands, the OBDF of $\bar{P}_0$ can be conveniently expressed as $I(x_0') = b_0 + b_1\ C(2\pi v x_0')$. Mathematically, the coordinate frame $(x'y'z')_0$ can be obtained by translating the coordinate frame $(xyz)_0$ by the vector $P_{0x}\bar{i} + P_{0y}\bar{j} + P_{0z}\bar{k}$ and then rotating the translated y-axis through an angle $-\mu$. Meanwhile, the frame $(x'y'z')_n$ is obtained by the translating frame $(xyz)_n$ by a vector $x_{n/chief}\bar{i} + z_{n/chief}\bar{k}$ and then rotating the translated y-axis through an angle $-\mu$. As a result, the pose matrix of $(x'y'z')_n$ with respect to $(xyz)_n$ is given by ${}^{n}\bar{A}_{n'} = \text{tran}(x_{n/\text{chief}},\ 0,\ z_{n/\text{chief}})\ \text{rot}(\bar{y},\ -\mu)$, and that of $(xyz)_n$ with respect to $(x'y'z')_n$ by ${}^{n'}\bar{A}_n = \text{rot}(\bar{y}, \mu)\text{tran}(-x_{n/\text{chief}},\ 0,\ -z_{n/\text{chief}})$. The following coordinate transformation exists between $(x'y'z')_n$ and $(xyz)_n$ for coordinates $[x_n\ \ 0\ \ z_n\ \ 1]^T$ on the image plane:

$$\begin{bmatrix} x_n' \\ 0 \\ z_n' \\ 1 \end{bmatrix} = {}^{n'}\bar{A}_n \begin{bmatrix} x_n \\ 0 \\ z_n \\ 1 \end{bmatrix} = \begin{bmatrix} C\mu & 0 & S\mu & -x_{n/\text{chief}}C\mu - z_{n/\text{chief}}S\mu \\ 0 & 1 & 0 & 0 \\ -S\mu & 0 & C\mu & x_{n/\text{chief}}S\mu - z_{n/\text{chief}}C\mu \\ 0 & 0 & 0 & 1 \end{bmatrix} \begin{bmatrix} x_n \\ 0 \\ z_n \\ 1 \end{bmatrix}$$
$$= \begin{bmatrix} (x_n - x_{n/\text{chief}})C\mu + (z_n - z_{n/\text{chief}})S\mu \\ 0 \\ -(x_n - x_{n/\text{chief}})S\mu + (z_n - z_{n/\text{chief}})C\mu \\ 1 \end{bmatrix}. \tag{6.13}$$

Since the $x'_0$ axis of $(x'y'z')_0$ is the direction of propagation of the OBDF, the equations derived in Sect. 3.6 for the on-axis case are also valid for the off-axis case considered here provided that the OBDF and energy intensity function on the image plane are expressed, respectively, as

$$\mathrm{I}(\mathrm{x}'_0) = \mathrm{b}_0 + \mathrm{b}_1\ \mathrm{C}(2\pi\nu \mathrm{x}'_0), \tag{6.14}$$

$$\mathrm{I}(\mathrm{x}'_\mathrm{n}) = \int \mathrm{L}(\delta', \mu)\mathrm{I}(\mathrm{x}'_\mathrm{n} - \delta')\mathrm{d}\delta', \tag{6.15}$$

where $\nu$ is the frequency of the brightness variation in cycles per unit length. Therefore, one has the two equations

$$\mathrm{L_c}(\nu, \mu) = \int \mathrm{L}(x'_\mathrm{n}, \mu)\ \mathrm{C}(2\pi\nu x'_\mathrm{n})\mathrm{d}x'_\mathrm{n}, \tag{6.16}$$

$$\mathrm{L_s}(\nu, \mu) = \int \mathrm{L}(x'_\mathrm{n}, \mu)\ \mathrm{S}(2\pi\nu x'_\mathrm{n})\mathrm{d}x'_\mathrm{n}, \tag{6.17}$$

where $x'_\mathrm{n} = (\mathrm{x_n} - \mathrm{x_{n/chief}})\mathrm{C}\mu + (\mathrm{z_n} - \mathrm{z_{n/chief}})\mathrm{S}\mu$ is obtained from Eq. (6.13). The MTF and phase shift are

$$\mathrm{MTF}(\nu, \mu) = \sqrt{\mathrm{L}^2_\mathrm{c}(\nu, \mu) + \mathrm{L}^2_\mathrm{s}(\nu, \mu)}, \tag{6.18}$$

and

$$\varpi(\nu, \mu) = \mathrm{a}\tan 2(\mathrm{L_s}(\nu, \mu), \mathrm{L_c}(\nu, \mu)), \tag{6.19}$$

respectively, for the propagation direction of the OBDF defined by angle $\mu$. The following two theorems are proposed to clarify the effects of a translation of $(x'y'z')_\mathrm{n}$ on the $\mathrm{MTF}(\nu, \mu)$ and $\varpi(\nu, \mu)$.

**Theorem 6.1** The MTF value of a source point $\bar{\mathrm{P}}_0$ is stationary when the origin of the coordinate frame $(x'y'z')_\mathrm{n}$ is located at $\begin{bmatrix}\mathrm{x_{n/chief}} & 0 & \mathrm{z_{n/chief}} & 1\end{bmatrix}^\mathrm{T}$ without rotation, since its gradients, $\partial\mathrm{MTF}(\nu, 0)/\partial\mathrm{x_{n/chief}}$ and $\partial\mathrm{MTF}(\nu, 0)/\partial\mathrm{z_{n/chief}}$, are zero. The proof is provided in Appendix A. Therefore, in Eq. (6.13), the point $\begin{bmatrix}\mathrm{x_{n/chief}} & 0 & \mathrm{z_{n/chief}} & 1\end{bmatrix}^\mathrm{T}$, i.e. the incident point of the chief ray on the image plane, is taken as the origin of the coordinate frame $(x'y'z')_\mathrm{n}$ since it usually has the highest irradiance point in the neighborhood.

**Theorem 6.2** The phase shift $\varpi(\nu, 0)$ is non-stationary along the $x'_\mathrm{n}$ axis when the origin of $(x'y'z')_\mathrm{n}$ is located at $\begin{bmatrix}\mathrm{x_{n/chief}} & 0 & \mathrm{z_{n/chief}} & 1\end{bmatrix}^T$ without rotation, since $\partial\varpi(\nu, 0)/\partial\mathrm{x_{n/chief}} = -1$. However, $\varpi(\nu, 0)$ is stationary along the $z'_\mathrm{n}$ axis when the origin of $(x'y'z')_\mathrm{n}$ is located at $\begin{bmatrix}\mathrm{x_{n/chief}} & 0 & \mathrm{z_{n/chief}} & 1\end{bmatrix}^\mathrm{T}$ without rotation, since its gradient with respect to $\mathrm{z_{n/chief}}$ is zero (i.e., $\partial\varpi(\nu, 0)/\partial\mathrm{z_{n/chief}} = 0$). The proof is provided in Appendix B.

# 6.5 MTF for any Arbitrary Direction of OBDF from Ray-Counting and Irradiance Methods

This section extends two methods, ray-counting and irradiance methods, for determining the LSF and MTF of an on-axis source point to the case of an off-axis source point with an arbitrary propagation direction of the OBDF.

## 6.5.1 *Ray-Counting Method*

In the ray-counting method [12], the PSF and the LSF are expressed in terms of the ray density (i.e. the number of rays intercepted by each grid meshed on the image plane). One example is shown in Fig. 6.6. Thus, the solutions are significantly dependent upon both the number of rays traced and the size of the grid used to mesh the image plane. As a result, the ray-counting method can provide only a qualitative estimation of the PSF and the LSF. In [12], the approximate value of the LSF, $L(x_n)$, was estimated simply by counting the number of rays in each increment $\Delta x_n$ (see Fig. 6.6). Consequently, the estimated value of the integration of LSF along $x_n$ axis was not precisely equal to one (i.e. $\int L(x_n)dx_n \neq 1$). Therefore, an alternative equation, Eq. (3.45), for the energy distribution at $x_n$ was proposed.

To determine the LSF and MTF along any arbitrary direction $x'_n$ shown in Fig. 6.7, it is necessary to transfer all the traced incident points ${}^n\bar{P}_n = [{}^nP_{nx} \quad 0 \quad {}^nP_{nz} \quad 1]^T = [x_n \quad 0 \quad z_n \quad 1]^T$ (note that ${}^nP_{ny} = y_n = 0$, since that incidence point is expressed with respect to $(xyz)_n$) to coordinate frame $(x'y'z')_n$ by means of Eq. (6.13). The LSF (denoted now as $L(x'_n, \mu)$ for the propagation direction defined by angle $\mu$) can then be approximately determined by counting

Zn

| | | | | | | | | | | |
|---|---|---|---|---|---|---|---|---|---|---|
| 0 | 0 | 0 | 0 | 0 | 0 | 0 | 0 | 0 | 0 | 0 |
| 0 | 0 | 0 | 6 | 2872 | 3774 | 2872 | 6 | 0 | 0 | 0 |
| 0 | 0 | 0 | 1252 | 3031 | 2674 | 3031 | 1252 | 0 | 0 | 0 |
| 0 | 0 | 0 | 3 | 1174 | 5551 | 1174 | 3 | 0 | 0 | 0 |
| 0 | 0 | 0 | 0 | 756 | 5125 | 756 | 0 | 0 | 0 | 0 |
| 0 | 0 | 16 | 664 | 1177 | 2095 | 1177 | 664 | 16 | 0 | 0 |
| 0 | 0 | 373 | 597 | 767 | 1218 | 767 | 597 | 373 | 0 | 0 |
| 0 | 69 | 401 | 448 | 609 | 795 | 609 | 448 | 401 | 69 | 0 |
| 0 | 152 | 317 | 374 | 482 | 576 | 482 | 374 | 317 | 152 | 0 |
| 0 | 37 | 223 | 319 | 391 | 435 | 391 | 319 | 223 | 37 | 0 |
| 0 | 0 | 13 | 143 | 294 | 325 | 294 | 143 | 13 | 0 | 0 |
| 0 | 0 | 0 | 0 | 0 | 41 | 0 | 0 | 0 | 0 | 0 |
| ≈0 | 0 | 0 | 0 | 0 | 0 | 0 | 0 | 0 | 0 | 0≈ |

Xn

$\Delta$Xn

**Fig. 6.6** The ray density of the source point $\bar{P}_0 = [0 \quad -507 \quad 150 \quad 1]^T$ on the image plane from the ray-counting method with grid size 0.1 × 0.1 mm and 56529 rays distributed uniformly over $\psi_0$

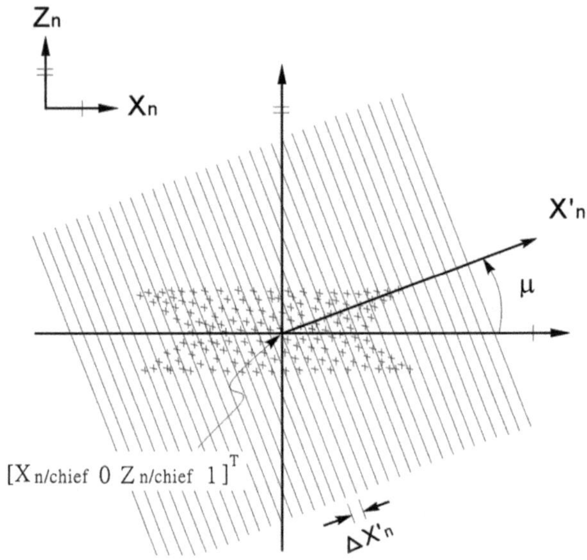


**Fig. 6.7** The ray-counting method in which the ray density (i.e. the number of rays intercepted by each grid on the image plane) is taken as a measure of the PSF and LSF

the number of rays in each increment $\Delta x'_n$ (see Fig. 6.7). $L^*_c(v)$ and $L^*_s(v)$ of Eqs. (3.46) and (3.47), now denoted as $L^*_c(v,\mu)$ and $L^*_s(v,\mu)$ respectively, are computed approximately via the following summation equations

$$L^*_c(v,\mu) = \frac{\sum L(x'_n,\mu)C(2\pi v x'_n)\Delta x'_n}{\sum L(x'_n,\mu)\Delta x'_n}, \tag{6.20}$$

$$L^*_s(v,\mu) = \frac{\sum L(x'_n,\mu)S(2\pi v x'_n)\Delta x'_n}{\sum L(x'_n,\mu)\Delta x'_n}. \tag{6.21}$$

The phase shift $\varpi^*(v,\mu)$ and $MTF^*(v,\mu)$ along the direction defined by angle μ are then obtained, respectively, as

$$\varpi^*(v,\mu) = \text{atan2}(L^*_s(v,\mu), L^*_c(v,\mu)), \tag{6.22}$$

and

$$MTF^*(v,\mu) = \sqrt{[L^*_c(v,\mu)]^2 + [L^*_s(v,\mu)]^2}. \tag{6.23}$$

### 6.5.2 Irradiance Method

Referring to Fig. 6.8, a general source ray $\bar{R}_0$ originates at the off-axis source point $\bar{P}_0 = [P_{0x} \quad P_{0y} \quad P_{0z} \quad 1]^T$ and travels along the unit directional vector $\bar{\ell}_0 = [C\beta_0 C(90^\circ + \alpha_0) \quad C\beta_0 S(90^\circ + \alpha_0) \quad S\beta_0 \quad 0]^T$. The refracted ray of the general source ray $\bar{R}_0$ intersects the image plane at coordinates $[x_n \quad 0 \quad z_n \quad 1]^T$, where $x_n = x_n(\alpha_0, \beta_0)$ and $z_n = z_n(\alpha_0, \beta_0)$. Defining $B_n(x'_n, z'_n)$ as the irradiance on an infinitesimal area $d\pi'_n = dx'_n dz'_n$ centered at the incident point on the image plane, the energy flux $dF'_n$ received by $d\pi'_n$ is equal to $dF'_n = B(x'_n, z'_n) d\pi'_n$. Assuming no transmission losses (i.e. $dF_0 = dF'_n$) and applying the principle of energy flux conservation along the ray path, the following equation is obtained for the energy flux emitted over the infinitesimal solid angle $d\psi_0$ as

$$dF_0 = I_0 d\psi_0 = \frac{C\beta_0}{\psi_0} d\alpha_0 d\beta_0 = dF'_n = B(x'_n, z'_n)\ dx'_n dz'_n. \tag{6.24}$$

Further, one can have the Jacobian matrix $\partial(x'_n, z'_n)/\partial(\alpha_0, \beta_0)$ to describe the extent that a point $(x'_n, z'_n)$ on the image plane is stretched in different directions in the neighborhood of $(\alpha_0, \beta_0)$. With this, the infinitesimal area $d\pi'_n = dx'_n dz'_n$ centered at the point of incidence point on the image plane can be determined by

$$d\pi'_n = dx'_n dz'_n = \left|\det\left(\frac{\partial(x'_n, z'_n)}{\partial(\alpha_0, \beta_0)}\right)\right| d\alpha_0 d\beta_0. \tag{6.25}$$

Substituting Eq. (6.25) into Eq. (6.24), the expression for the point spread function $B(x'_n, z'_n)$ on the image plane is obtained as

$$B(x'_n, z'_n) = \frac{C\beta_0}{\psi_0 \left|\det\left(\frac{\partial(x'_n, z'_n)}{\partial(\alpha_0, \beta_0)}\right)\right|}. \tag{6.26}$$

The LSF, $L(x'_n, \mu)$, of the source point $\bar{P}_0$ on the image plane can be obtained by integrating the PSF, $B(x'_n, z'_n)$, along $z'_n$ to give

$$L(x'_n, \mu) = \int B(x'_n, z'_n) dz'_n = \frac{1}{\psi_0} \int \frac{C\beta_0}{\left|\det\left(\frac{\partial(x'_n, z'_n)}{\partial(\alpha_0, \beta_0)}\right)\right|} dz'_n. \tag{6.27}$$

Note that an integration problem may arise as the Jacobian determinant approaches zero. A typical example of this problem occurs when calculating the LSF $L(x'_n, \mu)$ for an optical system in which the image plane is located at the cusp of a caustic surface [13–23]. In determining the MTF this problem can be avoided by substituting Eq. (6.27) into Eqs. (6.16) and (6.17) and then using Eq. (6.25) to obtain the following equations

$$\begin{aligned} L_c(\nu, \mu) &= \frac{1}{\psi_0} \int \int \frac{C\beta_0 C(2\pi\nu x'_n)}{\left|\det\left(\frac{\partial(x'_n, z'_n)}{\partial(\alpha_0, \beta_0)}\right)\right|} dz'_n dx'_n = \frac{1}{\psi_0} \iint C(2\pi\nu x'_n) C\beta_0 d\alpha_0 d\beta_0 \\ &= \frac{1}{\psi_0} \iint C(2\pi\nu[(x_n - x_{n/chief})C\mu + (z_n - z_{n/chief})S\mu]) C\beta_0 d\alpha_0 d\beta_0, \end{aligned} \tag{6.28}$$

$$L_s(\nu,\mu) = \frac{1}{\psi_0}\int\int \frac{C\beta_0 S(2\pi\nu x'_n)}{\left|\det\left(\frac{\partial(x'_n,z'_n)}{\partial(\alpha_0,\beta_0)}\right)\right|} dz'_n dx'_n = \frac{1}{\psi_0}\iint S(2\pi\nu x'_n)C\beta_0 d\alpha_0 d\beta_0$$
$$= \frac{1}{\psi_0}\iint S(2\pi\nu[(x_n - x_{n/chief})C\mu + (z_n - z_{n/chief})S\mu)])C\beta_0 d\alpha_0 d\overleftrightarrow{\beta}_0. \tag{6.29}$$

The MTF and phase shift can then be computed directly from

$$MTF(\nu,\mu) = \sqrt{[L_c(\nu,\mu)]^2 + [L_s(\nu,\mu)]^2}, \tag{6.30}$$

and

$$\varpi(\nu,\mu) = \text{atan2}(L_s(\nu,\mu), L_c(\nu,\mu)). \tag{6.31}$$

In other words, the LSF is not required, thus the integration problem in computing LSF $L(x'_n,\mu)$ is avoided. The following two theorems are provided to clarify the variations of the $MTF(\nu,\mu)$ and the phase shift $\varpi(\nu,\mu)$ over the range $0° \le \mu \le 360°$.

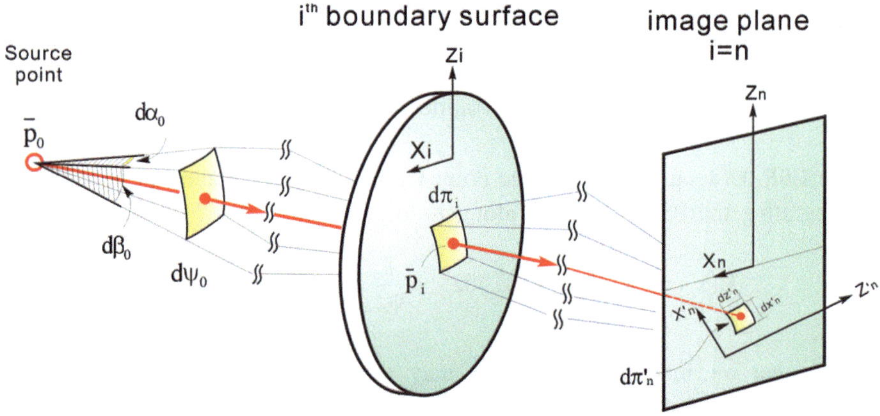


**Fig. 6.8** The infinitesimal area $d\pi'_n = dx'_n dz'_n$ centered at the incident point on the image plane receives the energy flux from the source point

**Theorem 6.3** For the axis-symmetrical system shown in Fig. 6.5, $L_c(\nu,\mu+180°) = L_c(\nu,\mu)$ and $L_s(\nu,\mu+180°) = -L_s(\nu,\mu)$, for which the $MTF(\nu,\mu+180°) = MTF(\nu,\mu)$ and $\varpi(\nu,\mu+180°) = -\varpi(\nu,\mu)$. The proof is provided in Appendix C.

**Theorem 6.4** If the source point $\bar{P}_0$ in Fig. 6.5 lies on the $y_0z_0$ plane (i.e. $\bar{P}_0 = [0 \quad P_{0y} \quad P_{0z} \quad 1]^T$), it follows that $L_c(\nu,90°+\mu) = L_c(\nu,90°-\mu)$ and $L_s(\nu,90°+\mu) = L_s(\nu,90°-\mu)$. As a result, $MTF(\nu,90°+\mu) = MTF(\nu,90°-\mu)$ and $\varpi(\nu,90°+\mu) = \varpi(\nu,90°-\mu)$. The proof is presented in Appendix D.

Due to symmetry of the system about its optical axis (i.e. the $y_0$ axis in Fig. 6.5) of an axis-symmetrical optical axis, it is possible with no loss of generality to define a source point $\bar{P}_0$ as lying on the $y_0z_0$ plane. In other words, Theorem 6.4 is always applicable for an axis-symmetrical optical system provided that the source points are located on the $y_0z_0$ plane. Significantly, Theorem 6.4 indicates that the MTF and phase shift curves are both symmetrical with respect to $\mu = 90°$ for a given frequency $\nu$. Meanwhile, Theorem 6.3 curves are anti-symmetrical with respect to $\mu = 180°$ for a given $\nu$. Therefore, provided that the source point $\bar{P}_0$ is located on the $y_0z_0$ plane of $(xyz)_0$, a complete understanding of the MTF and phase shift can be obtained by computing the variations of $\text{MTF}(\nu, \mu)$ and $\varpi(\nu, \mu)$ merely over the range $0° \leq \mu \leq 90°$. In other words, the extreme values of the MTF and phase shift can be determined without the need to explore the full range of $0° \leq \mu \leq 360°$.

The following examples demonstrate the validity of the proposed MTF computation methods using the system in Fig. 2.1, in which the 5th boundary surface is the aperture and n = 11 is the image plane. The integrations in Eqs. (6.28) and (6.29) are performed using Simpson's rule. In addition, the off-axis source points are assumed to be located on the $y_0z_0$ plane such that Theorem 6.4 can be applied.

*Example 6.5* Figures 6.9 and 6.10 present the variations of LSF, MTF, and phase shift values of a source point $\bar{P}_0 = [0 \quad -507 \quad 150 \quad 1]^T$ traveling in the system of Fig. 2.1. In general, the results show that when using the ray-counting method, the LSF, MTF and phase shift are all dependent on the number of rays traced and the grid size meshed on the image plane. In other words, if the grid mesh is not sufficiently fine, the LSF will have poor resolution and the accuracy of the MTF

**Fig. 6.9** The LSF of $\bar{P}_0 = [0 \quad -507 \quad 1501]^T$as determined using the ray-counting method. It is seen that the LSF depends on both the number of rays traced and the grid size on the image plane

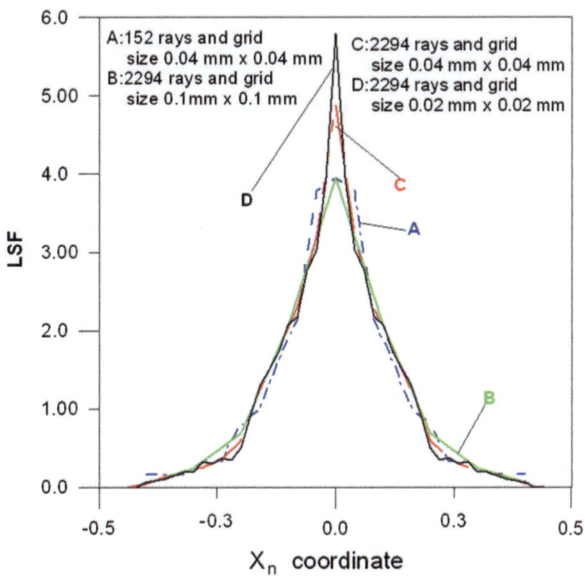

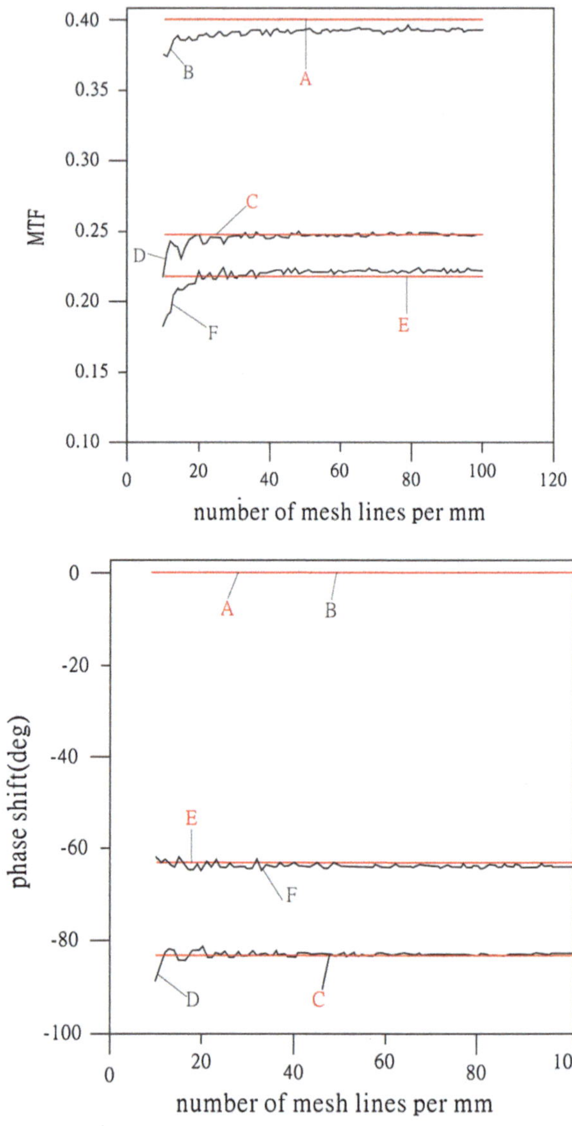


**Fig. 6.10** Variations of the MTF and phase shift of a source point $\bar{P}_0 = [0 \quad -507 \quad 1501]^T$ when tracing 2294 rays and using $v = 2$. It is shown that the MTF computed using the ray-counting method $(B: \mu = 0^\circ, D: \mu = 45^\circ, F: \mu = 90^\circ)$ is sensitive to the grid size meshed on the image plane. However, the MTF computed using the irradiance method $(A: \mu = 0^\circ, D: \mu = 45^\circ, F: \mu = 90^\circ)$ does not need grids

will be degraded. However, Fig. 6.10 shows that the value of the MTF computed using the irradiance method is constant since it is based solely on the use of irradiance, without the need of meshing and the LSF.

*Example 6.6* Figure 6.11 shows the variation of the LSF when the OBDF of $\bar{P}_0$ in Example 6.5 is rotated through different angles. It is seen that the shape of the LSF for $\mu = 0^\circ$ is significantly different from that for $\mu = 90^\circ$. Thus, it is inferred that coma and astigmatism aberrations exist in the system of Fig. 2.1 for this off-axis

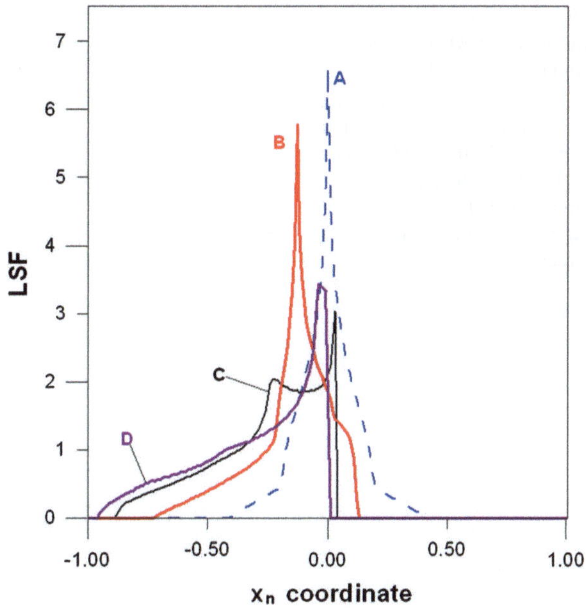


**Fig. 6.11** Effects of coma and astigmatism aberrations in changing the shape of the LSF when the OBDF is rotated through different angles ($\bar{P}_0 = [0 \quad -507 \quad 1501]^T$, 25781 rays, $A : \mu = 0°, B : \mu = 30°, C : \mu = 60°, D : \mu = 90°$)

source point. Consequently, the ray-counting method requires an appropriate choice of grid mesh size for improved LSF resolution and ensured MTF accuracy.

*Example 6.7* Figure 6.12 illustrates the variation of the MTF of the off-axis source point $\bar{P}_0 = [0 \quad -507 \quad 150 \quad 1]^T$ when the OBDF is rotated from $\mu = 0°$ to $\mu = 90°$. The results clearly show that the MTF varies with the rotation angle $\mu$ of the OBDF. Significantly, the MTF neither increases nor decreases monotonically in the domain $0° \leq \mu \leq 90°$. Therefore, it is impossible to determine the extreme values of the MTF and phase shift by considering only the sagittal and meridonial directions.

*Example 6.8* Figure 6.13 presents the MTF values of $\bar{P}_0$ in Example 6.5 obtained by the ray-counting and irradiance methods when different numbers of rays are traced. It is observed that for both methods, the MTF value oscillates if the number of traced rays is less than a certain threshold value ($\sim$500 rays in the present case). The main factor causing this drawback for irradiance method is that Eqs. (6.28) and (6.29) involve a numerical integration. The accuracy of any numerical integration method significantly depends on step size. In addition it is seen that in the ray-counting method, even when a sufficient number of rays are traced to ensure a constant MTF, the estimated value of the MTF still depends on the grid size on the image plane.

Overall, the results presented in Fig. 6.9 through Fig. 6.13 confirm that the ray-counting and irradiance methods both provide a feasible means of computing the MTF and phase shift of an off-axis source point for any arbitrary propagation direction of the OBDF. For both methods, a sufficient number of traced rays are

**Fig. 6.12** Non-monotonic variations of MTF in the domain $0° \leq \mu \leq 90°$ ($\bar{P}_0 = [0 \quad -507 \quad 150 \quad 1]^T$, 2294 rays, $v = 2$). It is noted that the extreme MTF values occur at intermediate directions between the sagittal and meridional directions

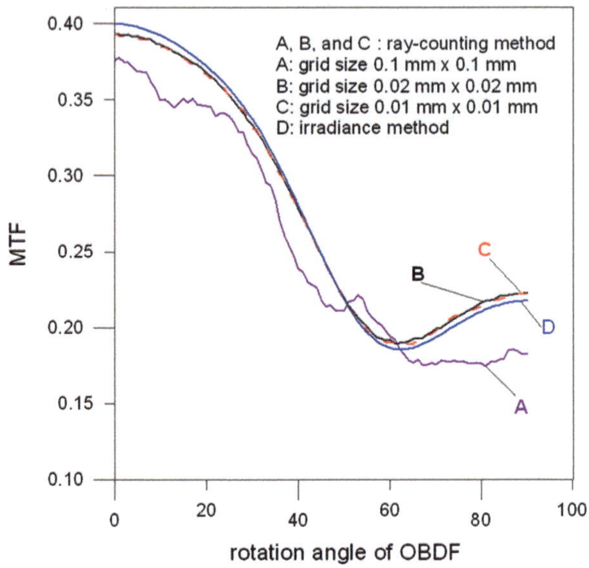


**Fig. 6.13** Variations of MTF with the number of traced rays for $v = 2$ and $\mu = 45°$. It is noted that for both the ray-counting and irradiance methods, the MTF values oscillate significantly unless a sufficient number of rays are traced

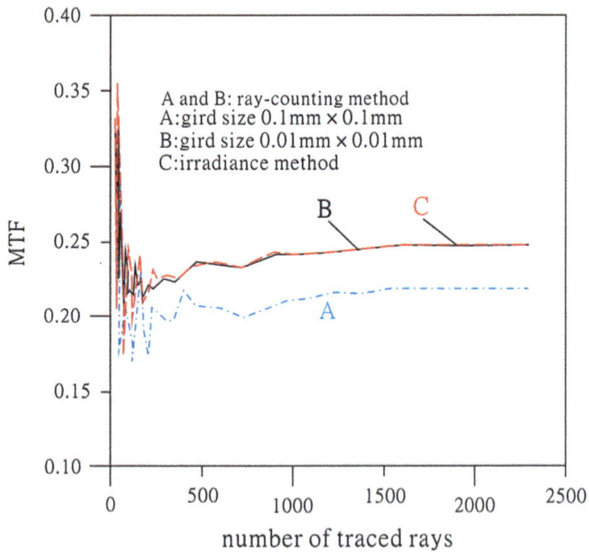


required to ensure the accuracy of the estimated MTF. The ray-counting method has two drawbacks: (1) the LSF must be computed before the MTF can be obtained; (2) the accuracy of the MTF and phase shift estimates is dependent upon the size of the grid on the image plane.

# Appendix A

For the case of zero rotation, the coordinate transformation between coordinate frames $(x'y'z')_n$ and $(xyz)_n$ can be obtained simply by setting $\mu = 0$ in Eq. (6.13). The MTF is therefore denoted as $\mathrm{MTF}(v,0)$. Since $\mathrm{MTF}^2(v,0) = L_c^2(v,0) + L_s^2(v,0)$, the gradient of $\mathrm{MTF}^2(v,0)$ with respect to $x_{n/chief}$ can be obtained from

$$\begin{aligned}\frac{\partial \mathrm{MTF}^2(v,0)}{\partial x_{n/chief}} &= 2\mathrm{MTF}(v,0)\frac{\partial \mathrm{MTF}(v,0)}{\partial x_{n/chief}} = \frac{\partial \mathrm{MTF}^2(v,0)}{\partial x'_n}\frac{\partial x'_n}{\partial x_{n/chief}} + \frac{\partial \mathrm{MTF}^2(v,0)}{\partial z'_n}\frac{\partial z'_n}{\partial x_{n/chief}} \\ &= 2\left(L_c(v,0)\frac{\partial L_c(v,0)}{\partial x'_n} + L_s(v,0)\frac{\partial L_s(v,0)}{\partial x'_n}\right)\frac{\partial x'_n}{\partial x_{n/chief}} \\ &+ 2\left(L_c(v,0)\frac{\partial L_c(v,0)}{\partial z'_n} + L_s(v,0)\frac{\partial L_s(v,0)}{\partial z'_n}\right)\frac{\partial z'_n}{\partial x_{n/chief}} \\ &= 0\,. \end{aligned} \tag{A.1}$$

Similarly, one has

$$\begin{aligned}\frac{\partial \mathrm{MTF}^2(v,0)}{\partial z_{n/chief}} &= 2\mathrm{MTF}(v,0)\frac{\partial \mathrm{MTF}(v,0)}{\partial z_{n/chief}} \\ &= \frac{\partial \mathrm{MTF}^2(v,\mu)}{\partial x'_n}\frac{\partial x'_n}{\partial z_{n/chief}} + \frac{\partial \mathrm{MTF}^2(v,\mu)}{\partial z'_n}\frac{\partial z'_n}{\partial z_{n/chief}} = 0. \end{aligned} \tag{A.2}$$

Equations (A.1) and (A.2) show $\partial \mathrm{MTF}(v,0)/\partial x_{n/chief} = \partial \mathrm{MTF}(v,0)/\partial z_{n/chief} = 0$, indicating that the $\mathrm{MTF}(v,0)$ value is stationary in the neighborhood of $\begin{bmatrix} x_{n/chief} & 0 & z_{n/chief} & 1\end{bmatrix}^T$, proving Theorem 6.1.

# Appendix B

The gradients of phase shift $\varpi(v,0)$ with respect to $x_{n/chief}$ and $z_{n/chief}$ can be obtained respectively as

$$\begin{aligned}\frac{\partial \varpi(v,0)}{\partial x_{n/chief}} &= \frac{\partial \varpi(v,0)}{\partial x'_n}\frac{\partial x'_n}{\partial x_{n/chief}} + \frac{\partial \varpi(v,0)}{\partial z'_n}\frac{\partial z'_n}{\partial x_{n/chief}} \\ &= \frac{L_c(v,0)}{L_c(v,0)^2 + L_s(v,0)^2}\frac{\partial L_s(v,0)}{\partial x'_n}\frac{\partial x'_n}{\partial x_{n/chief}} - \frac{L_s(v,0)}{L_c(v,0)^2 + L_s(v,0)^2}\frac{\partial L_c(v,0)}{\partial x'_n}\frac{\partial x'_n}{\partial x_{n/chief}} \\ &= \frac{\partial x'_n}{\partial x_{nc}} = -1\,, \end{aligned} \tag{B.1}$$

$$\begin{aligned}\frac{\partial\varpi(v,0)}{\partial z_{n/\mathrm{chief}}} &= \frac{\partial\varpi(v,0)}{\partial x_n'}\frac{\partial x'}{\partial z_{n/\mathrm{chief}}} + \frac{\partial\varpi(v,0)}{\partial z_n'}\frac{\partial z_n'}{\partial z_{n/\mathrm{chief}}}\\ &= \frac{L_c(v,0)}{L_c(v,0)^2+L_s(v,0)^2}\frac{\partial L_s(v,0)}{\partial x_n'}\frac{\partial x_n'}{\partial z_{n/\mathrm{chief}}} - \frac{L_s(v,0)}{L_c(v,0)^2+L_s(v,0)^2}\frac{\partial L_c(v,0)}{\partial x_n'}\frac{\partial x_n'}{\partial z_{n/\mathrm{chief}}}\\ &= \frac{\partial x_n'}{\partial z_{n/\mathrm{chief}}} = 0.\end{aligned} \tag{B.2}$$

Equation (B.1) shows that $\varpi(v,0)$ changes with a change of $x_{n/\mathrm{chief}}$. However, Eq. (B.2) shows that $\varpi(v,0)$ is unchanged with small changes of $z_{n/\mathrm{chief}}$. As discussed in Sect. 3.6, a phase shift of $\varpi = 180°$ yields a reversal of contrast.

## Appendix C

$$\begin{aligned}&L_c(v,\mu+180°)\\ &= \frac{1}{\psi_0}\iint C\left[2\pi v(x_n - x_{n/\mathrm{chief}})C(\mu+180°) + 2\pi v(z_n - z_{n/\mathrm{chief}})S(\mu+180°)\right]C\beta_0 d\alpha_0 d\beta_0\\ &= \frac{1}{\psi_0}\iint C\left[2\pi v(x_n - x_{n/\mathrm{chief}})C\mu + 2\pi v(z_n - z_{n/\mathrm{chief}})S\mu\right]C\beta_0 d\alpha_0 d\beta_0 = L_c(v,\mu).\end{aligned} \tag{C.1}$$

$$\begin{aligned}&L_s(v,\mu+180°)\\ &= \frac{1}{\psi_0}\iint S\left[2\pi v(x_n - x_{n/\mathrm{chief}})C(\mu+180°) + 2\pi v(z_n - z_{n/\mathrm{chief}})S(\mu+180°)\right]C\beta_0 d\alpha_0 d\beta_0\\ &= \frac{-1}{\psi_0}\iint S\left[2\pi v(x_n - x_{n/\mathrm{chief}})C(\mu) + 2\pi v(z_n - z_{n/\mathrm{chief}})S\mu\right]C\beta_0 d\alpha_0 d\beta_0 = -L_s(v,\mu).\end{aligned} \tag{C.2}$$

As a result, one has the following two equations from Eqs. (C.1) and (C.2):

$$\begin{aligned}\mathrm{MTF}(v,\mu+180°) &= \sqrt{[L_c(v,\mu+180°)]^2 + [L_s(v,\mu+180°)]^2}\\ &= \sqrt{[L_c(v,\mu)]^2 + [L_s(v,\mu)]^2} = \mathrm{MTF}(v,\mu),\end{aligned}$$

$$\begin{aligned}\varpi(v,\mu+180°) &= \mathrm{atan2}(L_s(v,\mu+180°), L_c(v,\mu+180°))\\ &= \mathrm{atan2}(-L_s(v,\mu), L_c(v,\mu)) = -\varpi(v,\mu).\end{aligned}$$

# Appendix D

Recall that in the irradiance method of Sect. 6.1, we indicated that a general source ray $\bar{\mathrm{R}}_0$ intersects the image plane at coordinates $[\mathrm{x_n} \quad 0 \quad \mathrm{z_n} \quad 1]^T$, where $\mathrm{x_n} = \mathrm{x_n}(\alpha_0, \beta_0)$ and $\mathrm{z_n} = \mathrm{z_n}(\alpha_0, \beta_0)$. As shown in Fig. 6.5, and with no loss in generality, the general source point can always be defined as lying on the $y_0 z_0$ plane and expressed as $\bar{\mathrm{P}}_0 = [0 \quad \mathrm{P_{0y}} \quad \mathrm{P_{0z}} \quad 1]^{\mathrm{T}}$ due to symmetry of the optical system. As a result, its chief ray intersects $z_n$ axis of the image coordinate frame $(xyz)_{\mathrm{n}}$, thus

$$\mathrm{x_{n/chief}} = \mathrm{y_{n/chief}} = 0. \tag{D.1}$$

Furthermore, two general rays $[\bar{\mathrm{P}}_0 \quad \bar{\ell}_{0/1}]^{\mathrm{T}}$ and $[\bar{\mathrm{P}}_0 \quad \bar{\ell}_{0/2}]^{\mathrm{T}}$, where $\bar{\ell}_{0/1} = [\mathrm{C}\beta_0\mathrm{C}(90^\circ - \alpha_0) \quad \mathrm{C}\beta_0\mathrm{S}(90^\circ - \alpha_0) \quad \mathrm{S}\beta_0 \quad 0]^{\mathrm{T}}$ and $\bar{\ell}_{0/2} = [\mathrm{C}\beta_0\mathrm{C}(90^\circ + \alpha_0) \quad \mathrm{C}\beta_0\mathrm{S}(90^\circ + \alpha_0) \quad \mathrm{S}\beta_0 \quad 0]^{\mathrm{T}}$, always have identical $\mathrm{z_n}$ intersections and oppositely signed $\mathrm{x_n}$ values in this axis-symmetrical optical system, i.e.

$$\mathrm{x_n}(90^\circ - \alpha_0, \beta_0) = -\mathrm{x_n}(90^\circ + \alpha_0, \beta_0), \tag{D.2}$$

$$\mathrm{z_n}(90^\circ - \alpha_0, \beta_0) = \mathrm{z_n}(90^\circ + \alpha_0, \beta_0). \tag{D.3}$$

Now we consider the calculation of $\mathrm{L_c}(\nu, 90^\circ + \mu)$ by substituting Eq. (D.1) into Eq. (6.28), yielding

$$\begin{aligned}
\mathrm{L_c}(\nu, 90^\circ + \mu) &= \frac{1}{\psi_0}\iint \mathrm{C}\left(2\pi\nu\left[\mathrm{x_n}\mathrm{C}(90^\circ + \mu) + \left(\mathrm{z_n} - \mathrm{z_{n/chief}}\right)\mathrm{S}(90^\circ + \mu)\right]\right)\mathrm{C}\beta_0 d\alpha_0 d\beta_0 \\
&= \frac{\mathrm{C}\left(2\pi\nu \mathrm{z_{n/chief}}\mathrm{C}\mu\right)}{\psi_0}\iint \mathrm{C}(2\pi\nu(\mathrm{x_n}\mathrm{S}\mu - \mathrm{z_n}\mathrm{C}\mu))\mathrm{C}\beta_0 d\alpha_0 d\beta_0 \\
&\quad - \frac{\mathrm{S}\left(2\pi\nu \mathrm{z_{n/chief}}\mathrm{C}\mu\right)}{\psi_0}\iint \mathrm{S}(2\pi\nu(\mathrm{x_n}\mathrm{S}\mu - \mathrm{z_n}\mathrm{C}\mu))\mathrm{C}\beta_0 d\alpha_0 d\beta_0 \\
&= \frac{\mathrm{C}\left(2\pi\nu \mathrm{z_{n/chief}}\mathrm{C}\mu\right)}{\psi_0}\left[\iint \mathrm{C}(2\pi\nu \mathrm{x_n}\mathrm{S}\mu)\mathrm{C}(2\pi\nu \mathrm{z_n}\mathrm{C}\mu)\mathrm{C}\beta_0 d\alpha_0 d\beta_0\right. \\
&\qquad \left. + \iint \mathrm{S}(2\pi\nu \mathrm{x_n}\mathrm{S}\mu)\mathrm{S}(2\pi\nu \mathrm{z_n}\mathrm{C}\mu)\mathrm{C}\beta_0 d\alpha_0 d\beta_0\right] \\
&\quad - \frac{\mathrm{S}\left(2\pi\nu \mathrm{z_{n/chief}}\mathrm{C}\mu\right)}{\psi_0}\left[\iint \mathrm{S}(2\pi\nu \mathrm{x_n}\mathrm{S}\mu)\mathrm{C}(2\pi\nu \mathrm{z_n}\mathrm{C}\mu)\mathrm{C}\beta_0 d\alpha_0 d\beta_0\right. \\
&\qquad \left. - \iint \mathrm{C}(2\pi\nu \mathrm{x_n}\mathrm{S}\mu)\mathrm{S}(2\pi\nu \mathrm{z_n}\mathrm{C}\mu)\mathrm{C}\beta_0 d\alpha_0 d\beta_0\right].
\end{aligned} \tag{D.4}$$

If the upper and lower integration limits of $\alpha_0$ are given by $90^\circ + \alpha_{0/\text{limit}}$ and $90^\circ - \alpha_{0/\text{limit}}$, respectively, the integrand of the first term in Eq. (D.4) can be written in the form shown in Eq. (D.5) when Eqs. (D.2) and (D.3) are used.

$$
\begin{aligned}
&\iint C(2\pi\nu x_n S\mu)C(2\pi\nu z_n C\mu)C\beta_0 d\alpha_0 d\beta_0 \\
&= \int\left[\int_{90^\circ-\alpha_{0/\text{limit}}}^{90^\circ+\alpha_{0/\text{limit}}} C(2\pi\nu x_n S\mu)C(2\pi\nu z_n C\mu)d(\alpha_0-90^\circ)\right]C\beta_0 d\beta_0 \\
&= \int\left[\int_{-\alpha_{0/\text{limit}}}^{\alpha_{0/\text{limit}}} C(2\pi\nu x_n S\mu)C(2\pi\nu z_n C\mu)d\alpha_0\right]C\beta_0 d\beta_0 \\
&= \int\left[\int_{-\alpha_{0/\text{limit}}}^{0} C(2\pi\nu x_n S\mu)C(2\pi\nu z_n C\mu)d\alpha_0 + \int_{0}^{\alpha_{0/\text{limit}}} C(2\pi\nu x_n S\mu)C(2\pi\nu z_n C\mu)d\alpha_0\right]C\beta_0 d\beta_0 \\
&= 2\int\left[\int_{0}^{\alpha_{0/\text{limit}}} C(2\pi\nu x_n S\mu)C(2\pi\nu z_n C\mu)d\alpha_0\right]C\beta_0 d\beta_0.
\end{aligned}
\tag{D.5}
$$

Similarly, the second, third, and fourth terms in Eq. (D.4) can be rewritten in the forms shown in Eqs. (D.6), (D.7) and (D.8), respectively, i.e.

$$\iint S(2\pi\nu x_n S\mu)S(2\pi\nu z_n C\mu)C\beta_0 d\alpha_0 d\beta_0 = 0, \tag{D.6}$$

$$\iint S(2\pi\nu x_n S\mu)C(2\pi\nu z_n C\mu)C\beta_0 d\alpha_0 d\beta_0 = 0, \tag{D.7}$$

$$
\begin{aligned}
&\iint C(2\pi\nu x_n S\mu)S(2\pi\nu z_n C\mu)C\beta_0 d\alpha_0 d\beta_0 \\
&= 2\int\left[\int_{0}^{\alpha_{0/\text{limit}}} C(2\pi\nu x_n S\mu)S(2\pi\nu z_n C\mu)d\alpha_0\right]C\beta_0 d\beta_0.
\end{aligned}
\tag{D.8}
$$

Substituting Eqs. (D.5), (D.6), (D.7) and (D.8) into Eq. (D.4) gives

$$
\begin{aligned}
L_c(\nu, 90^\circ + \mu) = &\frac{2C(2\pi\nu z_{n/\text{chief}} C\mu)}{\psi_0}\int\left[\int_{0}^{\alpha_{0\text{limit}}} C(2\pi\nu x_n S\mu)C(2\pi\nu z_n C\mu)d\alpha_0\right]C\beta_0 d\beta_0 \\
&+ \frac{2S(2\pi\nu z_{n/\text{chief}} C\mu)}{\psi_0}\int\left[\int_{0}^{\alpha_{0/\text{limit}}} C(2\pi\nu x_n S\mu)S(2\pi\nu z_n C\mu)d\alpha_0\right]C\beta_0 d\beta_0.
\end{aligned}
\tag{D.9}
$$

Since Eq. (D.9) indicates that $L_c(\nu, 90^\circ + \mu)$ is an even function with respect to $\mu$, we have

$$\mathrm{L_c}(v, 90^\circ + \mu) = \mathrm{L_c}(v, 90^\circ - \mu). \tag{D.10}$$

Similarly, by substituting $\mu = 90^\circ + \mu$ and Eq. (D.1) into Eq. (6.29), $\mathrm{L_s}(v, 90^\circ + \mu)$ can be rewritten as

$$\begin{aligned}
\mathrm{L_s}(v, 90^\circ + \mu) &= \frac{1}{\psi_0}\iint \mathrm{S}\left(2\pi v\left[\mathrm{x_n C}(90^\circ + \mu) + (\mathrm{z_n} - \mathrm{z_{n/chief}})\mathrm{C}(90^\circ + \mu)\right]\right)\mathrm{C}\beta_0 \mathrm{d}\alpha_0 \mathrm{d}\beta_0 \\
&= \frac{-\mathrm{S}(2\pi v \mathrm{C}\mu \mathrm{z_{n/chief}})}{\psi_0}\iint \mathrm{C}(2\pi v(\mathrm{x_n S}\mu - \mathrm{z_n C}\mu))\mathrm{C}\beta_0 \mathrm{d}\alpha_0 \mathrm{d}\beta_0 \\
&\quad - \frac{\mathrm{C}(2\pi v \mathrm{z_{n/chief}} \mathrm{C}\mu)}{\psi_0}\iint \mathrm{S}(2\pi v(\mathrm{x_n S}\mu - \mathrm{z_n C}\mu))\mathrm{C}\beta_0 \mathrm{d}\alpha_0 \mathrm{d}\beta_0 \\
&= \frac{-\mathrm{S}(2\pi v \mathrm{z_{n/chief}} \mathrm{C}\mu)}{\psi_0}\left[\iint \mathrm{C}(2\pi v \mathrm{x_n S}\mu)\mathrm{C}(2\pi v \mathrm{z_n C}\mu)\mathrm{C}\beta_0 \mathrm{d}\alpha_0 \mathrm{d}\beta_0\right. \\
&\quad \left. + \iint \mathrm{S}(2\pi v \mathrm{x_n S}\mu)\mathrm{S}(2\pi v \mathrm{z_n C}\mu)\mathrm{C}\beta_0 \mathrm{d}\alpha_0 \mathrm{d}\beta_0\right] \\
&\quad - \frac{\mathrm{C}(2\pi v \mathrm{z_{n/chief}} \mathrm{C}\mu)}{\psi_0}\left[\iint \mathrm{S}(2\pi v \mathrm{x_n S}\mu)\mathrm{C}(2\pi v \mathrm{z_n C}\mu)\mathrm{C}\beta_0 \mathrm{d}\alpha_0 \mathrm{d}\beta_0\right. \\
&\quad \left. - \iint \mathrm{C}(2\pi v \mathrm{x_n S}\mu)\mathrm{S}(2\pi v \mathrm{z_n C}\mu)\mathrm{C}\beta_0 \mathrm{d}\alpha_0 \mathrm{d}\beta_0\right] \\
&= \frac{-2\mathrm{S}(2\pi v \mathrm{z_{n/chief}} \mathrm{C}\mu)}{\psi_0}\int\left[\int_0^{\alpha_{0/\mathrm{limit}}} \mathrm{C}(2\pi v \mathrm{x_n S}\mu)\mathrm{C}(2\pi v \mathrm{z_n C}\mu)\mathrm{d}\alpha_0\right]\mathrm{C}\beta_0 \mathrm{d}\beta_0 \\
&\quad + \frac{2\mathrm{C}(2\pi v \mathrm{z_{n/chief}} \mathrm{C}\mu)}{\psi_0}\int\left[\int_0^{\alpha_{0/\mathrm{limit}}} \mathrm{C}(2\pi v \mathrm{x_n S}\mu)\mathrm{S}(2\pi v \mathrm{z_n C}\mu)\mathrm{d}\alpha_0\right]\mathrm{C}\beta_0 \mathrm{d}\beta_0.
\end{aligned} \tag{D.11}$$

Again, since $\mathrm{L_s}(v, 90^\circ + \mu)$ is an even function with respect to $\mu$, one has

$$\mathrm{L_s}(v, 90^\circ + \mu) = \mathrm{L_s}(v, 90^\circ - \mu). \tag{D.12}$$

Then, from Eqs. (D.10) and (D.12), we have

$$\begin{aligned}
\mathrm{MTF}(v, 90^\circ + \mu) &= \sqrt{[\mathrm{L_c}(v, 90^\circ + \mu)]^2 + [\mathrm{L_s}(v, 90^\circ + \mu)]^2} \\
&= \sqrt{[\mathrm{L_c}(v, 90^\circ - \mu)]^2 + [\mathrm{L_s}(v, 90^\circ - \mu)]^2} = \mathrm{MTF}(v, 90^\circ - \mu),
\end{aligned} \tag{D.13}$$

and

$$\begin{aligned}
\varpi(v, 90^\circ + \mu) &= \mathrm{atan2}(\mathrm{L_s}(v, 90^\circ + \mu), \mathrm{L_c}(v, 90^\circ + \mu)) \\
&= \mathrm{atan2}(\mathrm{L_s}(v, 90^\circ - \mu), \mathrm{L_c}(v, 90^\circ - \mu)) = \varpi(v, 90^\circ - \mu).
\end{aligned} \tag{D.14}$$

## References

1. V.N. Mahajan, *Optical Imaging and Aberrations Part I Ray Geometrical Optics* (SPIE—The Interational Society for Optical Engineering, Bellingham, 1998)
2. K.H. Tseng, C. Kung, T.T. Liao, H.P. Chang, Calculation of modulation transfer function of an optical system by using skew ray tracing. Trans. Can. Soc. Mech. Eng. (J. Mech. Eng.) **33**, 429–442 (2009)
3. S. Inoue, N. Tsumura, Y. Miyake, Measuring MTF of paper by sinusoidal test pattern projection. J. Imaging Sci. Technol. **41**, 657–661 (1997)
4. G.D. Boreman, S. Yang, Modulation transfer function measurement using three- and four-bar targets. Appl. Opt. **34**, 8050–8052 (1995)
5. D.N. Sitter, J.S. Goddard, R.K. Ferrell, Method for the measurement of the modulation transfer function of sampled imaging systems from bar-target patterns. Appl. Opt. **34**, 746–751 (1995)
6. R. Barakat, Determination of the optical transfer function directly from the edge spread function. J. Opt. Soc. Am. **55**, 1217–1221 (1965)
7. G.L. Rogers, Measurement of the modulation transfer function of paper. Appl. Opt. **37**, 7235–7240 (1998)
8. S.K. Park, R. Schowengerdt, M. Kaczynski, Modulation-transfer-function analysis for sampled image system. Appl. Opt. **23**, 2572–2582 (1984)
9. S. Inoue, N. Tsumura, Y. Miyake, Measuring MTF of paper by sinusoidal test pattern projection. J. Imaging Sci. Technol. **41**, 657–661 (1997)
10. K.H. Tseng, C. Kung, T.T. Liao, H.P. Chang, Calculation of modulation transfer function of an optical system by using skew ray tracing. Trans. Can. Soc. Mech. Eng. (J. Mech. Eng.) **33**, 429–442 (2009)
11. E. Giakoumakis, M.C. Katsarioti, G.S. Panayiotakis, Modulation transfer function of thin transparent foils in radiographic cassettes Appl. Phys. Solids Surf. **52**, 210–212 (1991)
12. W.J. Smith, *Modern Optical Engineering*, 3rd edn. (Edmund Industrial Optics, Barrington, 2001)
13. O.N. Stavroudis, *The Optics of Rays, Wavefronts, and Caustics* (Academic press, New York, 1972)
14. O.N. Stavroudis, *The Mathematics of Geometrical and Physical Optics* (Wiley-VCH Verlag, Weinheim, 2006)
15. J.A. Hoffnagle, D.L. Shealy, Refracting the k-function: Stavroudis's solution to the eikonal equation for multielement optical systems. J. Opt. Soc. Am. A **28**, 1312–1321 (2011)
16. D.L. Shealy, D.G. Burkhard, Caustic surfaces and irradiance for reflection and refraction from an ellipsoid, elliptic parabolid and elliptic cone. Appl. Opt. **12**, 2955–2959 (1973)
17. D.L. Shealy, D.G. Burkhard, Caustic surface merit functions in optical design. J. Opt. Soc. Am. **66**, 1122 (1976)
18. D.L. Shealy, Analytical illuminance and caustic surface calculations in geometrical optics. Appl. Opt. **15**, 2588–2596 (1976)
19. T.B. Andersen, Optical aberration functions: computation of caustic surfaces and illuminance in symmetrical systems. Appl. Opt. **20**, 3723–3728 (1981)
20. A.M. Kassim, D.L. Shealy, Wave front equation, caustics, and wave aberration function of simple lenses and mirrors. Appl. Opt. **21**, 516–522 (1988)
21. A.M. Kassim, D.L. Shealy, D.G. Burkhard, Caustic merit function for optical design. Appl. Opt. **28**, 601–606 (1989)
22. G. Silva-Orthigoza, M. Marciano-Melchor, O. Carvente-Munoz, R. Silva-Ortigoza, Exact computation of the caustic associated with the evolution of an aberrated wavefront. J. Opt. A Pure Appl. Opt. **4**, 358–365 (2002)
23. D.L. Shealy, J.A. Hoffnagle, Wavefront and caustics of a plane wave refracted by an arbitrary surface. J. Opt. Soc Am. A **25**, 2370–2382 (2008)

# Chapter 7
# Optical Path Length and Its Jacobian Matrix with Respect to System Variable Vector

Optical path length (OPL) is important because it determines the phase of the light and governs interference and diffraction of ray as it propagates. This chapter presents a mathematical approach for determining the Jacobian matrix of wavefront aberration with respect to system variables by using OPL. Wavefront aberration is one of the important optical quantities. Thus, determining the Jacobian matrix of wavefront aberration is fundamental when one designs an optical system for fine tuning the performance of an existing system. The chapter not only resolves the error inherent in finite difference methods as a result of the denominator being far smaller than the numerator, but also avoids the requirement of multiple raytracing operations; hence it yields a substantial improvement in computational accuracy. This developed approach is important to an optical system designed by optimization methods when a merit function is formulated in terms of the wavefront aberration.

## 7.1 The Jacobian Matrix of $OPL_i$ Between (i−1)th and ith Boundary Surfaces

As illustrated in Fig. 7.1, in a medium of constant refractive index $\xi_{i-1}$ the OPL between points $\bar{P}_{i-1}$ and $\bar{P}_i$ (denoted as $OPL(\bar{P}_{i-1}, \bar{P}_i) = OPL_i = \xi_{i-1}\lambda_i$) is defined as the product of the geometric length $\lambda_i$ between these two points and $\xi_{i-1}$. It has been noted from Figs. 5.1, 5.2, 5.3, 5.4 and 5.5 that geometrical length $\lambda_i$ between points $\bar{P}_{i-1}$ and $\bar{P}_i$ is a function of not only the incoming ray $\bar{R}_{i-1}$, but also the boundary variable vector $\bar{X}_i$. Therefore, this section formulates the Jacobian matrix of $OPL_i = \xi_{i-1}\lambda_i$ with respect to the incoming ray $\bar{R}_{i-1}$ and boundary variable vector $\bar{X}_i$ separately.

PD Lin, *New Computation Methods for Geometrical Optics*,
Springer Series in Optical Sciences 178, DOI: 10.1007/978-981-4451-79-6_7,

**Fig. 7.1** The $OPL_i$ between points $\bar{P}_{i-1}$ and $\bar{P}_i$ is defined as the product of the geometric length $\lambda_i$ between these two points and $\xi_{i-1}$

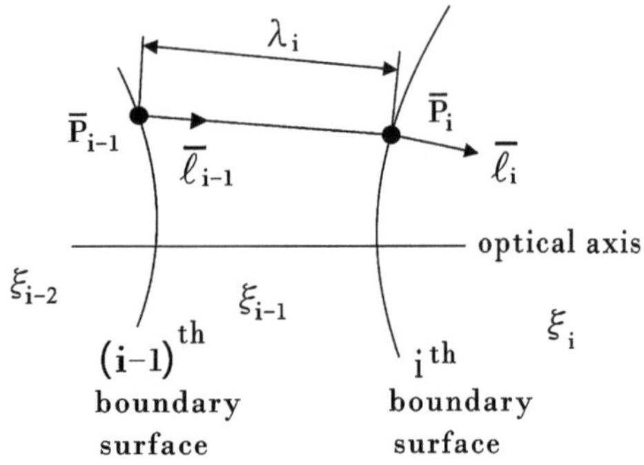


### 7.1.1 The Jacobian Matrix of $OPL_i$ with Respect to its Incoming Ray $\bar{R}_{i-1}$

Equations (2.15) and (2.38) respectively show that the $\lambda_i$ of a spherical and flat boundary surface varies as a function of the incoming ray $\bar{R}_{i-1}$. Therefore, the $OPL_i = \xi_{i-1}\lambda_i$ is influenced by the incoming ray $\bar{R}_{i-1}$. The Jacobian matrix of $OPL_i$ with respect to the incoming ray $\bar{R}_{i-1}$ (i.e. $\partial OPL_i/\partial \bar{R}_{i-1}$) can be obtained directly by differentiating $OPL_i = \xi_{i-1}\lambda_i$ with respect to its incoming ray $\bar{R}_{i-1}$,

$$\frac{\partial OPL_i}{\partial \bar{R}_{i-1}} = \xi_{i-1}\frac{\partial \lambda_i}{\partial \bar{R}_{i-1}}. \tag{7.1}$$

$\partial\lambda_i/\partial\bar{R}_{i-1}$ of Eq. (7.1) for flat and spherical boundary surface is respectively given by Eqs. (5.4) and (5.19). Now the change in $OPL_i$, $\Delta OPL_i$, induced by the change in the incoming ray, $\Delta\bar{R}_{i-1}$, can be computed by matrix product of the Jacobian matrix of $OPL_i$ and $\Delta\bar{R}_{i-1}$, i.e.

$$\Delta OPL_i = \frac{\partial OPL_i}{\partial \bar{R}_{i-1}}\Delta\bar{R}_{i-1}. \tag{7.2}$$

We will investigate the wavefront shape, irradiance, and caustic surface along the ray path in Chap. 8 for a given optical system by using $\partial OPL_i/\partial\bar{X}_0$, the Jacobian of $OPL_i$ with respect to $\bar{X}_0$. This can be obtained by substituting $\Delta\bar{R}_{i-1} = \bar{M}_{i-1}\bar{M}_{i-2}\ldots\bar{M}_2\bar{M}_1\bar{S}_0\Delta\bar{X}_0$ (obtained from Eq. (5.33) by setting i to be i−1) into Eq. (7.2) to give

$$\Delta OPL_i = \frac{\partial OPL_i}{\partial \bar{R}_{i-1}}\bar{M}_{i-1}\bar{M}_{i-2}\ldots\bar{M}_2\bar{M}_1\bar{S}_0\Delta\bar{X}_0. \tag{7.3}$$

The expression of $\partial OPL_i/\partial\bar{X}_0$ can be obtained as follows if $\Delta OPL_i$ is defined as $\Delta OPL_i = \partial OPL_i/\partial\bar{X}_0\Delta\bar{X}_0$,

$$\frac{\partial \mathrm{OPL}_i}{\partial \bar{X}_0} = \frac{\partial \mathrm{OPL}_i}{\partial \bar{R}_{i-1}} \bar{\mathrm{M}}_{i-1}\bar{\mathrm{M}}_{i-2}\ldots\bar{\mathrm{M}}_2\bar{\mathrm{M}}_1\bar{\mathrm{S}}_0. \tag{7.4}$$

One can determine the Jacobian matrix $\partial \mathrm{OPL}_i/\partial \bar{X}_0$ for a given optical system to investigate the variations of wavefront when any variable in variable vector $\bar{\mathrm{X}}_0$ of source ray is changing.

### *7.1.2 The Jacobian Matrix of OPL$_i$ with Respect to Boundary Variable Vector* $\bar{\mathbf{X}}_\mathbf{i}$

Changes of boundary variables in $\bar{\mathrm{X}}_i$ may also cause changes of OPL$_i$. The Jacobian matrix of OPL$_i$ with respect to boundary variables $\bar{\mathrm{X}}_i$ can be obtained by taking the first-order differentiation of $\mathrm{OPL}_i = \xi_{i-1}\lambda_i$ by

$$\frac{\partial \mathrm{OPL}_i}{\partial \bar{\mathrm{X}}_i} = \lambda_i \frac{\partial \xi_{i-1}}{\partial \bar{\mathrm{X}}_i} + \xi_{i-1}\frac{\partial \lambda_i}{\partial \bar{\mathrm{X}}_i} = \lambda_i\left[\bar{0}_{1\times 6} \quad 1 \quad \bar{0}_{1\times 2}\right] + \xi_{i-1}\frac{\partial \lambda_i}{\partial \bar{\mathrm{X}}_i}, \tag{7.5}$$

where $\partial\lambda_i/\partial\bar{X}_i$ is given in Eqs. (5.36) and (5.48) for flat and spherical boundary surfaces, respectively. From Eq. (7.5) we can determine the change in OPL$_i$ induced by any change of boundary variables in X$_i$ by

$$\Delta \mathrm{OPL}_i = \frac{\partial \mathrm{OPL}_i}{\partial \bar{\mathrm{X}}_i}\Delta\bar{\mathrm{X}}_i. \tag{7.6}$$

The sum of Eq. (7.2) and Eq. (7.6) gives the total change in OPL$_i$ caused by changes in the incoming ray $\bar{R}_{i-1}$ and boundary variable vector $\bar{\mathrm{X}}_i$ (see Fig. 7.2).

$$\Delta \mathrm{OPL}_i = \frac{\partial \mathrm{OPL}_i}{\partial \bar{\mathrm{R}}_{i-1}}\Delta\bar{\mathrm{R}}_{i-1} + \frac{\partial \mathrm{OPL}_i}{\partial \bar{\mathrm{X}}_i}\Delta\bar{\mathrm{X}}_i. \tag{7.7}$$

Note that, as stated in Sect. 5.7, not all variables in $\bar{\mathrm{X}}_i$ (i = 0 to i = n) are independent variables for an optical system, since an element may contain multiply boundary surfaces and a boundary variable may repeatedly appear in different $\bar{\mathrm{X}}_i$. Therefore, we can transfer $\Delta\bar{\mathrm{X}}_i$ in Eq. (7.7) into $\Delta\bar{\mathrm{X}}_{sys}$, which contains the

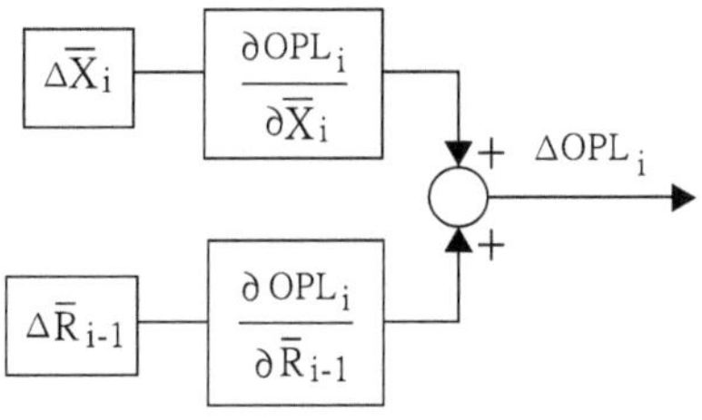


**Fig. 7.2** The $\Delta$OPL$_i$ is the sum of the changes caused by $\Delta\bar{\mathrm{R}}_{i-1}$ and $\Delta\bar{\mathrm{X}}_i$

variables of a system, by $\Delta\bar{X}_i = \partial\bar{X}_i/\partial\bar{X}_{sys}\Delta\bar{X}_{sys}$, where $\partial\bar{X}_i/\partial\bar{X}_{sys}$ is computed by Eq. (5.69), to give

$$\Delta OPL_i = \frac{\partial OPL_i}{\partial\bar{R}_{i-1}}\Delta\bar{R}_{i-1} + \frac{\partial OPL_i}{\partial\bar{X}_i}\frac{\partial\bar{X}_i}{\partial\bar{X}_{sys}}\Delta\bar{X}_{sys}. \tag{7.8}$$

## 7.2 The Jacobian Matrix of OPL between any Two Incidence Points

It is convenient to use $OPL(\bar{P}_g, \bar{P}_h)$, where g and h are two positive integers and $0 \le g < h \le n$, to represent the OPL between two incidence points $\bar{P}_g$ and $\bar{P}_h$ of a ray at the gth and hth boundary surfaces, respectively. This section constructs the Jacobian matrix of $OPL(\bar{P}_g, \bar{P}_h)$ with respect to system variables $\bar{X}_{sys}$ of an optical system. $OPL(\bar{P}_g, \bar{P}_h)$ is obtained by summing the $OPL_i$ associated with each of its boundary surfaces, i.e.

$$OPL(\bar{P}_g, \bar{P}_h) = \sum_{i=g+1}^{i=h} OPL_i. \tag{7.9}$$

The total change in $OPL(\bar{P}_g, \bar{P}_h)$ is given by

$$\Delta OPL(\bar{P}_g, \bar{P}_h) = \sum_{i=g+1}^{i=h} \Delta OPL_i. \tag{7.10}$$

After substituting Eq. (7.7) into Eq. (7.10), one has the following expression for $\Delta OPL(\bar{P}_g, \bar{P}_h)$,

$$\Delta OPL(\bar{P}_g, \bar{P}_h) = \sum_{i=g+1}^{i=h} \frac{\partial OPL_i}{\partial\bar{R}_{i-1}}\Delta\bar{R}_{i-1} + \sum_{i=g+1}^{i=h} \frac{\partial OPL_i}{\partial\bar{X}_i}\Delta\bar{X}_i. \tag{7.11}$$

The second term of Eq. (7.11) gives the effect of any changes of boundary variables in $\bar{X}_i$ ($i = g + 1$ to $i = h$) on $\Delta OPL(\bar{P}_g, \bar{P}_h)$. The changes of incoming rays $\bar{R}_{i-1}$ of Eq. (7.11) is given by Eq. (5.63) by setting $g = i - 1$. Now Eq. (7.11) can be re-written as (see Fig. 7.3):

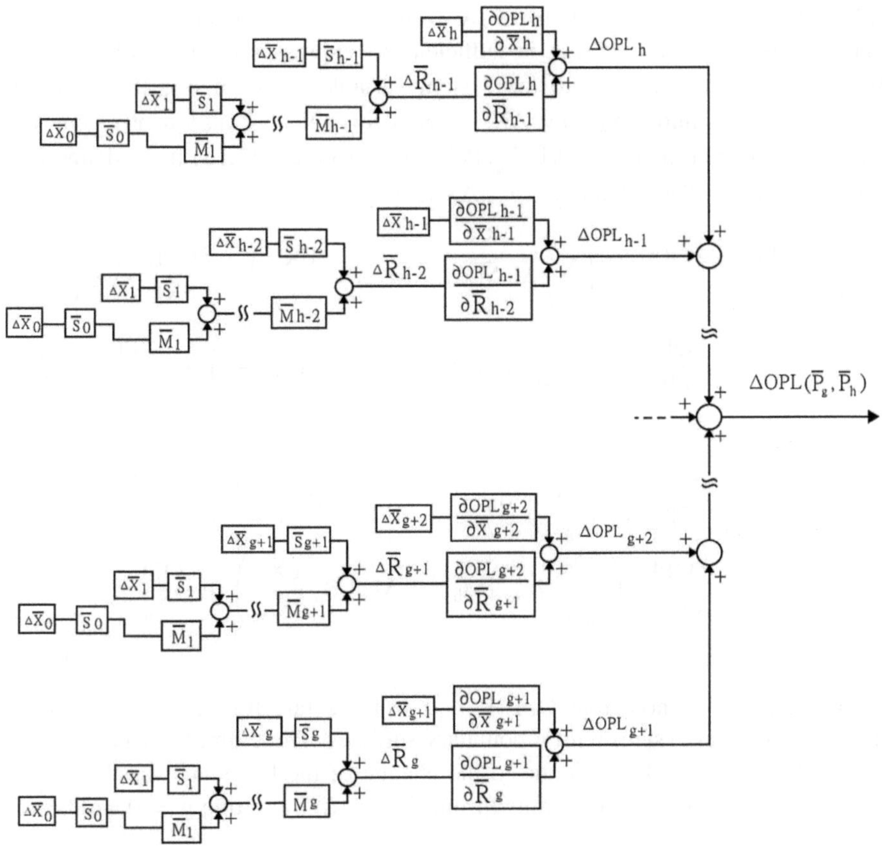

**Fig. 7.3** Determination of $\Delta OPL(\bar{P}_g, \bar{P}_h)$ when a ray travels through an optical system, where $OPL(\bar{P}_g, \bar{P}_h)$ is the optical path length from point $\bar{P}_g$ to point $\bar{P}_h$

$$
\begin{aligned}
\Delta OPL(\bar{P}_g, \bar{P}_h) &= \sum_{i=g+1}^{i=h} \frac{\partial OPL_i}{\partial \bar{R}_{i-1}} \left( \sum_{u=0}^{u=i-1} \bar{M}_{i-1}\bar{M}_{i-2}\ldots\bar{M}_{u+1}\bar{S}_u \Delta\bar{X}_u \right) + \sum_{i=g+1}^{i=h} \frac{\partial OPL_i}{\partial \bar{X}_i} \Delta\bar{X}_i \\
&= \frac{\partial OPL_{g+1}}{\partial \bar{R}_g} \left( \sum_{u=0}^{u=g} \bar{M}_g\bar{M}_{g-1}\ldots\bar{M}_{u+1}\bar{S}_u \Delta\bar{X}_u \right) + \frac{\partial OPL_{g+1}}{\partial \bar{X}_{g+1}} \Delta\bar{X}_{g+1} \\
&+ \frac{\partial OPL_{g+2}}{\partial \bar{R}_{g+1}} \left( \sum_{u=0}^{u=g+1} \bar{M}_{g+1}\bar{M}_{g+2}\ldots\bar{M}_{u+1}\bar{S}_u \Delta\bar{X}_u \right) + \frac{\partial OPL_{g+2}}{\partial \bar{X}_{g+2}} \Delta\bar{X}_{g+2} \\
&+ \ldots \\
&+ \frac{\partial OPL_{h-1}}{\partial \bar{R}_{h-2}} \left( \sum_{u=0}^{u=h-2} \bar{M}_{h-2}\bar{M}_{h-3}\ldots\bar{M}_{u+1}\bar{S}_u \Delta\bar{X}_u \right) + \frac{\partial OPL_{h-1}}{\partial \bar{X}_{h-1}} \Delta\bar{X}_{h-1} \\
&+ \frac{\partial OPL_h}{\partial \bar{R}_{h-1}} \left( \sum_{u=0}^{u=h-1} \bar{M}_{h-1}\bar{M}_{h-2}\ldots\bar{M}_{u+1}\bar{S}_u \Delta\bar{X}_u \right) + \frac{\partial OPL_h}{\partial \bar{X}_h} \Delta\bar{X}_h.
\end{aligned}
\tag{7.12}
$$

Note that some terms are deliberately excluded in Eq. (7.12) for simplicity. The final expression in Eq. (7.12) is sufficient to express the total OPL change, $\Delta OPL(\bar{P}_g, \bar{P}_h)$, with respect to the system variable vector if one transfers $\Delta\bar{X}_i$ ($i = 0$ to $i = h$) into $\Delta\bar{X}_{sys}$ by $\Delta\bar{X}_i = \partial\bar{X}_i/\partial\bar{X}_{sys}\Delta\bar{X}_{sys}$. Consequently, one can obtain the Jacobian matrix $\partial OPL(\bar{P}_g, \bar{P}_h)/\partial\bar{X}_{sys}$ from Eq. (7.12) if we define it as $\Delta OPL(\bar{P}_g, \bar{P}_h) = \partial OPL(\bar{P}_g, \bar{P}_h)/\partial\bar{X}_{sys}\Delta\bar{X}_{sys}$ to give

$$\begin{aligned}\frac{\partial OPL(\bar{P}_g,\bar{P}_h)}{\partial\bar{X}_{sys}} &= \frac{\partial OPL_{g+1}}{\partial\bar{R}_g}\left(\sum_{u=0}^{u=g}\bar{M}_g\bar{M}_{g-1}\ldots\bar{M}_{u+1}\bar{S}_u\frac{\partial\bar{X}_u}{\partial\bar{X}_{sys}}\right)+\frac{\partial OPL_{g+1}}{\partial\bar{X}_{g+1}}\frac{\partial\bar{X}_{g+1}}{\partial\bar{X}_{sys}}\\ &+\frac{\partial OPL_{g+2}}{\partial\bar{R}_{g+1}}\left(\sum_{u=0}^{u=g+1}\bar{M}_{g+1}\bar{M}_{g+2}\ldots\bar{M}_{u+1}\bar{S}_u\frac{\partial\bar{X}_u}{\partial\bar{X}_{sys}}\right)+\frac{\partial OPL_{g+2}}{\partial\bar{X}_{g+2}}\frac{\partial\bar{X}_{g+2}}{\partial\bar{X}_{sys}}\\ &+\ldots\\ &+\frac{\partial OPL_{h-1}}{\partial\bar{R}_{h-2}}\left(\sum_{u=0}^{u=h-2}\bar{M}_{h-2}\bar{M}_{h-3}\ldots\bar{M}_{u+1}\bar{S}_u\frac{\partial\bar{X}_u}{\partial\bar{X}_{sys}}\right)+\frac{\partial OPL_{h-1}}{\partial\bar{X}_{h-1}}\frac{\partial\bar{X}_{h-1}}{\partial\bar{X}_{sys}}\\ &+\frac{\partial OPL_{h}}{\partial\bar{R}_{h-1}}\left(\sum_{u=0}^{u=h-1}\bar{M}_{h-1}\bar{M}_{h-2}\ldots\bar{M}_{u+1}\bar{S}_u\frac{\partial\bar{X}_u}{\partial\bar{X}_{sys}}\right)+\frac{\partial OPL_{h}}{\partial\bar{X}_{h}}\frac{\partial\bar{X}_{h}}{\partial\bar{X}_{sys}}\end{aligned} \tag{7.13}$$

One also has to note that $OPL(\bar{P}_0, \bar{P}_n)$ is the total OPL traversed by a ray through an optical system with n boundary surfaces. Therefore, the change of the total OPL (i.e $\Delta OPL(\bar{P}_0, \bar{P}_n)$) due to the change of the boundary variable vector $\Delta\bar{X}_i$ ($i = 0$ to $i = n$) can be obtained from Eq. (7.12) by setting $g = 0$ and $h = n$. That is

$$\begin{aligned}\Delta OPL(\bar{P}_0,\bar{P}_n) &= \frac{\partial OPL_1}{\partial\bar{R}_0}(\bar{S}_0\Delta\bar{X}_0)+\frac{\partial OPL_1}{\partial\bar{X}_1}\Delta\bar{X}_1+\frac{\partial OPL_2}{\partial\bar{R}_1}(\bar{M}_1\bar{S}_0\Delta\bar{X}_0+\bar{S}_1\Delta\bar{X}_1)\\ &+\frac{\partial OPL_2}{\partial\bar{X}_2}\Delta\bar{X}_2+\ldots+\frac{\partial OPL_{n-1}}{\partial\bar{R}_{n-2}}(\bar{M}_{n-2}\ldots\bar{M}_1\bar{S}_0\Delta\bar{X}_0\\ &+\bar{M}_{n-2}\ldots\bar{M}_2\bar{S}_1\Delta\bar{X}_1+\ldots+\bar{M}_{n-2}\bar{S}_{n-3}\Delta\bar{X}_{n-3}+\bar{S}_{n-2}\Delta\bar{X}_{n-2})\\ &+\frac{\partial OPL_{n-1}}{\partial\bar{X}_{n-1}}\Delta\bar{X}_{n-1}+\frac{\partial OPL_n}{\partial\bar{R}_{n-1}}(\bar{M}_{n-1}\ldots\bar{M}_1\bar{S}_0\Delta\bar{X}_0\\ &+\bar{M}_{n-1}\bar{M}_{n-2}\ldots\bar{M}_2\bar{S}_1\Delta\bar{X}_1+\ldots+\bar{M}_{n-1}\bar{S}_{n-2}\Delta\bar{X}_{n-2}\\ &+\bar{S}_{n-1}\Delta\bar{X}_{n-1})+\frac{\partial OPL_n}{\partial\bar{X}_n}\Delta\bar{X}_n.\end{aligned} \tag{7.14}$$

The Jacobian matrix $\partial OPL(\bar{P}_0, \bar{P}_n)/\partial\bar{X}_{sys}$ can be obtained from Eq. (7.14), as

$$\begin{aligned}
\frac{\partial \mathrm{OPL}(\bar{\mathrm{P}}_0,\bar{\mathrm{P}}_n)}{\partial \bar{\mathrm{X}}_{\mathrm{sys}}} &= \frac{\partial \mathrm{OPL}_1}{\partial \bar{\mathrm{R}}_0}\left(\bar{S}_0\frac{\partial \bar{\mathrm{X}}_0}{\partial \bar{\mathrm{X}}_{\mathrm{sys}}}\right) + \frac{\partial \mathrm{OPL}_1}{\partial \bar{\mathrm{X}}_1}\frac{\partial \bar{\mathrm{X}}_1}{\partial \bar{\mathrm{X}}_{\mathrm{sys}}} + \frac{\partial \mathrm{OPL}_2}{\partial \bar{\mathrm{R}}_1}\left(\bar{\mathrm{M}}_1\bar{\mathrm{S}}_0\frac{\partial \bar{\mathrm{X}}_0}{\partial \bar{\mathrm{X}}_{\mathrm{sys}}} + \bar{S}_1\frac{\partial \bar{\mathrm{X}}_1}{\partial \bar{\mathrm{X}}_{\mathrm{sys}}}\right)\\
&+\frac{\partial \mathrm{OPL}_2}{\partial \bar{\mathrm{X}}_2}\frac{\partial \bar{\mathrm{X}}_2}{\partial \bar{\mathrm{X}}_{\mathrm{sys}}} + \ldots\\
&+\frac{\partial \mathrm{OPL}_{\mathrm{n}-1}}{\partial \bar{\mathrm{R}}_{\mathrm{n}-2}}\left(\bar{\mathrm{M}}_{\mathrm{n}-2}\ldots\bar{\mathrm{M}}_1\bar{\mathrm{S}}_0\frac{\partial \bar{\mathrm{X}}_0}{\partial \bar{\mathrm{X}}_{\mathrm{sys}}} + \bar{\mathrm{M}}_{\mathrm{n}-2}\ldots\bar{\mathrm{M}}_2\bar{\mathrm{S}}_1\frac{\partial \bar{\mathrm{X}}_1}{\partial \bar{\mathrm{X}}_{\mathrm{sys}}}\right.\\
&\left.+\ldots+\bar{\mathrm{M}}_{\mathrm{n}-2}\bar{\mathrm{S}}_{\mathrm{n}-3}\frac{\partial \bar{\mathrm{X}}_{\mathrm{n}-3}}{\partial \bar{\mathrm{X}}_{\mathrm{sys}}} + \bar{\mathrm{S}}_{\mathrm{n}-2}\frac{\partial \bar{\mathrm{X}}_{\mathrm{n}-2}}{\partial \bar{\mathrm{X}}_{\mathrm{sys}}}\right) + \frac{\partial \mathrm{OPL}_{\mathrm{n}-1}}{\partial \bar{\mathrm{X}}_{\mathrm{n}-1}}\frac{\partial \bar{\mathrm{X}}_{\mathrm{n}-1}}{\partial \bar{\mathrm{X}}_{\mathrm{sys}}}\\
&+\frac{\partial \mathrm{OPL}_{\mathrm{n}}}{\partial \bar{\mathrm{R}}_{\mathrm{n}-1}}\left(\bar{\mathrm{M}}_{\mathrm{n}-1}\ldots\bar{\mathrm{M}}_1\bar{\mathrm{S}}_0\frac{\partial \bar{\mathrm{X}}_0}{\partial \bar{\mathrm{X}}_{\mathrm{sys}}} + \bar{\mathrm{M}}_{\mathrm{n}-1}\ldots\bar{\mathrm{M}}_2\bar{\mathrm{S}}_1\frac{\partial \bar{\mathrm{X}}_1}{\partial \bar{\mathrm{X}}_{\mathrm{sys}}}\right.\\
&\left.+\ldots+\bar{\mathrm{M}}_{\mathrm{n}-1}\bar{\mathrm{S}}_{\mathrm{n}-2}\frac{\partial \bar{\mathrm{X}}_{\mathrm{n}-2}}{\partial \bar{\mathrm{X}}_{\mathrm{sys}}} + \bar{\mathrm{S}}_{\mathrm{n}-1}\frac{\partial \bar{\mathrm{X}}_{\mathrm{n}-1}}{\partial \bar{X}_{\mathrm{sys}}}\right) + \frac{\partial OPL_n}{\partial \bar{\mathrm{X}}_n}\frac{\partial \bar{\mathrm{X}}_{\mathrm{n}}}{\partial \bar{\mathrm{X}}_{\mathrm{sys}}}.
\end{aligned} \tag{7.15}$$

Sometimes, we have to compute the change of $\mathrm{OPL}(\bar{\mathrm{P}}_0,\bar{\mathrm{P}}_{\mathrm{n}})$ after a ray traveled through an existing optical system (this means that all $\Delta\bar{\mathrm{X}}_{\mathrm{i}}$, $\mathrm{i}=1$ to $\mathrm{i}=\mathrm{n}$, are zero matrices) for analyzing the wavefront aberration. This can be obtained from Eq. (7.14) by setting $\Delta\bar{\mathrm{X}}_{\mathrm{i}}$, $\mathrm{i}=1$ to $\mathrm{i}=\mathrm{n}$, as zero matrix, to give

$$\begin{aligned}
\Delta\mathrm{OPL}(\bar{\mathrm{P}}_0,\bar{\mathrm{P}}_{\mathrm{n}}) &= \left(\frac{\partial \mathrm{OPL}_1}{\partial \bar{\mathrm{R}}_0} + \frac{\partial \mathrm{OPL}_2}{\partial \bar{\mathrm{R}}_1}\bar{\mathrm{M}}_1 + \ldots + \frac{\partial \mathrm{OPL}_{\mathrm{n}-1}}{\partial \bar{\mathrm{R}}_{\mathrm{n}-2}}\bar{\mathrm{M}}_{\mathrm{n}-2}\ldots\bar{\mathrm{M}}_1\right.\\
&\left.+\frac{\partial \mathrm{OPL}_{\mathrm{n}}}{\partial \bar{\mathrm{R}}_{\mathrm{n}-1}}\bar{\mathrm{M}}_{\mathrm{n}-1}\ldots\bar{\mathrm{M}}_1\right)\bar{\mathrm{S}}_0\Delta\bar{\mathrm{X}}_0.
\end{aligned} \tag{7.16}$$

Therefore, the Jacobian matrix of $\mathrm{OPL}(\bar{\mathrm{P}}_0,\bar{\mathrm{P}}_n)$ with respect to $\bar{\mathrm{X}}_0$ is obtained from Eq. (7.16), as

$$\begin{aligned}
\frac{\partial \mathrm{OPL}(\bar{\mathrm{P}}_0,\bar{\mathrm{P}}_{\mathrm{n}})}{\partial \bar{\mathrm{X}}_0} &= \left(\frac{\partial \mathrm{OPL}_1}{\partial \bar{\mathrm{R}}_0} + \frac{\partial \mathrm{OPL}_2}{\partial \bar{\mathrm{R}}_1}\bar{\mathrm{M}}_1 + \ldots + \frac{\partial \mathrm{OPL}_{\mathrm{n}-1}}{\partial \bar{\mathrm{R}}_{\mathrm{n}-2}}\bar{\mathrm{M}}_{\mathrm{n}-2}\ldots\bar{\mathrm{M}}_1\right.\\
&\left.+\frac{\partial \mathrm{OPL}_{\mathrm{n}}}{\partial \bar{\mathrm{R}}_{\mathrm{n}-1}}\bar{\mathrm{M}}_{\mathrm{n}-1}\ldots\bar{\mathrm{M}}_1\right)\bar{\mathrm{S}}_0\frac{\partial \bar{\mathrm{X}}_0}{\partial \bar{\mathrm{X}}_{\mathrm{sys}}}.
\end{aligned} \tag{7.17}$$

*Example 7.1* Referring to the system shown in Fig. 2.1, the expression of $\Delta\mathrm{OPL}_{\mathrm{e}1}$ due to $\Delta\bar{\mathrm{X}}_{\mathrm{sys}}$,, where $\mathrm{OPL}_{\mathrm{e}1} = \mathrm{OPL}(\bar{\mathrm{P}}_1,\bar{\mathrm{P}}_2)$ is the OPL when ray travelling in the first element, is given by (see Fig. 7.4)

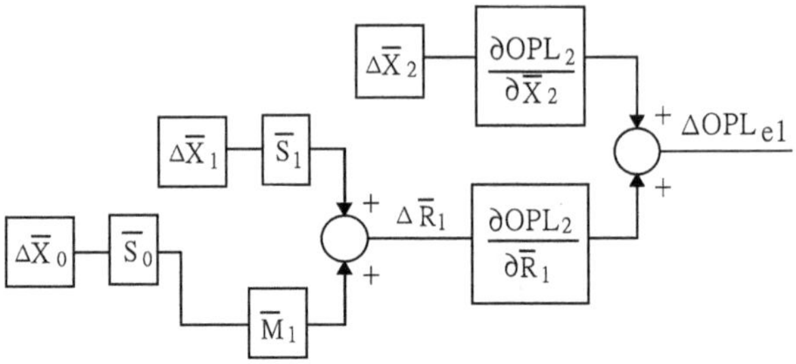


**Fig. 7.4** Determination of $\Delta OPL_{e1}$ for the system in Fig. 2.1, where $OPL_{e1} = OPL(\bar{P}_1, \bar{P}_2)$ is the OPL measured from incidence point $\bar{P}_1$ to the exit point $\bar{P}_2$ of the first element

$$\Delta OPL_{e1} = \Delta OPL_2 = \partial OPL_{e1}/\partial\bar{X}_{sys}\Delta\bar{X}_{sys} = \frac{\partial OPL_2}{\partial\bar{R}_1}(\bar{M}_1\bar{S}_0\Delta\bar{X}_0 + \bar{S}_1\Delta\bar{X}_1) + \frac{\partial OPL_2}{\partial\bar{X}_2}\Delta\bar{X}_2$$
$$= \frac{\partial OPL_2}{\partial\bar{R}_1}\left(\bar{M}_1\bar{S}_0\frac{\partial\bar{X}_0}{\partial\bar{X}_{sys}}\Delta\bar{X}_{sys} + \bar{S}_1\frac{\partial\bar{X}_1}{\partial\bar{X}_{sys}}\Delta\bar{X}_{sys}\right) + \frac{\partial OPL_2}{\partial\bar{X}_2}\frac{\partial\bar{X}_2}{\partial\bar{X}_{sys}}\Delta\bar{X}_{sys}. \tag{7.18a}$$

The Jacobian matrix $\partial OPL_{e1}/\partial\bar{X}_{sys}$ can be obtained from Eq. (7.18a) to give

$$\frac{\partial OPL_{e1}}{\partial\bar{X}_{sys}} = \frac{\partial OPL_2}{\partial\bar{R}_1}\left(\bar{M}_1\bar{S}_0\frac{\partial\bar{X}_0}{\partial\bar{X}_{sys}} + \bar{S}_1\frac{\partial\bar{X}_1}{\partial\bar{X}_{sys}}\right) + \frac{\partial OPL_2}{\partial\bar{X}_2}\frac{\partial\bar{X}_2}{\partial\bar{X}_{sys}}. \tag{7.18b}$$

When Eq. (7.18a) is evaluated by the chief ray originated from $\bar{P}_0 = [0 \quad -507 \quad 170 \quad 1]^T$, its numerical result is given by

$$\Delta OPL_{e1} = -0.098205\Delta P_{0y} - 0.305797\Delta P_{0z} - 176.286003\Delta\beta_0 + 0.098205\Delta v_1$$
$$+ 0.305797\Delta t_{e1z} - 7.135859\Delta\omega_{e1x} + 0.644910\Delta\xi_{air} + 15.268118\Delta\xi_{e1}$$
$$+ 1.731846\Delta q_{e1} + 0.011776\Delta R_1 + 0.002387\Delta R_2. \tag{7.18c}$$

*Example 7.2* Referring the system shown in Fig. 2.1, the change in $OPL(\bar{P}_8, \bar{P}_{10})$ is given by (see Fig. 7.5):

$$\Delta OPL(\bar{P}_8, \bar{P}_{10}) = \Delta OPL_9 + \Delta OPL_{10}$$
$$= \frac{\partial OPL_9}{\partial\bar{R}_8}\Delta\bar{R}_8 + \frac{\partial OPL_9}{\partial\bar{X}_9}\Delta\bar{X}_9 + \frac{\partial OPL_{10}}{\partial\bar{R}_9}\Delta\bar{R}_9 + \frac{\partial OPL_{10}}{\partial\bar{X}_{10}}\Delta\bar{X}_{10}$$
$$= \frac{\partial OPL_9}{\partial\bar{R}_8}\Delta\bar{R}_8 + \frac{\partial OPL_9}{\partial\bar{X}_9}\Delta\bar{X}_9 + \frac{\partial OPL_{10}}{\partial\bar{R}_9}(\bar{M}_9\Delta\bar{R}_8 + \bar{S}_9\Delta\bar{X}_9)$$
$$+ \frac{\partial OPL_{10}}{\partial\bar{X}_{10}}\Delta\bar{X}_{10}. \tag{7.19a}$$

Note that Eq. (7.19a) is derived using Eq. (5.61). The term $\Delta\bar{R}_8$ of Eq. (7.19a) is given by Eq. (5.63) by setting g = 8, to give

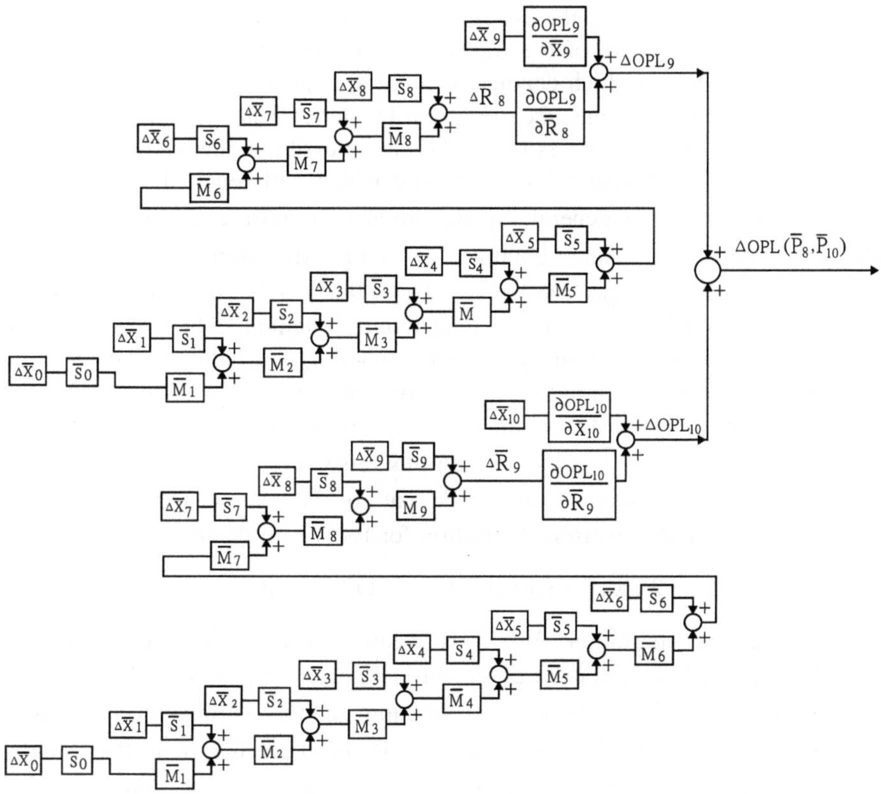


**Fig. 7.5** Determination of $\Delta\mathrm{OPL}(\bar{\mathrm{P}}_8, \bar{\mathrm{P}}_{10})$ for the system shown in Fig. 2.1

$$\begin{aligned}\Delta\bar{\mathrm{R}}_8 = {} & \bar{\mathrm{M}}_8\bar{\mathrm{M}}_7\bar{\mathrm{M}}_6\bar{\mathrm{M}}_5\bar{\mathrm{M}}_4\bar{\mathrm{M}}_3\bar{\mathrm{M}}_2\bar{\mathrm{M}}_1\bar{\mathrm{S}}_0\Delta\bar{\mathrm{X}}_0 + \bar{\mathrm{M}}_8\bar{\mathrm{M}}_7\bar{\mathrm{M}}_6\bar{\mathrm{M}}_5\bar{\mathrm{M}}_4\bar{\mathrm{M}}_3\bar{\mathrm{M}}_2\bar{\mathrm{S}}_1\Delta\bar{\mathrm{X}}_1 \\ & + \bar{\mathrm{M}}_8\bar{\mathrm{M}}_7\bar{\mathrm{M}}_6\bar{\mathrm{M}}_5\bar{\mathrm{M}}_4\bar{\mathrm{M}}_3\bar{\mathrm{S}}_2\Delta\bar{\mathrm{X}}_2 + \bar{\mathrm{M}}_8\bar{\mathrm{M}}_7\bar{\mathrm{M}}_6\bar{\mathrm{M}}_5\bar{\mathrm{M}}_4\bar{\mathrm{S}}_3\Delta\bar{\mathrm{X}}_3 + \bar{\mathrm{M}}_8\bar{\mathrm{M}}_7\bar{\mathrm{M}}_6\bar{\mathrm{M}}_5\bar{\mathrm{S}}_4\Delta\bar{\mathrm{X}}_4 \\ & + \bar{\mathrm{M}}_8\bar{\mathrm{M}}_7\bar{\mathrm{M}}_6\bar{\mathrm{S}}_5\Delta\bar{\mathrm{X}}_5 + \bar{\mathrm{M}}_8\bar{\mathrm{M}}_7\bar{\mathrm{S}}_6\Delta\bar{\mathrm{X}}_6 + \bar{\mathrm{M}}_8\bar{\mathrm{S}}_7\Delta\bar{\mathrm{X}}_7 + \bar{\mathrm{S}}_8\Delta\bar{\mathrm{X}}_8.\end{aligned} \tag{7.19b}$$

The Jacobian matrix $\partial OPL(\bar{P}_8, \bar{P}_{10})/\partial \bar{X}_{sys}$ can be obtained from Eqs. (7.19a) and (7.19b) if $\Delta\bar{\mathrm{X}}_\mathrm{i} = \partial\bar{\mathrm{X}}_\mathrm{i}/\partial\bar{\mathrm{X}}_\mathrm{sys}\Delta\bar{\mathrm{X}}_\mathrm{sys}$ (i = 0 to i = 10) is used.

## 7.3 Computation of Wavefront Aberration

In geometrical optics, a wavefront is defined by the locus of points in space having the same OPL from a common source point $\bar{\mathrm{P}}_0$. In paraxial optics, the emerging wavefront is essentially spherical and will form a perfect image at its center. However, the wavefront propagating through an actual optical system, is often no longer perfectly spherical by the time it exits the system. Therefore, it is important to determine the departure of the wavefront shape traveling along a ray from a perfect spherical form (which denoted as reference sphere $\bar{r}_{ref}$ in this book).

In wavefront computations, the wavefront $\bar{\Omega}$ and reference sphere $\bar{r}_{ref.}$ may not pass through the center of the exit pupil [1]. It is possible to define the wavefront aberration by using another reference sphere with a different radius. Figure 7.6 illustrates the path of a general ray $\bar{R}_0$ originating from a source point $\bar{P}_0 = [P_{0x} \quad P_{0y} \quad P_{0z} \quad 1]^T$ and having a unit directional vector $\bar{\ell}_0 = [\ell_{0x} \quad \ell_{0y} \quad \ell_{0z} \quad 0]^T = [C\beta_0 C(90^\circ + \alpha_0) \quad C\beta_0 S(90^\circ + \alpha_0) \quad S\beta_0 \quad 0]^T$ as it propagates through an axis-symmetrical optical system containing n boundary surfaces. Figure 7.6 also shows an arbitrary reference sphere $\bar{r}_{ref}$ of radius $R_{ref}$ centered at $\bar{P}_{n/chief}$ (i.e., the image point of the chief ray originating from $\bar{P}_0$) used to define the wavefront aberration. As shown, the chief ray originating from the source point $\bar{P}_0 = [P_{0x} \quad P_{0y} \quad P_{0z} \quad 1]^T$ is incident on this reference sphere at point $\bar{P}_{ref/chief}$. Let $\bar{P}_\Omega$ denote the intersection point of the path of the general source ray $\bar{R}_0$ and the wavefront $\bar{\Omega}$. The wavefront aberration for $\bar{R}_0$ is then determined as

$$W(\bar{X}_0) = \mathrm{OPL}(\bar{P}_0, \bar{P}_{ref}) - \mathrm{OPL}(\bar{P}_0, \bar{P}_\Omega). \tag{7.20}$$

The wavefront aberration $W(\bar{X}_0)$ is positive if the reference sphere $\bar{\mathbf{r}}_{ref}$ leads the aberrated wavefront $\bar{\Omega}$ as shown in Fig. 7.6. Equation (7.20) indicates that the wavefront aberration is a function of the variable vector $\bar{X}_0 = [P_{0x} \quad P_{0y} \quad P_{0z} \quad \alpha_0 \quad \beta_0]^T$ of the source ray $\bar{R}_0$ for a given system. Particularly, if we are discussing the wavefront aberration of a fixed source point (e.g., $\bar{P}_0$),

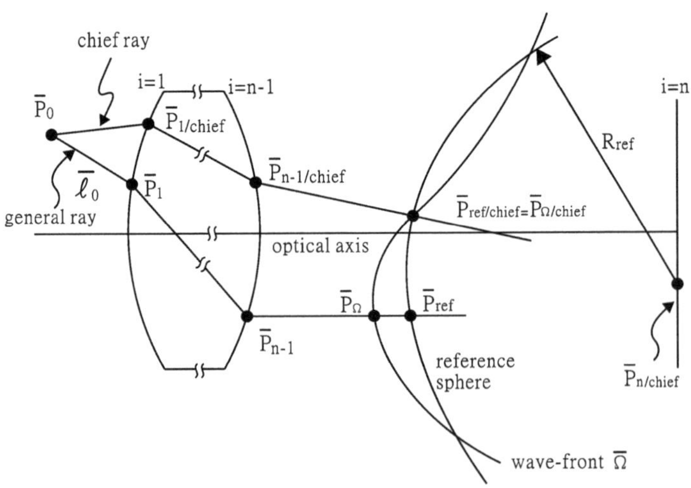


**Fig. 7.6** The determination of wavefront aberration of an axis-symmetrical optical system possessing n boundary surfaces

$W(\bar{X}_0)$ can be further simplified as a function of $\alpha_0$ and $\beta_0$ only. One can use $W(\alpha_0, \beta_0)$ to represent the wavefront aberration of the fixed source point $\bar{P}_0$ of a given system as:

$$W(\alpha_0, \beta_0) = OPL(\bar{P}_0, \bar{P}_{ref}) - OPL(\bar{P}_0, \bar{P}_\Omega). \tag{7.21}$$

Significantly, Eq. (7.21) expresses the wavefront aberration in terms of the two polar coordinates $\alpha_0$ and $\beta_0$ of the unit directional vector $\bar{\ell}_0$ of a general ray $\bar{R}_0$ for a given system. By contrast, most methods on wavefront aberration calculation take the entrance or exit pupil coordinates (e.g., $x_5$ and $z_5$ of Fig. 2.1) as the variables. Thus, the wavefront aberration function becomes a composite function, $W(x_5(\alpha_0, \beta_0), y_5(\alpha_0, \beta_0))$. A composite function is a function which operates in turn on other functions. The manipulation of such composite function is, more or less, awkward in computations. Consequently, Eq. (7.21) has two potential advantages: 1) the shape of the wavefront aberration of a ray following reflection/ or refraction can be numerically determined without the need to compute the corresponding coordinates of the entrance or exit pupil (note that these two pupils are virtual surfaces obtained from the images of the aperture); and 2) the local principal curvatures of the wavefront aberration can be investigated by differentiating Eq. (7.18) with respect to $\alpha_0$ and $\beta_0$ for a given system.

Equation (7.20) can be rewritten as

$$W(\bar{X}_0) = OPL(\bar{P}_0, \bar{P}_{ref}) - OPL\left(\bar{P}_0, \bar{P}_{ref/chief}\right), \tag{7.22}$$

since both points $\bar{P}_\Omega$ and $\bar{P}_{ref/chief}$ lie on the wavefront $\bar{\Omega}$; leading to $OPL(\bar{P}_0, \bar{P}_\Omega) = OPL\left(\bar{P}_0, \bar{P}_{ref/chief}\right)$. Equation (7.22) indicates that the wavefront aberration $W(\bar{X}_0)$ of a general source ray $\bar{R}_0$ is equal to the difference between $OPL(\bar{P}_0, \bar{P}_{ref})$ and $OPL\left(\bar{P}_0, \bar{P}_{ref/chief}\right)$, where $\bar{P}_{ref}$ and $\bar{P}_{ref/chief}$ are the incident points of the source ray $\bar{R}_0$ and the chief ray $\bar{R}_{0/chief}$, respectively, on the reference sphere $\bar{\mathbf{r}}_{ref}$.

It should be noted that the reference sphere $\bar{\mathbf{r}}_{ref}$ is centered at the imaging point $\bar{P}_{n/chief}$ of the chief ray originating from source point $\bar{P}_0$ (see Fig. 7.6). However, $\bar{P}_{n/chief}$ does not lie on the optical axis when $\bar{P}_0$ is an off-axis point. In other words, it is actually necessary to compute the wavefront aberration in a non-axially symmetrical system, even though the system is in fact an axis-symmetrical system. In Sect. 7.2, a methodology was presented for determining the Jacobain matrix of the OPL with respect to the system variable vector $\bar{X}_{sys}$ for non-axially symmetrical optical systems. As stated before, to analyze a system possessing n optical boundary surfaces, it is first necessary to label the surfaces sequentially from $i = 0$ to the last boundary surface with $i = n$ (Fig. 2.1). Figure 7.7 shows the non-axially symmetrical system obtained by including the reference sphere $\bar{\mathbf{r}}_{ref}$ as one of the spherical boundary surfaces when applying the approach presented in the Chap. 3 to the system shown in Fig. 7.6. One has to note that in applying this approach, $\bar{\mathbf{r}}_{ref}$ in Fig. 7.6 has to be labeled as the nth boundary surface. Thus, the image plane, which was originally labeled as $i = n$, has to be labeled as the $(n + 1)$th boundary

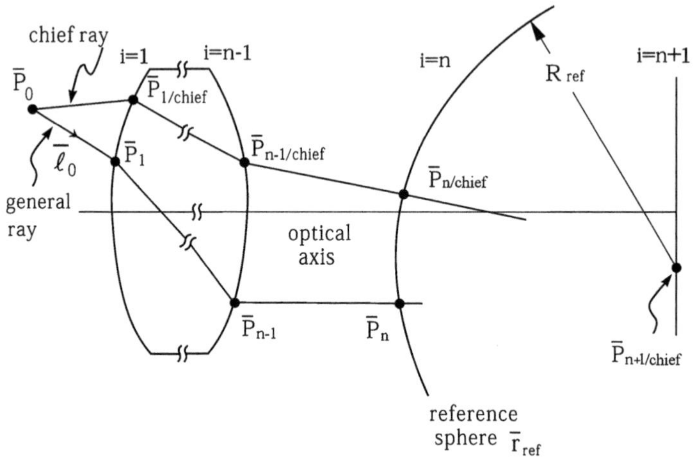


**Fig. 7.7** The system of Fig. 7.6 becomes a non-axially symmetrical system when the reference sphere $\bar{r}_{ref}$ is considered as a virtual boundary surface with radius $R_{ref}$ centered at the incidence point of the chief ray on the imaging plane

surface. Using the modified system shown in Fig. 7.7, the wavefront aberration given in Eq. (7.22) becomes

$$W(\bar{X}_0) = OPL(\bar{P}_0, \bar{P}_n) - OPL(\bar{P}_0, \bar{P}_{n/chief}). \tag{7.23}$$

The first and second terms of Eq. (7.23) can be written respectively as

$$OPL(\bar{P}_0, \bar{P}_n) = \sum_{i=1}^{i=n} OPL_i, \tag{7.24}$$

$$OPL(\bar{P}_0, \bar{P}_{n/chief}) = \sum_{i=1}^{i=n} OPL_{i/chief}. \tag{7.25}$$

The $OPL_i$ term in Eq. (7.21) and $OPL_{i/chief}$ term in Eq. (7.25) are the OPLs of a general ray $\bar{R}_0 = [\bar{P}_0 \quad \bar{\ell}_0]^T$ and the chief ray $\bar{R}_{0/chief} = [\bar{P}_0 \quad \bar{\ell}_{0/chief}]^T$, respectively, between the (i−1)th and ith boundary surfaces for a given source point $\bar{P}_0$. Substituting Eqs. (7.24) and (7.25) into Eq. (7.23), one obtains the following equation for the wavefront aberration:

$$W(\bar{X}_0) = \sum_{i=1}^{i=n} OPL_i - \sum_{i=1}^{i=n} OPL_{i/chief} = \sum_{i=1}^{i=n} (OPL_i - OPL_{i/chief}). \tag{7.26}$$

Equation (7.26) shows that the wavefront aberration of a system with multiple boundary surfaces is additive, i.e., the wavefront aberration of a ray for the entire system is equal to the sum of the wavefront aberrations at each of the boundary surfaces. (Note that this property does not hold true for ray aberrations.)

Using the modified system shown in Fig. 7.7, the Jacobain matrix of the wavefront aberration given in Eq. (7.26) becomes

$$\frac{\partial W(\bar{X}_0)}{\partial \bar{X}_{sys}}=\frac{\partial OPL(\bar{P}_0,\bar{P}_n)}{\partial \bar{X}_{sys}}-\frac{\partial OPL(\bar{P}_0,\bar{P}_{n/chief})}{\partial \bar{X}_{sys}}=\sum_{i=1}^{i=n}\frac{\partial OPL_i}{\partial \bar{X}_{sys}}-\sum_{i=1}^{i=n}\frac{\partial OPL_{i/chief}}{\partial \bar{X}_{sys}}. \tag{7.27}$$

Note that $\bar{P}_n$ and $\bar{P}_{n/chief}$ in Eq. (7.27) are the incident points of the general ray $\bar{R}_0$ and chief ray, respectively, on the reference sphere $\bar{\mathbf{r}}_{ref}$ (i.e., the nth boundary surface of the modified system in Fig. 7.7).

Figure 7.8 shows two rays originating from a source point $\bar{P}_0$, namely, a reference source ray $\bar{R}_0=\left[\bar{P}_0\ \bar{\ell}_0\right]^T$ with variable vector $\bar{X}_0=\left[P_{0x}\quad P_{0y}\quad P_{0z}\quad \alpha_0\quad \beta_0\right]^T$ and a neighboring source ray $\bar{R}'_0=\left[\bar{P}_0\quad \bar{\ell}'_0\right]^T$ with variable vector $\bar{X}'_0=\left[P_{0x}\quad P_{0y}\quad P_{0z}\quad \alpha'_0\quad \beta'_0\right]^T$. It is possible to estimate the wavefront aberration $W(\bar{X}'_0)=W(\bar{X}_0)+\Delta W(\bar{X}_0)$ of the ray $\bar{R}'_0$ by first-order Taylor series expansion if the wavefront aberration $W(\bar{X}_0)$ of the reference ray $\bar{R}_0$ is given for an existing optical system. The change of the wavefront aberration $\Delta W(\bar{X}_0)$ is given by:

$$\Delta W(\bar{X}_0)=\left(\sum_{i=1}^{i=n}\frac{\partial OPL_i}{\partial \bar{X}_0}\right)\Delta \bar{X}_0, \tag{7.28}$$

where $\Delta\bar{X}_0=\bar{X}'_0-\bar{X}_0=\left[0\quad 0\quad 0\quad \alpha'_0-\alpha_0\quad \beta'_0-\beta_0\right]^T$. The Jacobian matrix of Eq. (7.28) is evaluated by the reference ray $\bar{R}_0$. The wavefront aberration $W(\bar{X}'_0)$ of the neighboring ray $\bar{R}'_0$ is the sum of $W(\bar{X}_0)$ and $\Delta W(\bar{X}_0)$.

*Example 7.3* As shown by Eq. (7.22), in principle, it is possible to trace a general ray $\bar{R}_0=\left[\bar{P}_0\quad \bar{\ell}_0\right]^T$ and the chief ray $\bar{R}_{0/chief}=\left[\bar{P}_0\quad \bar{\ell}_{0/chief}\right]^T$, and then subtract

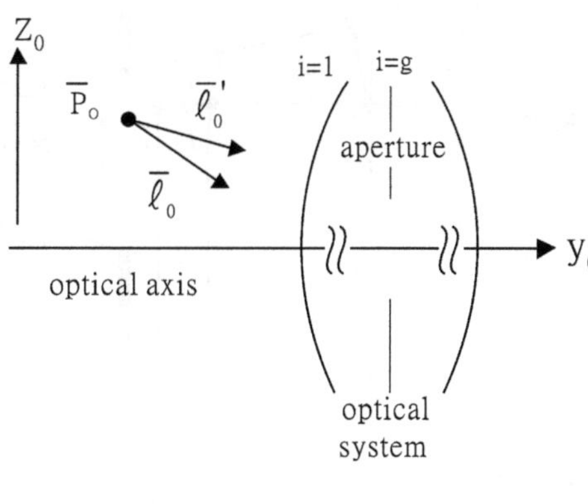


**Fig. 7.8** Feasibility of estimating wavefront aberration of neighboring ray $\bar{R}'_0$ via Taylor series expansion if wavefront aberration of reference ray $\bar{R}_0$ is given

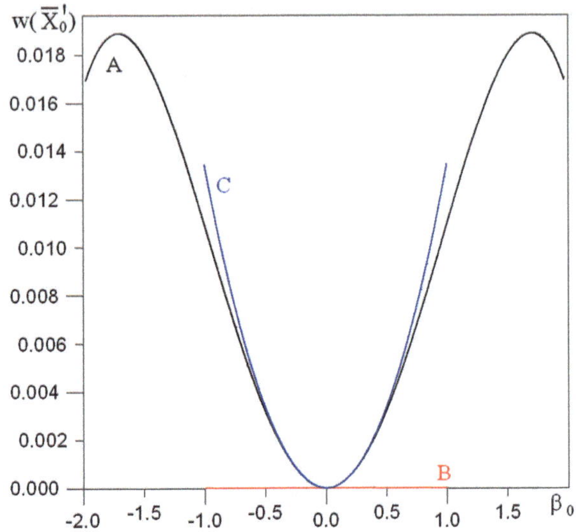


**Fig. 7.9** *Curve A* is the wavefront aberration of on-axis source point $\bar{P}_0 = [0 \quad -507 \quad 0 \quad 1]^T$ of system in Fig. 2.1 as computed by raytracing method. *Curves B* and *C* will be discussed in example 8.1

their optical path lengths to obtain the wavefront aberration $W(\bar{X}_0)$ of $\bar{R}_0$. Curve A in Fig. 7.9 shows the wavefront aberration $W(\bar{X}_0)$ obtained via raytracing for an on-axis source point $\bar{P}_0 = [0 \quad -507 \quad 0 \quad 1]^T$ in the system shown in Fig. 2.1 given a reference sphere $\bar{r}_{ref}$ with a radius of $R_{ref} = 45$ mm (see Fig. 7.10, the modified figure of Fig. 2.1). Note that in this particular example, $W(\bar{X}_0)$ is an axis-symmetric function since the system shown in Fig. 2.1 (or Fig. 7.10) is axis-symmetric and $\bar{P}_0$ lies on the optical axis. As a result, Curve A of Fig. 7.9 shows only a cross-sectional view of $W(\bar{X}_0)$ rather than its entire distribution. It is observed from the curve A that $\bar{P}_0$ is well focused and has only a small wavefront

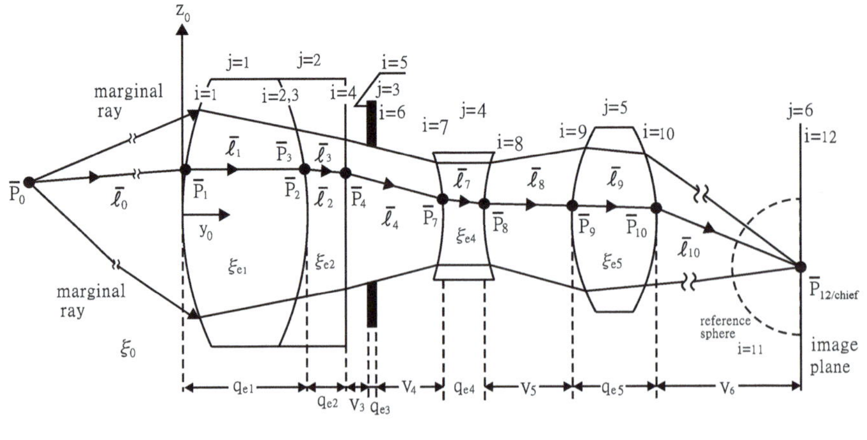

**Fig. 7.10** The modified system of Fig. 2.1 when the reference sphere $\bar{r}_{ref}$ is considered as a virtual boundary surface with radius $R_{ref}$

aberration since it lies within the depth of field of the system. The root mean square, mean value and variance [2] of the wavefront aberration are found to be $\text{rms}(W(\bar{X}_0)) = 0.0002198$, $\text{mean}(W(\bar{X}_0)) = 0.0135739$, and $\text{variance}(W(\bar{X}_0)) = 0.0000355$, respectively.

*Example 7.4* The maximum positive and negative wavefront aberration determine the peak-to-valley (P-V) wavefront aberration and represent the maximum departure of the actual wavefront from the desired wavefront in the positive and negative directions. When utilizing raytracing methods, it is generally necessary to trace many rays in order to determine accurately the maximum positive and negative values of $W(\bar{X}_0)$. Of course, a small amount of traced rays are enough if only an estimated P-V value with lower accuracy is required. An alternative is to use some form of numerical methods to obtain the roots of $\sum_{i=1}^{i=n} \partial OPL_i/\partial \bar{X}_0 = 0$. Clearly, however, it is also necessary to compute the wavefront aberration $W(\bar{X}_0)$for all the marginal rays since it is possible that the local maximum and minimum values of the wavefront aberration may exist. In the case of Example 7.3, the local minimum value is $W(\bar{X}_0) = 0$ and is located at $\beta_0 = 0$, while the local maximum value is $W(\bar{X}_0) = 0.0188926$ and is located at $\beta_0 = \pm 1.7053^o$. The wavefront aberration of the marginal ray is $W(\bar{X}_0) = 0.0169624$. The P-V wavefront aberration is therefore equal to 0.0188926. However, the P-V property simply states the maximum deviation of the wavefront aberration. In other words, it provides no indication of the area over which the aberration occurs. In practice, an optical system with a large P-V error may actually perform better than a system with a small P-V error. As a result, it is generally more meaningful to specify the wavefront quality using the root mean square, mean value, and variance of the wavefront aberration, as given in Example 7.3.

## 7.4 The Merit Function Based on Wavefront Aberration

The Jacobian matrix of the wavefront aberration presented in Sect. 7.3 provides an ideal basis for the optimization process performed in automatic optical system design applications. The two merit functions proposed by Meiron [2] (i.e., mean square value of OPD and variance of OPD) require the tracing of a large number of rays since they involve the integration of Eq. (7.26) over all portions of the reference sphere filled by the wavefront. To resolve this problem, We can use the sum of square wavefront aberration as a merit function. Assume that $\bar{P}_0$ is a single source point which emits a large number of rays into space. To calculate the wavefront aberration, a subset of these rays (say, m) must be traced from the source point $\bar{P}_0$ through the system. In general, the merit function is evaluated for several source points (say, $q$) in the field of view. The following merit function is adopted to indicate the amount of image defects:

$$\Phi=\sum_{1}^{q}\sum_{1}^{m}W^2(\bar{X}_0)=\sum_{1}^{q}\sum_{1}^{m}\left[OPL(\bar{P}_0,\bar{P}_n)-OPL\left(\bar{P}_0,\bar{P}_{n/chief}\right)\right]^2. \tag{7.29}$$

The Jacobian matrix $\partial\Phi/\partial\bar{X}_{sys}$, which is necessary for the determination of search direction for optimization programs, can be obtained by differentiating Eq. (7.29) to give

$$\begin{aligned}\frac{\partial\Phi}{\partial\bar{X}_{sys}} &= 2\sum_{1}^{q}\sum_{1}^{m}W(\bar{X}_0)\frac{\partial W(\bar{X}_0)}{\partial\bar{X}_{sys}}\\ &= 2\sum_{1}^{q}\sum_{1}^{m}\left[OPL(\bar{P}_0,\bar{P}_n)-OPL\left(\bar{P}_0,\bar{P}_{n/chief}\right)\right]\\ &\quad\left[\frac{\partial OPL(\bar{P}_0,\bar{P}_n)}{\partial\bar{X}_{sys}}-\frac{\partial OPL\left(\bar{P}_0,\bar{P}_{n/chief}\right)}{\partial\bar{X}_{sys}}\right].\end{aligned} \tag{7.30}$$

Note that Eq. (7.30) is derived using Eq. (7.27). In the simplest sense, optimizing an optical system involves minimizing $\Phi$, where the numerical value of $\Phi$ depends on the system variable vector $\bar{X}_{sys}$.

## References


1. A. Milkŝ, Dependence of the wave-front aberration on the radius of the reference sphere. J. Opt. Soc. Am. **19**, 1187–1190 (2002)
2. J. Meiron, The use of merit functions on wave-front aberrations in automatic lens design. Appl. Opt. **7**, 667–672 (1968)

# Chapter 8
# The Wavefront Shape, Irradiance, and Caustic Surface in an Optical System

The wavefront shape, illuminance, and caustic surface at any point along the ray path can be computed either by k-function method [1, 2] or by the differential geometry-based method [3]. The former approach was further extended by Shealy and his colleagues [4–8] for systems with multiple optical elements illuminated by a plane wavefront propagating parallel to the optical axis. However, how to generate the k-function for non-axially symmetrical systems with reflecting surfaces for a skew source ray is still highly challenging. This chapter will show that the differential geometry-based approach based on the first and second fundamental forms of the wavefront is a more general approach for discussing the shape of wavefront. Implementing this method requires (1) equations which accurately describe the reflection/or refraction processes each time the ray encounters a boundary surface (addressed in Chap. 3); (2) expressions for the Jacobian matrices of the ray $\bar{R}_i$ and OPL with respect to the variable vector $\bar{X}_0$ of the source ray (i.e., $\partial\bar{R}_i/\partial\bar{X}_0$ and $\partial OPL_i/\partial\bar{X}_0$, given respectively in Chaps. 5 and 7); and (3) expressions for the Hessian matrices of the ray $\bar{R}_i$ and OPL with respect to $\bar{X}_0$ (i.e., $\partial^2\bar{R}_i/\partial\bar{X}_0^2$ and $\partial^2 OPL_i/\partial\bar{X}_0^2$). However, due to the high degree of nonlinearity of the reflection/refraction equations, deriving these two Hessian matrices is extremely cumbersome. Various authors have used the local Gaussian curvature of the wavefront to compute the wavefront shape or irradiance along the ray path [1–8]. Curiously, in our scan of the literature, we were not able to find a reference to provide any formal means of computing the required matrices. In order to implement the differential geometry approach to discuss these topics, this chapter will address the required Hessian matrices, $\partial^2\bar{R}_i/\partial\bar{X}_0^2$ and $\partial^2 OPL_i/\partial\bar{X}_0^2$. Prior to determining these two Hessian matrices, one needs the matrix $\partial^2\bar{R}_i/\partial\bar{R}_{i-1^2}$, which will be addressed in Sects. 8.1 and 8.2.

PD Lin, *New Computation Methods for Geometrical Optics*,
Springer Series in Optical Sciences 178, DOI: 10.1007/978-981-4451-79-6_8,

## 8.1 The Hessian Matrix of the Ray $\overline{\mathbf{R}}_{\mathbf{i}}$ with Respect to its Incoming Ray $\overline{\mathbf{R}}_{\mathbf{i}-1}$ at a Flat Boundary Surface

As addressed in Sect. 5.2, any change of the incoming ray, $\Delta\overline{R}_{i-1}$, from last boundary surface $\bar{r}_{i-1}$ may cause a corresponding change in the refracted/or reflected ray, $\Delta\overline{R}_i$, at the current boundary surface $\bar{r}_i$. The ith ray $\overline{R}_i$ includes the incidence point $\overline{P}_i$ and its unit directional vector $\overline{\ell}_i$. Therefore, the Hessian matrix $\partial^2\overline{R}_i/\partial\overline{R}_{i-1}^2$comprises two components, namely, $\partial^2\overline{P}_i\Big/\partial\overline{R}_{i-1}^2$ and $\partial^2\overline{\ell}_i\Big/\partial\overline{R}_{i-1}^2$, as discussed in the following.

(a) $\partial^2\overline{P}_i\Big/\partial\overline{R}_{i-1}^2 = \left[\partial^2\overline{P}_i/\partial x_w\partial x_v\right]$, $x_w \in \overline{R}_{i-1}$ and $x_v \in \overline{R}_{i-1}$, at a flat boundary surface can be obtained by differentiating Eq. (5.3) to give

$$\frac{\partial^2\overline{P}_i}{\partial\overline{R}_{i-1}^2} = \begin{bmatrix} \partial^2 P_{ix}/\partial x_w\partial x_v \\ \partial^2 P_{iy}/\partial x_w\partial x_v \\ \partial^2 P_{iz}/\partial x_w\partial x_v \end{bmatrix}, \tag{8.1}$$

where

$$\left[\frac{\partial^2 P_{ix}}{\partial x_w\partial x_v}\right] = \left[\frac{\partial\lambda_i}{\partial x_w}\right]^T \begin{bmatrix}0 & 0 & 0 & 1 & 0 & 0\end{bmatrix} + \begin{bmatrix}0 & 0 & 0 & 1 & 0 & 0\end{bmatrix}^T \left[\frac{\partial\lambda_i}{\partial x_v}\right] + \ell_{i-1x}\left[\frac{\partial^2\lambda_i}{\partial x_w\partial x_v}\right], \tag{8.1a}$$

$$\left[\frac{\partial^2 P_{iy}}{\partial x_w\partial x_v}\right] = \left[\frac{\partial\lambda_i}{\partial x_w}\right]^T \begin{bmatrix}0 & 0 & 0 & 0 & 1 & 0\end{bmatrix} + \begin{bmatrix}0 & 0 & 0 & 0 & 1 & 0\end{bmatrix}^T \left[\frac{\partial\lambda_i}{\partial x_v}\right] + \ell_{i-1y}\left[\frac{\partial^2\lambda_i}{\partial x_w\partial x_v}\right], \tag{8.1b}$$

$$\left[\frac{\partial^2 P_{iz}}{\partial x_w\partial x_v}\right] = \left[\frac{\partial\lambda_i}{\partial x_w}\right]^T \begin{bmatrix}0 & 0 & 0 & 0 & 0 & 1\end{bmatrix} + \begin{bmatrix}0 & 0 & 0 & 0 & 0 & 1\end{bmatrix}^T \left[\frac{\partial\lambda_i}{\partial x_v}\right] + \ell_{i-1z}\left[\frac{\partial^2\lambda_i}{\partial x_w\partial x_v}\right]. \tag{8.1c}$$

$\partial\lambda_i/\partial x_w$ and $\partial\lambda_i/\partial x_v$, respectively, are the wth and vth components of $\partial\lambda_i/\partial\overline{R}_{i-1}$ given in Eq. (5.4), and

$$\frac{\partial^2\lambda_i}{\partial\overline{R}_{i-1}^2} = \left[\frac{\partial^2\lambda_i}{\partial x_w \partial x_v}\right]$$

$$= \frac{1}{E_i^2}\begin{bmatrix} 0 & 0 & 0 & J_{ix}J_{ix} & J_{ix}J_{iy} & J_{ix}J_{iz} \\ & 0 & 0 & J_{iy}J_{ix} & J_{iy}J_{iy} & J_{iy}J_{iz} \\ & & 0 & J_{iz}J_{ix} & J_{iz}J_{iy} & J_{iz}J_{iz} \\ & & & 0 & 0 & 0 \\ & \text{symm.} & & & 0 & 0 \\ & & & & & 0 \end{bmatrix} - \frac{2D_i}{E_i^3}\begin{bmatrix} 0 & 0 & 0 & 0 & 0 & 0 \\ & 0 & 0 & 0 & 0 & 0 \\ & & 0 & 0 & 0 & 0 \\ & & & J_{ix}J_{ix} & J_{ix}J_{iy} & J_{ix}J_{iz} \\ & \text{symm.} & & & J_{iy}J_{iy} & J_{iy}J_{iz} \\ & & & & & J_{iz}J_{iz} \end{bmatrix}. \tag{8.2}$$

The annotation "symm." in Eq. (8.2) indicates that $\partial^2\lambda_i/\partial\overline{R}_{i-1^2}$ is a symmetrical matrix. The explicit expressions of $[\partial^2 P_{ix}/\partial x_w \partial x_v]$, $[\partial^2 P_{iy}/\partial x_w \partial x_v]$, and $[\partial^2 P_{iz}/\partial x_w \partial x_v]$ are respectively given in Eqs. (A.1), (A.2), and (A.2) of the Appendix A.

(b) When $\overline{\ell}_i$ is the unit directional vector of the reflected ray at a flat boundary surface, $\partial^2\overline{\ell}_i/\partial^2\overline{\ell}_i\partial\overline{R}_{i-1^2}\partial\overline{R}_{i-1^2} = [\partial^2\overline{\ell}_i/\partial x_w \partial x_v]$, $x_w \in \overline{R}_{i-1}$ and $x_v \in \overline{R}_{i-1}$, can be obtained by differentiating Eq. (5.5) to give

$$\left[\frac{\partial^2\overline{\ell}_i}{\partial\overline{R}_{i-1}^2}\right] = \begin{bmatrix} \partial^2\ell_{ix}/\partial x_w \partial x_v \\ \partial^2\ell_{iy}/\partial x_w \partial x_v \\ \partial^2\ell_{iz}/\partial x_w \partial x_v \end{bmatrix} = \begin{bmatrix} \overline{0}_{6\times 6} \\ \overline{0}_{6\times 6} \\ \overline{0}_{6\times 6} \end{bmatrix} = \overline{0}_{3\times 6\times 6}. \tag{8.3}$$

Equation (5.5) shows that $\partial\overline{\ell}_i/\partial\overline{R}_{i-1}$ is a function of the unit normal vector $\overline{n}_i$ of boundary surface, not of $\overline{R}_{i-1}$, yielding $\partial^2\overline{\ell}_i\big/\partial\overline{R}_{i-1}^2 = \overline{0}_{3\times 6\times 6}$.

(c) When $\overline{\ell}_i$ is the unit directional vector of the refracted ray at a flat boundary surface, $\partial^2\overline{\ell}_i\big/\partial\overline{R}_{i-1}^2 = [\partial^2\overline{\ell}_i/\partial x_w \partial x_v]$, $x_w \in \overline{R}_{i-1}$ and $x_v \in \overline{R}_{i-1}$, can be obtained by differentiating Eq. (5.6) to give

$$\left[\frac{\partial^2\overline{\ell}_i}{\partial\overline{R}_{i-1}^2}\right] = \begin{bmatrix} \partial^2\ell_{ix}/\partial x_w \partial x_v \\ \partial^2\ell_{iy}/\partial x_w \partial x_v \\ \partial^2\ell_{iz}/\partial x_w \partial x_v \end{bmatrix}, \tag{8.4}$$

where the explicit expression of $[\partial^2\ell_{ix}/\partial x_w \partial x_v]$, $[\partial^2\ell_{iy}/\partial x_w \partial x_v]$, and $[\partial^2\ell_{iz}/\partial x_w \partial x_v]$ are respectively given in Eqs. (A.4), (A.5), and (A.6) of Appendix A.

If the ith boundary surface of the optical system is flat, then by combining Eqs. (8.1) and (8.3) (when the ray $\overline{R}_i$ is reflected), or Eqs. (8.1) and (8.4) (when the ray $\overline{R}_i$ is refracted), the Hessian matrix $\partial^2\overline{R}_i\big/\partial\overline{R}_{i-1}^2$ is obtained as

$$\frac{\partial^2 \overline{R}_i}{\partial \overline{R}_{i-1}^2} = \begin{bmatrix} \partial^2 P_{ix} \Big/ \partial \overline{R}_{i-1}^2 \\ \partial^2 P_{iy} \Big/ \partial \overline{R}_{i-1}^2 \\ \partial^2 P_{iz} \Big/ \partial \overline{R}_{i-1}^2 \\ \partial^2 \ell_{ix} \Big/ \partial \overline{R}_{i-1}^2 \\ \partial^2 \ell_{iy} \Big/ \partial \overline{R}_{i-1}^2 \\ \partial^2 \ell_{iz} \Big/ \partial \overline{R}_{i-1}^2 \end{bmatrix}_{6\times6\times6.} \tag{8.5}$$

## 8.2 The Hessian Matrix of the Ray $\overline{R}_i$ with Respect to Its Incoming Ray $\overline{R}_{i-1}$ at a Spherical Boundary Surface

Equations (5.18), (5.22) and (5.30) express respectively the Jacobian matricx of the incidence point $\overline{P}_i$, refracted ray $\overline{\ell}_i$ and reflected ray $\overline{\ell}_i$ with respect to the incoming ray $\overline{R}_{i-1}$ at a spherical boundary surface. They enable the matrix $\partial^2 \overline{R}_i \Big/ \partial \overline{R}_{i-1}^2$ (which is a $6 \times 6 \times 6$ matrix) to be determined by further differentiating themselves with respect to the incoming ray $\overline{R}_{i-1}$:

(a) $\partial^2 \overline{P}_i \Big/ \partial \overline{R}_{i-1}^2 = \left[\partial^2 \overline{P}_i / \partial x_w \partial x_v\right]$, $x_w \in \overline{R}_{i-1}$ and $x_v \in \overline{R}_{i-1}$, can be obtained by differentiating Eq. (5.18) to give

$$\frac{\partial^2 \overline{P}_i}{\partial \overline{R}_{i-1}^2} = \begin{bmatrix} \partial^2 P_{ix} / \partial x_w \partial x_v \\ \partial^2 P_{iy} / \partial x_w \partial x_v \\ \partial^2 P_{iz} / \partial x_w \partial x_v \end{bmatrix}, \tag{8.6}$$

where

$$\left[\frac{\partial^2 P_{ix}}{\partial x_w \partial x_v}\right] = \left[\frac{\partial \lambda_i}{\partial x_w}\right]^T \begin{bmatrix} 0 & 0 & 0 & 1 & 0 & 0 \end{bmatrix} + \begin{bmatrix} 0 & 0 & 0 & 1 & 0 & 0 \end{bmatrix}^T \left[\frac{\partial \lambda_i}{\partial x_v}\right] + \ell_{i-1x} \left[\frac{\partial^2 \lambda_i}{\partial x_w \partial x_v}\right], \tag{8.6a}$$

$$\left[\frac{\partial^2 P_{iy}}{\partial x_w \partial x_v}\right] = \left[\frac{\partial \lambda_i}{\partial x_w}\right]^T \begin{bmatrix} 0 & 0 & 0 & 0 & 1 & 0 \end{bmatrix} + \begin{bmatrix} 0 & 0 & 0 & 0 & 1 & 0 \end{bmatrix}^T \left[\frac{\partial \lambda_i}{\partial x_v}\right] + \ell_{i-1y} \left[\frac{\partial^2 \lambda_i}{\partial x_w \partial x_v}\right], \tag{8.6b}$$

$$\left[\frac{\partial^2 P_{iz}}{\partial x_w \partial x_v}\right] = \left[\frac{\partial \lambda_i}{\partial x_w}\right]^T [0 \quad 0 \quad 0 \quad 0 \quad 0 \quad 1] + [0 \quad 0 \quad 0 \quad 0 \quad 0 \quad 1]^T \left[\frac{\partial \lambda_i}{\partial x_v}\right] + \ell_{i-1z}\left[\frac{\partial^2 \lambda_i}{\partial x_w \partial x_v}\right], \tag{8.6c}$$

$\partial\lambda_i/\partial x_w$ and $\partial\lambda_i/\partial x_v$, respectively, are the wth and vth components of $\partial\lambda_i/\partial\overline{R}_{i-1}$ given in Eq. (5.19), and

$$\frac{\partial^2 \lambda_i}{\partial \overline{R}_{i-1}^2} = \left[\frac{\partial^2 \lambda_i}{\partial x_w \partial x_v}\right] = \begin{bmatrix} \lambda_{i,11} & \lambda_{i,21} & \lambda_{i,31} & \lambda_{i,41} & \lambda_{i,51} & \lambda_{i,61} \\ & \lambda_{i,22} & \lambda_{i,32} & \lambda_{i,42} & \lambda_{i,52} & \lambda_{i,62} \\ & & \lambda_{i,33} & \lambda_{i,43} & \lambda_{i,53} & \lambda_{i,63} \\ & & & \lambda_{i,44} & \lambda_{i,54} & \lambda_{i,64} \\ & \text{symm.} & & & \lambda_{i,55} & \lambda_{i,65} \\ & & & & & \lambda_{i,66} \end{bmatrix}. \tag{8.7}$$

The explicit expressions of $\left[\partial^2 P_{ix}/\partial x_w \partial x_v\right]$, $\left[\partial^2 P_{iy}/\partial x_w \partial x_v\right]$, $\left[\partial^2 P_{iz}/\partial x_w \partial x_v\right]$, and $\left[\partial^2 \lambda_i/\partial x_w \partial x_v\right]$ are respectively given in Eqs. (B.1), (B.2), (B.3), and (B.4) of Appendix B.

(b) When $\overline{\ell}_i$ is the unit directional vector of the refracted ray at a spherical boundary surface, $\partial^2\overline{\ell}_i/\partial\overline{R}_{i-1}^2 = \left[\partial^2\overline{\ell}_i/\partial x_w \partial x_v\right]$, $x_w \in \overline{R}_{i-1}$ and $x_v \in \overline{R}_{i-1}$, can be obtained by differentiating Eq. (5.22) to give

$$\left[\frac{\partial^2 \overline{\ell}_i}{\partial \overline{R}_{i-1}^2}\right] = \begin{bmatrix} \partial^2 \ell_{ix}/\partial x_w \partial x_v \\ \partial^2 \ell_{iy}/\partial x_w \partial x_v \\ \partial^2 \ell_{iz}/\partial x_w \partial x_v \end{bmatrix}, \tag{8.8}$$

where $\left[\partial^2 \ell_{ix}/\partial x_w \partial x_v\right]$, $\left[\partial^2 \ell_{iy}/\partial x_w \partial x_v\right]$, and $\left[\partial^2 \ell_{iz}/\partial x_w \partial x_v\right]$ for a ray refracted at a spherical boundary surface are respectively given in Eqs. (B.5), (B.6), and (B.7) of Appendix B.

(c) When $\overline{\ell}_i$ is the unit directional vector of the reflected ray at a spherical boundary surface, $\partial^2\overline{\ell}_i/\partial\overline{R}_{i-1}^2 = \left[\partial^2\overline{\ell}_i/\partial x_w \partial x_v\right]$, $x_w \in \overline{R}_{i-1}$ and $x_v \in \overline{R}_{i-1}$, can be obtained by differentiating Eq. (5.30) to give

$$\left[\frac{\partial^2 \overline{\ell}_i}{\partial \overline{R}_{i-1}^2}\right] = \begin{bmatrix} \partial^2 \ell_{ix}/\partial x_w \partial x_v \\ \partial^2 \ell_{iy}/\partial x_w \partial x_v \\ \partial^2 \ell_{iz}/\partial x_w \partial x_v \end{bmatrix}, \tag{8.9}$$

where $\left[\partial^2 \ell_{ix}/\partial x_w \partial x_v\right]$, $\left[\partial^2 \ell_{iy}/\partial x_w \partial x_v\right]$, and $\left[\partial^2 \ell_{iz}/\partial x_w \partial x_v\right]$ for a ray reflected at a spherical boundary surface are given respectively in Eqs. (B.8), (B.9), and (B.10) of Appendix B.

If the ith boundary surface of the optical system is spherical, then by combining Eqs. (8.6) and (8.8) (when the ray $\overline{R}_i$ is reflected), or Eqs. (8.6) and (8.9) (when the ray $\overline{R}_i$ is refracted), the Hessian matrix $\partial^2\overline{R}_i/\partial\overline{R}_{i-1}^2$ is obtained as

$$\frac{\partial^2 \overline{R}_i}{\partial \overline{R}_{i-1}^2} = \begin{bmatrix} \partial^2 P_{ix} / \partial \overline{R}_{i-1}^2 \\ \partial^2 P_{iy} / \partial \overline{R}_{i-1}^2 \\ \partial^2 P_{iz} / \partial \overline{R}_{i-1}^2 \\ \partial^2 \ell_{ix} / \partial \overline{R}_{i-1}^2 \\ \partial^2 \ell_{iy} / \partial \overline{R}_{i-1}^2 \\ \partial^2 \ell_{iz} / \partial \overline{R}_{i-1}^2 \end{bmatrix}_{6\times 6\times 6.} \tag{8.10}$$

It is shown from Eq. (8.1) to Eq. (8.10) that $\partial^2 f / \partial \overline{R}_{i-1}^2$, where $f \in \overline{R}_i$, is a symmetrical matrix. This indicates that the Hessian matrix $\partial^2 f/\partial X_0^2$ computed in the following section is also a symmetrical matrix.

## 8.3 Computation of Hessian Matrix $\partial^2 \overline{\mathbf{R}}_i / \partial \overline{\mathbf{X}}_0^2$

The Hessian matrix $\partial^2 \overline{R}_i / \partial \overline{X}_0^2$ has the form

$$\frac{\partial^2 \overline{R}_i}{\partial \overline{X}_0^2} = \left[\frac{\partial^2 \overline{R}_i}{\partial x_w \partial x_v}\right] = \begin{bmatrix} \frac{\partial^2 \overline{R}_i}{\partial P_{0x}\partial P_{0x}} & \frac{\partial^2 \overline{R}_i}{\partial P_{0x}\partial P_{0y}} & \frac{\partial^2 \overline{R}_i}{\partial P_{0x}\partial P_{0z}} & \frac{\partial^2 \overline{R}_i}{\partial P_{0x}\partial \alpha_0} & \frac{\partial^2 \overline{R}_i}{\partial P_{0x}\partial \beta_0} \\ & \frac{\partial^2 \overline{R}_i}{\partial P_{0y}\partial P_{0y}} & \frac{\partial^2 \overline{R}_i}{\partial P_{0y}\partial P_{0z}} & \frac{\partial^2 \overline{R}_i}{\partial P_{0y}\partial \alpha_0} & \frac{\partial^2 \overline{R}_i}{\partial P_{0y}\partial \beta_0} \\ & & \frac{\partial^2 \overline{R}_i}{\partial P_{0z}\partial P_{0z}} & \frac{\partial^2 \overline{R}_i}{\partial P_{0z}\partial \alpha_0} & \frac{\partial^2 \overline{R}_i}{\partial P_{0z}\partial \beta_0} \\ & \text{symm.} & & \frac{\partial^2 \overline{R}_i}{\partial \alpha_0\partial \alpha_0} & \frac{\partial^2 \overline{R}_i}{\partial \alpha_0\partial \beta_0} \\ & & & & \frac{\partial^2 \overline{R}_i}{\partial \beta_0\partial \beta_0} \end{bmatrix}, \tag{8.11}$$

where $x_w \in \overline{X}_0$ and $x_v \in \overline{X}_0$. Equation (8.11) can be determined from Eq. (5.105) by setting $\overline{X}_i$ to be a zero matrix to give

$$\frac{\partial^2 \overline{R}_i}{\partial \overline{X}_0^2} = \left[\frac{\partial^2 \overline{R}_i}{\partial x_w \partial x_v}\right] = \left[\frac{\partial \overline{R}_{i-1}}{\partial x_w}\right]^T \frac{\partial^2 \overline{R}_i}{\partial \overline{R}_{i-1}^2} \left[\frac{\partial \overline{R}_{i-1}}{\partial x_v}\right] + \frac{\partial \overline{R}_i}{\partial \overline{R}_{i-1}} \left[\frac{\partial^2 \overline{R}_{i-1}}{\partial x_w \partial x_v}\right]. \tag{8.12}$$

$\partial \overline{R}_{i-1}/\partial x_w$ and $\partial \overline{R}_{i-1}/\partial x_v$of 8.12 are respectively the wth and vth components of $\partial \overline{R}_{i-1}/\partial \overline{X}_0$ shown in Eq. (5.33) by setting i to be i − 1. $\partial \overline{R}_i/\partial \overline{R}_{i-1}$ and $\partial^2 \overline{R}_i / \partial \overline{R}_{i-1}^2$ of Eq. (8.12) were already addressed in Sects. 5.2 and 5.3 for flat and Sects. 8.1 and 8.2 for spherical boundary surface. Therefore, now it is possible to compute the Hessian matrix $\partial^2 \overline{R}_i/\partial \overline{X}_0^2$ sequentially from i = 1 to i = n by given $\partial^2 \overline{R}_0/\partial \overline{X}_0^2$, which is obtained by directly differentiating Eq. (5.10) to give

$$\frac{\partial^2 P_{0x}}{\partial \overline{X}_0^2} = \frac{\partial^2 P_{0y}}{\partial \overline{X}_0^2} = \frac{\partial^2 P_{0z}}{\partial \overline{X}_0^2} = \overline{0}_{5\times 5}, \tag{8.13}$$

$$\frac{\partial^2 \ell_{0x}}{\partial \overline{X}_0^2} = \begin{bmatrix} 0 & 0 & 0 & 0 & 0 \\ 0 & 0 & 0 & 0 & 0 \\ 0 & 0 & 0 & 0 & 0 \\ 0 & 0 & 0 & -C\beta_0 C\alpha_0 & S\beta_0 S\alpha_0 \\ 0 & 0 & 0 & S\beta_0 S\alpha_0 & -C\beta_0 C\alpha_0 \end{bmatrix}, \tag{8.14}$$

$$\frac{\partial^2 \ell_{0y}}{\partial \overline{X}_0^2} = \begin{bmatrix} 0 & 0 & 0 & 0 & 0 \\ 0 & 0 & 0 & 0 & 0 \\ 0 & 0 & 0 & 0 & 0 \\ 0 & 0 & 0 & -C\beta_0 S\alpha_0 & -S\beta_0 C\alpha_0 \\ 0 & 0 & 0 & -S\beta_0 C\alpha_0 & -C\beta_0 S\alpha_0 \end{bmatrix}, \tag{8.15}$$

$$\frac{\partial^2 \ell_{0z}}{\partial \overline{X}_0^2} = \begin{bmatrix} 0 & 0 & 0 & 0 & 0 \\ 0 & 0 & 0 & 0 & 0 \\ 0 & 0 & 0 & 0 & 0 \\ 0 & 0 & 0 & -S\beta_0 & 0 \\ 0 & 0 & 0 & 0 & 0 \end{bmatrix}. \tag{8.16}$$

The optical system shown in Fig. 2.1 is used as an illustrative example to demonstrate numerically the accuracy and calculation speed of the method in Eq. (8.12) for Hessian matrix $\partial^2 P_{11z}/\partial X_0^2$ computation verses traditional FD methodology. The source ray variables are given as

$$\begin{aligned}\overline{X}_{e0} &= [P_{0x} \quad P_{0y} \quad P_{0z} \quad \alpha_0 \quad \beta_0]^T \\ &= [0 \quad -507 \quad 170 \quad 89.69974^\circ \quad -17.94211^\circ]^T,\end{aligned} \tag{8.17}$$

yielding the source ray $\overline{R}_0$ that enters the system of Fig. 2.1 as

$$\overline{R}_0 = [\overline{P}_0 \quad \overline{\ell}_0]^T = [0 \quad -507 \quad 170 \quad 0.00498 \quad 0.95135 \quad -0.30806]^T. \tag{8.18}$$

In the comparative trials, the presented method and two traditional but alternative FD methods, denoted as FD#1 and FD#2, are used to calculate the Hessian matrix $\partial^2 P_{11z}/\partial^2 X_0$, where $P_{11z}$ is the z coordinate of the incidence point on the image plane, of the system shown in Fig. 2.1. FD#1 determines the components of the Hessian matrix $\partial^2 P_{11z}/\partial^2 \overline{X}_0$ by using the result $\partial P_{11z}/\partial \overline{X}_0$ from the Jacobian matrix as

$$\frac{\partial^2 P_{11z}}{\partial x_w \partial x_v} = \frac{\frac{\partial P_{11z}(x_w + \Delta x_w)}{\partial x_v} - \frac{\partial P_{11z}(x_w)}{\partial x_v}}{\Delta x_w}. \tag{8.19}$$

Notably, solving the Hessian matrix using the FD#1 method involves two ray-tracing operations (i.e. one operation to determine $\partial P_{11z}(x_w)/\partial x_v$ from Eq. (5.65) and a second operation using the same variables but with one variable increased by $\Delta x_w = 0.0000001$ to determine $\partial P_{11z}(x_w + \Delta x_w)/\partial x_v$. In contrast to FD#1's two raytracing operations, the method presented in Eq. (8.12) requires only one raytracing operation.

The alternative FD#2 method uses the result $P_{11z}$ from raytracing to determine the Hessian matrix $\partial^2 P_{11z}/\partial^2 \overline{X}_0$ as

$$\frac{\partial^2 P_{11z}}{\partial x_w \partial x_v} = \frac{\frac{P_{11z}(x_v+\Delta x_v, x_w+\Delta x_w)-P_{11z}(x_v, x_w+\Delta x_w)}{\Delta x_v} - \frac{P_{11z}(x_v+\Delta x_v, x_w)-P_{11z}(x_v, x_w)}{\Delta x_v}}{\Delta x_w}, \quad (8.20)$$

where $\Delta x_w = \Delta x_v = 0.0000001$. In contrast to the one raytracing operation required by the proposed method and the two raytracing operations required by FD#1, the FD#2 method involves four raytracing operations to determine the Hessian matrix. A special characteristic of the FD#2 method is that the results are particularly liable to error due to the far smaller value of the denominator than that of the numerator. FD methods hardly arrive at exact numeric solutions when performing complex calculations and yield many round off or truncated approximations. The results of the presented method (based on differential geometry) are more accurate than the results of either FD#1 or FD#2 methods. Table 8.1 summarizes the results obtained by the three methods (i.e. the developed approach, FD#1 and FD#2) for $\partial^2 P_{11z}/\partial \overline{X}_0^2$. The performances of the FD#1 and FD#2 methods are compared with the presented method by the percentage errors listed in Table 8.1, which take the results of Eq. (8.12) as refrences. From Table 8.1, one can see that, compared with FD#2 method, FD#1 provides relatively stable approximations, with random trivial errors. This is because FD#1 method uses a forward difference formula by taking exact mathematical values of first-order derivatives to approximate the second-order derivative, yielding truncated errors of order $\Delta x_w = 0.0000001$. The FD#2 method uses four raytracing operations, producing larger round-off error, since round-off error tends to accompany every floating point arithmetic operation (addition, subtraction, multiplication and division). We also use a series of values of $\Delta\beta_0$ to compare the accuracy in $\partial^2 P_{11z}/\partial\beta_0\partial\beta_0$ of FD#1 and FD#2 methods (see Table 8.2). The presented method produces an exact mathematical value of $\partial^2 P_{11z}/\partial\beta_0\partial\beta_0 = 13185.86375$. Table 8.2 shows that the FD#2 results range from 0.12 % to $\infty$ % as $\Delta\beta_0$ ranges from $5 \times 10^{-2}$ to $1 \times 10^{-15}$ rad. When using FD methods, the division operation is highly sensitive to the size of the denominator relative to the size of the numerator. Inappropriate choice of $\Delta\beta_0$ produces results with great error. It is noted from Table 8.2 that round-off errors generate serious mistakes in numerical calculations involving very small numerator. Therefore, since the precision of the FD#2 method is unreliable and subject to severe degradation, it is impractical to use FD#2 method with ray-tracing data to determine the Hessian matrix of a ray.

With regard to the time required to calculate the Hessian matrix for the illustrative system of Fig. 2.1, the CPU time for the FD#1, FD#2 methods and the developed method for computing $\partial^2 \overline{R}_{11}/\partial \overline{X}_0^2$ is 0.0081, 0.0035 and 0.0092, respectively. FD#1 method and the developed method consume approximately the

**Table 8.1** Results of $\partial^2\overline{R}_{11}/\partial^2\overline{X}_0$; the percentage errors are computed with the FD results relative to those of presented method

| Results/components | Methods | | | | |
|---|---|---|---|---|---|
| | Presented method | FD#1 | | FD#2 | |
| | | Value | Error (%) | Value | Error (%) |
| $\partial^2 P_{11z}/\partial P_{0x}\partial P_{0x}$ | −0.00187 | −0.00187 | 0.00 | −532.90706 | ∞ |
| $\partial^2 P_{11z}/\partial P_{0x}\partial P_{0y}$ | −0.00023 | −0.00023 | 0.007 | −532.90706 | ∞ |
| $\partial^2 P_{11z}/\partial P_{0x}\partial P_{0z}$ | −0.00075 | −0.00075 | 0.006 | −177.63569 | ∞ |
| $\partial^2 P_{11z}/\partial P_{0x}\partial \alpha_0$ | 0.74810 | 0.74810 | 0.00 | −1065.81412 | ∞ |
| $\partial^2 P_{11z}/\partial P_{0x}\partial \beta_0$ | −0.79356 | −0.79355 | 0.002 | −1065.81412 | ∞ |
| $\partial^2 P_{11z}/\partial P_{0y}\partial P_{0y}$ | 0.00030 | 0.00030 | 0.004 | 355.27137 | ∞ |
| $\partial^2 P_{11z}/\partial P_{0y}\partial P_{0z}$ | 0.00091 | 0.00091 | 0.004 | −1030.28698 | ∞ |
| $\partial^2 P_{11z}/\partial P_{0y}\partial \alpha_0$ | 0.16967 | 0.16967 | 0.001 | −248.68998 | ∞ |
| $\partial^2 P_{11z}/\partial P_{0y}\partial \beta_0$ | 2.80754 | 2.80759 | 0.001 | −284.21710 | ∞ |
| $\partial^2 P_{11z}/\partial P_{0z}\partial P_{0z}$ | 0.00279 | 0.00279 | 0.000 | −1669.77545 | ∞ |
| $\partial^2 P_{11z}/\partial P_{0z}\partial \alpha_0$ | 0.54705 | 0.54705 | 0.000 | −319.74423 | ∞ |
| $\partial^2 P_{11z}/\partial P_{0z}\partial \beta_0$ | 7.87667 | 7.87669 | 0.000 | 1065.81412 | ∞ |
| $\partial^2 P_{11z}/\partial \alpha_0\partial \alpha_0$ | −54.68989 | −54.69016 | 0.001 | −1669.77545 | 2953 |
| $\partial^2 P_{11z}/\partial \alpha_0\partial \beta_0$ | 638.41364 | 638.40747 | 0.001 | −1065.81412 | ∞ |
| $\partial^2 P_{11z}/\partial \beta_0\partial \beta_0$ | 13185.86375 | 13185.68066 | 0.001 | 13145.04077 | 0.31 |

**Table 8.2** $\partial^2 P_{11z}/\partial\beta_0^2$ from FD#1 and FD#2 methods for different $\Delta\beta_0$; percentage errors are their results relative to our invariant result $\partial^2 P_{11z}/\partial\beta_0^2 = 13185.86375$

| $\Delta\beta_0$ | FD#1 | | FD#2 | |
|---|---|---|---|---|
| | Value | Error (%) | Value | Error (%) |
| $5 \times 10^{-2}$ | 3891.33342 | 70.489 | −115.00217 | 100.872 |
| $1 \times 10^{-3}$ | 11513.15724 | 12.686 | 10047.93832 | 23.798 |
| $1 \times 10^{-5}$ | 13167.56273 | 0.319 | 13149.28680 | 0.277 |
| $1 \times 10^{-7}$ | 13185.68066 | 0.001 | 13169.554033 | 0.124 |
| $1 \times 10^{-9}$ | 13185.56360 | 0.000 | 174082.98012 | 1220.224 |
| $1 \times 10^{-11}$ | 13185.94519 | 0.001 | ∞ | ∞ |
| $1 \times 10^{-13}$ | 13193.92618 | 0.061 | ∞ | ∞ |
| $1 \times 10^{-15}$ | 8469.66938 | 35.767 | ∞ | ∞ |

same computation time. Moreover, because both FD#1 and the developed methods require additional computations, they spend more time than FD#2 method does to obtain the Hessian matrix. However, the proposed method determines the Hessian matrix from differential geometry with a single raytracing operation and tends to yield exact solutions. In contrast, the FD methods require multiple raytracing operations; thus, the accuracy of their results is vulnerable to rounding error and truncation error.

In conclusion, our numerical results indicate that the presented method is more accurate and reliable than traditional FD methodology for computation of $\partial^2\overline{R}_i/\partial\overline{X}_0^2$, and the computation speed of the presented method is quite acceptable. The methodology presented in Sects. 5.2 and 5.3 can be implemented easily in computer code. It also provides a basis for determination of the Hessian matrix $\partial^2\overline{R}_i/\partial\overline{X}_{sys}^2$ of a ray $\overline{R}_i$ with respect to the system variable vector $\overline{X}_{sys}$ of an optical system.

## 8.4 The Hessian Matrix of $OPL_i$ with Respect to the Variable Vector $\overline{X}_0$

The Hessian matrix $\partial^2 OPL_i/\partial\overline{X}_0^2$ of $OPL_i$ with respect to the variable vector $\overline{X}_0$ of source ray has the form

$$\frac{\partial^2 OPL_i}{\partial\overline{X}_0^2}=\left[\frac{\partial^2 OPL_i}{\partial x_w \partial x_v}\right]=\begin{bmatrix} \frac{\partial^2 OPL_i}{\partial P_{0x}\partial P_{0x}} & \frac{\partial^2 OPL_i}{\partial P_{0x}\partial P_{0y}} & \frac{\partial^2 OPL_i}{\partial P_{0x}\partial P_{0z}} & \frac{\partial^2 OPL_i}{\partial P_{0x}\partial \alpha_0} & \frac{\partial^2 OPL_i}{\partial P_{0x}\partial \beta_0} \\ & \frac{\partial^2 OPL_i}{\partial P_{0y}\partial P_{0y}} & \frac{\partial^2 OPL_i}{\partial P_{0y}\partial P_{0z}} & \frac{\partial^2 OPL_i}{\partial P_{0y}\partial \alpha_0} & \frac{\partial^2 OPL_i}{\partial P_{0y}\partial \beta_0} \\ & & \frac{\partial^2 OPL_i}{\partial P_{0z}\partial P_{0z}} & \frac{\partial^2 OPL_i}{\partial P_{0z}\partial \alpha_0} & \frac{\partial^2 OPL_i}{\partial P_{0z}\partial \beta_0} \\ & \text{symm.} & & \frac{\partial^2 OPL_i}{\partial \alpha_0\partial \alpha_0} & \frac{\partial^2 OPL_i}{\partial \alpha_0\partial \beta_0} \\ & & & & \frac{\partial^2 OPL_i}{\partial \beta_0\partial \beta_0} \end{bmatrix}, \tag{8.21}$$

where, $x_w \in \overline{X}_0$ and $x_v \in \overline{X}_0$. Equation (8.21) can be computed from Eq. (8.12) by replacing $\overline{R}_i$ with $OPL_i$ to give

$$\frac{\partial^2 OPL_i}{\partial\overline{X}_0^2}=\left[\frac{\partial^2 OPL_i}{\partial x_w \partial x_v}\right]=\left[\frac{\partial\overline{R}_{i-1}}{\partial x_w}\right]^T\frac{\partial^2 OPL_i}{\partial\overline{R}_{i-1}^2}\left[\frac{\partial\overline{R}_{i-1}}{\partial x_v}\right]+\frac{\partial OPL_i}{\partial\overline{R}_{i-1}}\left[\frac{\partial^2\overline{R}_{i-1}}{\partial x_w \partial x_v}\right]. \tag{8.22}$$

Again, $\partial\overline{R}_{i-1}/\partial x_w$ and $\partial\overline{R}_{i-1}/\partial x_v$ of Eq. (8.22) are respectively the wth and vth components of $\partial\overline{R}_{i-1}/\partial\overline{X}_0$ given in Eq. (5.33) by setting i to be i − 1. $\partial OPL_i/\partial\overline{R}_{i-1}$ was shown in Eq. (7.1), and $\partial^2\overline{R}_{i-1}/\partial x_w\partial x_v$ is the (w, v)th component of $\partial^2\overline{R}_{i-1}/\partial\overline{X}_0^2$, which can be obtained from Eq. (8.11) by setting i to be i − 1. Now we further need $\partial^2 OPL_i/\partial\overline{R}_{i-1}^2$ in order to determine $\partial^2 OPL_i/\partial\overline{X}_0^2$. $\partial^2 OPL_i/\partial\overline{R}_{i-1}^2$ can be obtained by differentiating Eq. (7.1) with respect to the incoming ray $\overline{R}_{i-1}$, to give

$$\frac{\partial^2 OPL_i}{\partial\overline{R}_{i-1}^2}=\xi_{i-1}\frac{\partial^2\lambda_i}{\partial\overline{R}_{i-1}^2}. \tag{8.23}$$

For a flat and a spherical boundary surface, $\partial^2\lambda_i/\partial\overline{R}_{i-1}^2$ are given respectively in Eqs. (8.2) and (8.7). The Hessian matrix $\partial^2 OPL_i/\partial\overline{X}_0^2$ given in Eq. (8.21) enables various wavefront aberration problems to be solved. The solution is discussed in the following and Sect. 8.5.

We already addressed the computation of wavefront aberration of an optical system by adding a reference sphere as a virtual boundary surface in an optical system (see Fig. 7.7). For the modified system as shown in Fig. 7.7, the wavefront aberration $W(\overline{X}_0)$ can be computed by Eq. (7.26), which can be rewritten as

$$W(\overline{X}_0) = OPL(\overline{P}_0, \overline{P}_n) - OPL(\overline{P}_0, \overline{P}_{n/chief}) = \sum_{i=1}^{i=n} OPL_i - \sum_{i=1}^{i=n} OPL_{i/chief}$$
$$= \sum_{i=1}^{i=n} \left(OPL_i - OPL_{i/chief}\right). \tag{8.24}$$

We also derived the first-order change of wavefront aberration, $\Delta W(\overline{X}_0)$, in Eq. (7.28). Now we are going to estimate the change of wavefront aberration, $\Delta W(\overline{X}_0)$, by Taylor series expansion with up to the quadratic term. As shown in Fig. 7.8, two rays originating from a same source point $\overline{P}_0$, namely a reference source ray $\overline{R}_0 = \begin{bmatrix}\overline{P}_0 & \overline{\ell}_0\end{bmatrix}^T$ with variable vector $\overline{X}_0 = [P_{0x} \quad P_{0y} \quad P_{0z} \quad \alpha_0 \quad \beta_0]^T$ and a neighboring source ray $\overline{R}'_0 = \begin{bmatrix}\overline{P}_0 & \overline{\ell}'_0\end{bmatrix}^T$ with variable vector $\overline{X}'_0 = [P_{0x} \quad P_{0y} \quad P_{0z} \quad \alpha'_0 \quad \beta'_0]^T$. If the wavefront aberration $W(\overline{X}'_0)$ of the reference ray $\overline{R}'_0$ is given, it is possible to estimate $W\left(\overline{X}'_0\right)$ of the ray $\overline{R}'_0$ with up to the quadratic term of the Taylor series expansion by $W\left(\overline{X}'_0\right) = W(\overline{X}_0) + \Delta W(\overline{X}_0)$, where

$$\Delta W(\overline{X}_0) = \left(\sum_{i=1}^{i=n} \frac{\partial OPL_i}{\partial \overline{X}_0}\right)\Delta\overline{X}_0 + \frac{1}{2}\Delta\overline{X}_0^T\left(\sum_{i=1}^{i=n} \frac{\partial^2 OPL_i}{\partial \overline{X}_0^2}\right)\Delta\overline{X}_0, \tag{8.25}$$

with $\Delta\overline{X}_0 = \overline{X}'_0 - \overline{X}_0 = \begin{bmatrix}0 & 0 & 0 & \alpha'_0 - \alpha_0 & \beta'_0 - \beta_0\end{bmatrix}^T$. The Jacobian and Hessian matrices of Eq. (8.25) is evaluated by the reference ray $\overline{R}_0$. Then the wavefront aberration $W\left(\overline{X}'_0\right)$ of the neighboring ray $\overline{R}'_0$ is the sum of $W(\overline{X}_0)$ and $\Delta W(\overline{X}_0)$.

*Example 8.1* As shown in Eqs. (7.28) and (8.25), given two rays originating from the same source point (denoted respectively as the reference ray $\overline{R}_0$ and the neighboring ray $\overline{R}'_0$, see Fig. 7.8), the wavefront of the neighboring ray $W\left(\overline{X}'_0\right)$ can be estimated with up to the quadratic term of the Taylor series expansion provided that the wavefront aberration of the reference ray $W(\overline{X}_0)$ is known. Curves B and C in Fig. 7.9 show the wavefront aberrations

$W(\overline{X}'_0) = W(\overline{X}_0) + \Delta W(\overline{X}_0)$, where $\Delta W(\overline{X}_0)$ is estimated by Eqs. (7.28) and (8.25), respectively, by setting the reference ray $\overline{R}_0$ as the chief ray $\overline{R}_{0/\text{chief}}$. It is noted from Curve B that since $\partial OPL_i/\partial \overline{X}_0 = \overline{0}_{1\times 5}$ (i = 1 to i = n) for the reference ray $\overline{R}_{0/\text{chief}}$, the value of $W(\overline{X}'_0) = W(\overline{X}_0) + \Delta W(\overline{X}_0)$ using Eq. (7.28) is equal to zero for all neighboring rays. The value of $W(\overline{X}'_0) = W(\overline{X}_0) + \Delta W(\overline{X}_0)$, where $\Delta W(\overline{X}_0)$ is estimated using Eq. (8.25), which uses up to the quadratic term of the Taylor series expansion, is in well agreement with the solution obtained via raytracing for close neighboring rays. However, the accuracy of the estimated solution reduces as the distance of the neighboring ray $\overline{R}'_0$ from the reference ray $\overline{R}_0$ increases.

## 8.5 Change of Wavefront Aberration due to Translation of Source Point

The chief ray of a source point $\overline{P}_0$ is a ray $\overline{R}_{0/\text{chief}} = \begin{bmatrix} \overline{P}_0 & \overline{\ell}_{0/\text{chief}} \end{bmatrix}^T$ with unit directional vector $\overline{\ell}_{0/\text{chief}} = \begin{bmatrix} C\beta_{0/\text{chief}} C(90^\circ + \alpha_{0/\text{chief}}) & C\beta_{0/\text{chief}} S(90^\circ + \alpha_{0/\text{chief}}) & S\beta_{0/\text{chief}} 0 \end{bmatrix}^T$, that is, the ray $\overline{R}_{0/\text{chief}}$ is incident on the center of the aperture (see Fig. 8.1). As the discussion after Eq. (5.32), when the aperture is labeled as the gth boundary surface of the optical system, the change in the chief ray at the aperture, $\Delta\overline{R}_g$, caused by any change in the variable vector of the chief ray, $\Delta\overline{X}_{0/\text{chief}} = \begin{bmatrix} \Delta P_{0x} & \Delta P_{0y} & \Delta P_{0z} & \Delta\alpha_{0/\text{chief}} & \Delta\beta_{0/\text{chief}} \end{bmatrix}^T$, can be obtained as

$$\begin{aligned} \Delta\overline{R}_g &= \begin{bmatrix} \Delta P_{gx} & \Delta P_{gy} & \Delta P_{gz} & \Delta\ell_{gx} & \Delta\ell_{gy} & \Delta\ell_{gz} \end{bmatrix}^T \\ &= \overline{M}_g \overline{M}_{g-1} \cdots \overline{M}_2 \overline{M}_1 \overline{S}_0 \, \Delta\overline{X}_{0/\text{chief}} \\ &= [c_{wv}] \Delta\overline{X}_{0/\text{chief}}. \end{aligned} \tag{8.26}$$

Taking the first and third components of Eq. (8.26) and setting $\Delta P_{gx} = \Delta P_{gz} = 0$, yields

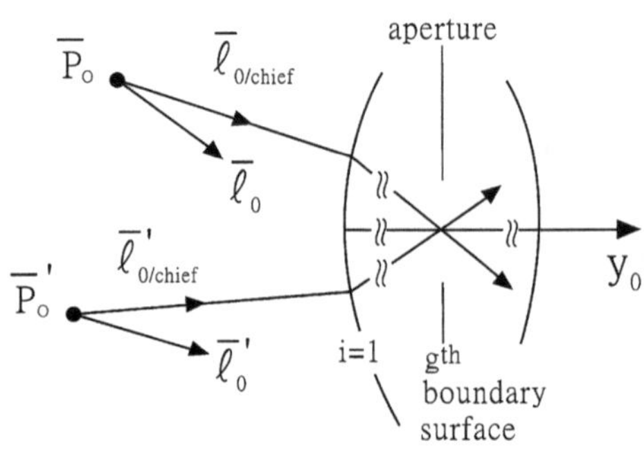


**Fig. 8.1** Feasibility of estimating wavefront aberration via Taylor series expansion when source point $\overline{P}_0$ is translated to a neighboring point $\overline{P}'_0$

$$\begin{bmatrix} \Delta P_{gx} \\ \Delta P_{gz} \end{bmatrix} = \begin{bmatrix} 0 \\ 0 \end{bmatrix} = \begin{bmatrix} c_{11} & c_{12} & c_{13} & c_{14} & c_{15} \\ c_{31} & c_{32} & c_{33} & c_{34} & c_{35} \end{bmatrix} \begin{bmatrix} \Delta P_{0x} \\ \Delta P_{0y} \\ \Delta P_{0z} \\ \Delta\alpha_{0/chief} \\ \Delta\beta_{0/chief} \end{bmatrix}. \tag{8.27}$$

From Eq. (8.27), the change in the polar coordinates $\alpha_{0/chief}$ and $\beta_{0/chief}$ as the reference source point translates from $\overline{P}_0 = [P_{0x} \quad P_{0x} \quad P_{0x}\ 1]^T$ to a neighboring source point $\overline{P}'_0 = [P'_{0x} \quad P'_{0x} \quad P'_{0x}\ 1]^T$ can be obtained as:

$$\begin{aligned} \begin{bmatrix} \Delta\alpha_{0/chief} \\ \Delta\beta_{0/chief} \end{bmatrix} &= \frac{-1}{(c_{14}c_{35} - c_{34}c_{15})} \begin{bmatrix} c_{35} & -c_{15} \\ -c_{34} & c_{14} \end{bmatrix} \begin{bmatrix} c_{11} & c_{12} & c_{13} \\ c_{31} & c_{32} & c_{33} \end{bmatrix} \begin{bmatrix} \Delta P_{0x} \\ \Delta P_{0y} \\ \Delta P_{0z} \end{bmatrix} \\ &= \begin{bmatrix} u_{41} & u_{42} & u_{43} \\ u_{51} & u_{52} & u_{53} \end{bmatrix} \begin{bmatrix} \Delta P_{0x} \\ \Delta P_{0y} \\ \Delta P_{0z} \end{bmatrix}, \end{aligned} \tag{8.28}$$

where $\Delta\overline{P}_0 = [\Delta P_{0x} \quad \Delta P_{0y} \quad \Delta P_{0z}]^T = \overline{P}'_0 - \overline{P}_0 = [P'_{0x} - P_{0x} \quad P'_{0y} - P_{0y} \quad P'_{0z} - P_{0z}]^T$. The change in the variable vector $\Delta\overline{X}_{0/chief}$ between the two chief rays $\overline{R}'_{0/chief}$ and $\overline{R}_{0/chief}$ originating from source points $\overline{P}'_0$ and$\overline{P}_0$, respectively, is given as

$$\begin{aligned} \Delta\overline{X}_{0/chief} &= \begin{bmatrix} \Delta P_{0x} \\ \Delta P_{0y} \\ \Delta P_{0z} \\ \Delta\alpha_{0/chief} \\ \Delta\beta_{0/chief} \end{bmatrix} = \begin{bmatrix} P'_{0x} - P_{0x} \\ P'_{0y} - P_{0y} \\ P'_{0z} - P_{0x} \\ \alpha'_{0/chief} - \alpha_{0/chief} \\ \beta'_{0/chief} - \beta_{0/chief} \end{bmatrix} \\ &= \begin{bmatrix} 1 & 0 & 0 \\ 0 & 1 & 0 \\ 0 & 0 & 1 \\ u_{41} & u_{42} & u_{43} \\ u_{51} & u_{52} & u_{53} \end{bmatrix} \begin{bmatrix} P'_{0x} - P_{0x} \\ P'_{0y} - P_{0y} \\ P'_{0x} - P_{0x} \end{bmatrix} = \begin{bmatrix} 1 & 0 & 0 \\ 0 & 1 & 0 \\ 0 & 0 & 1 \\ u_{41} & u_{42} & u_{43} \\ u_{51} & u_{52} & u_{53} \end{bmatrix} \Delta\overline{P}_0. \end{aligned} \tag{8.29}$$

From Eq. (8.29), $\Delta\overline{X}_{0/chief}$ can be computed using

$$\Delta\overline{X}_{0/chief} = \frac{\partial\overline{X}_{0/chief}}{\partial\overline{P}_0} \Delta\overline{P}_0 \tag{8.30}$$

with

$$\frac{\partial\overline{X}_{0/chief}}{\partial\overline{P}_0} = \begin{bmatrix} 1 & 0 & 0 \\ 0 & 1 & 0 \\ 0 & 0 & 1 \\ u_{41} & u_{42} & u_{43} \\ u_{51} & u_{52} & u_{53} \end{bmatrix}. \tag{8.31}$$

It should be noted that all existing discussions of wavefront aberrations in the literature assume that the source point $\overline{P}_0$ is fixed. Under this condition, the chief ray is unchanged (i.e., $\Delta\overline{X}_{0/chief} = \overline{0}_{5\times 1}$). Now it is possible to develop an algorithm for estimating the change in the wavefront aberration $W\left(\overline{X}'_0\right)$ of a neighboring ray when the source point $\overline{P}_0$ translates to a neighboring point $\overline{P}'_0$ using a Taylor series expansion. Consider the case shown in Fig. 8.1 involving four rays, namely two rays originating from the reference source point $\overline{P}_0$ (i.e., a general ray $\overline{R}_0 = \left[\overline{P}_0 \quad \overline{\ell}_0\right]^T$ with variable vector $\overline{X}_0 = \left[P_{0x} \quad P_{0y} \quad P_{0z} \quad \alpha_0 \quad \beta_0\right]^T$ and its chief ray $\overline{R}_{0/chief} = \left[\overline{P}_0 \quad \overline{\ell}_{0/chief}\right]^T$ with variable vector $\overline{X}_{0/chief} = \left[P_{0x} \quad P_{0y} \quad P_{0z} \quad \alpha_{0/chief} \quad \beta_{0/chief}\right]^T$) and two rays originating from a neighboring source point $\overline{P}'_0$ (i.e., a general ray $\overline{R'_0} = \left[\overline{P}'_0 \quad \overline{\ell}'_0\right]^T$ with variable vector $\overline{X}'_0 = \left[P'_{0x} \quad P'_{0y} \quad P'_{0z} \quad \alpha'_0 \quad \beta'_0\right]^T$ and its chief source ray $\overline{R}'_{0/chief} = \left[\overline{P}'_0 \quad \overline{\ell}'_{0/chief}\right]^T$ with variable vector $\overline{X}'_{0/chief} = \left[P'_{0x} \quad P'_{0y} \quad P'_{0z} \quad \alpha'_{0/chief} \quad \beta'_{0/chief}\right]^T$. When the source point translates from point $\overline{P}_0$ to a neighboring source point $\overline{P}'_0$, the changes in the wavefront aberration $\Delta W\left(\overline{X}_0\right)$ can be estimated by Taylor series expansion with either first-order or up to the quadratic term,

$$
\begin{aligned}
\Delta W(\overline{X}_0) &= \left(\frac{\partial OPL\left(\overline{P}_0, \overline{P}_n\right)}{\partial \overline{X}_0}\right)_{\overline{R}_0} \Delta\overline{X}_0 - \left(\frac{\partial OPL\left(\overline{P}_0, \overline{P}_n\right)}{\partial \overline{X}_0}\right)_{\overline{R}_{0/chief}} \Delta\overline{X}_{0/chief} \\
&= \left(\sum_{i=1}^{i=n} \frac{\partial OPL_i}{\partial \overline{X}_0}\right)_{\overline{R}_0} \Delta\overline{X}_0 - \left(\sum_{i=1}^{i=n} \frac{\partial OPL_i}{\partial \overline{X}_0}\right)_{\overline{R}_{0/chief}} \Delta\overline{X}_{0/chief},
\end{aligned}
\tag{8.32}
$$

$$
\begin{aligned}
\Delta W(\overline{X}_0) &= \left(\frac{\partial OPL\left(\overline{P}_0, \overline{P}_n\right)}{\partial \overline{X}_0}\right)_{\overline{R}_0} \Delta\overline{X}_0 - \left(\frac{\partial OPL\left(\overline{P}_0, \overline{P}_n\right)}{\partial \overline{X}_0}\right)_{\overline{R}_{0/chief}} \Delta\overline{X}_{0/chief} \\
&\quad + \frac{1}{2}\left(\Delta\overline{X}_0\right)^T \left(\frac{\partial^2 OPL\left(\overline{P}_0, \overline{P}_n\right)}{\partial \overline{X}_0^2}\right)_{\overline{R}_0} \Delta\overline{X}_0 \\
&\quad - \frac{1}{2}\left(\Delta\overline{X}_{0/chief}\right)^T \left(\frac{\partial^2 OPL\left(\overline{P}_0, \overline{P}_n\right)}{\partial \overline{X}_0^2}\right)_{\overline{R}_{0/chief}} \Delta\overline{X}_{0/chief} \\
&= \left(\sum_{i=1}^{i=n} \frac{\partial OPL_i}{\partial \overline{X}_0}\right)_{\overline{R}_0} \Delta\overline{X}_0 - \left(\sum_{i=1}^{i=n} \frac{\partial OPL_i}{\partial \overline{X}_0}\right)_{\overline{R}_{0/chief}} \Delta\overline{X}_{0/chief} \\
&\quad + \frac{1}{2}\left(\Delta\overline{X}_0\right)^T \left(\sum_{i=1}^{i=n} \frac{\partial OPL_i}{\partial \overline{X}_0^2}\right)_{\overline{R}_0} \Delta\overline{X}_0 \\
&\quad - \frac{1}{2}\left(\Delta\overline{X}_{0/chief}\right)^T \left(\sum_{i=1}^{i=n} \frac{\partial^2 OPL_i}{\partial \overline{X}_0^2}\right)_{\overline{R}_{0/chief}} \Delta\overline{X}_{0/chief},
\end{aligned}
\tag{8.33}
$$

where $()_{\overline{R}_0}$ and $()_{\overline{R}_{0/chief}}$ indicate that the matrices within () are evaluated in terms of the reference ray $\overline{R}_0$ and the chief ray $\overline{R}_{0/chief}$, respectively, and

$$\Delta\overline{X}_0 = \begin{bmatrix} P'_{0x} - P_{0x} \\ P'_{0y} - P_{0y} \\ P'_{0x} - P_{0x} \\ \alpha'_0 - \alpha_0 \\ \beta'_0 - \beta_0 \end{bmatrix}. \tag{8.34}$$

Equations (8.32) and (8.33) can be rewritten as Eqs. (8.35) and (8.36) when Eq. (8.31) is used, i.e.,

$$\Delta W(\overline{X}_0) = \left(\sum_{i=1}^{i=n} \frac{\partial OPL_i}{\partial \overline{X}_0}\right)_{\overline{R}_0} \Delta\overline{X}_0 - \left(\sum_{i=1}^{i=n} \frac{\partial OPL_i}{\partial \overline{X}_0}\right)_{\overline{R}_{0/chief}} \frac{\partial \overline{X}_{0/chief}}{\partial \overline{P}_0} \Delta\overline{P}_0, \tag{8.35}$$

$$\begin{aligned} \Delta W(\overline{X}_0) = & \left(\sum_{i=1}^{i=n} \frac{\partial OPL_i}{\partial \overline{X}_0}\right)_{\overline{R}_0} \Delta\overline{X}_0 - \left(\sum_{i=1}^{i=n} \frac{\partial OPL_i}{\partial \overline{X}_0}\right)_{\overline{R}_{0/chief}} \frac{\partial \overline{X}_{0/chief}}{\partial \overline{P}_0} \Delta\overline{P}_0 \\ & + \frac{1}{2}(\Delta\overline{X}_0)^T \left(\sum_{i=1}^{i=n} \frac{\partial OPL_i}{\partial \overline{X}_0^2}\right)_{\overline{R}_0} \Delta\overline{X}_0 \\ & - \frac{1}{2}(\Delta\overline{P}_0)^T \left(\frac{\partial \overline{X}_{0/chief}}{\partial \overline{P}_0}\right)^T \left(\sum_{i=1}^{i=n} \frac{\partial OPL_i}{\partial \overline{X}_0^2}\right)_{\overline{R}_{0/chief}} \frac{\partial \overline{X}_{0/chief}}{\partial \overline{P}_0} \Delta\overline{P}_0, \end{aligned} \tag{8.36}$$

where $\Delta\overline{P}_0 = \begin{bmatrix} P'_{0x} - P_{0x} & P'_{0y} - P_{0y} & P'_{0z} - P_{0z} \end{bmatrix}^T$ and $\Delta\overline{X}_0$ are given in Eq. (8.34). The wavefront aberration $W\left(\overline{X}'_0\right)$ of the neighboring ray $\overline{R}'_0$ can then be estimated as the sum of the wavefront aberration $W(\overline{X}_0)$ of the reference ray $\overline{R}_0 = \begin{bmatrix} \overline{P}_0 & \overline{\ell}_0 \end{bmatrix}^T$ and $\Delta W(\overline{X}_0)$ given in Eq. (8.35) or (8.36).

*Example 8.2* Table 8.3 compares the results obtained via raytracing and Eqs. (7.28), (8.25), (8.35) and (8.36), respectively, for $W\left(\overline{X}'_0\right)$ given a reference ray with a variable vector of $\overline{X}_0 = [0 \quad -507 \quad 0 \quad 0.7° \quad -0.4°]^T$ and a translated source ray with $\Delta\overline{X}_0 = [0 \quad \Delta P_{0y} \quad 0 \quad 0.2° \quad 0.2°]^T$, where $\Delta P_{0y} = 6$, 12 or 18, respectively. The chief ray $\overline{R}'_{0/chief}$ is evaluated by $\overline{X}'_{0/chief} = \overline{X}_{0/chief} + \Delta\overline{X}_{0/chief}$ using Eq. (8.30). In general, the results in Table 8.3 show that Eqs. (8.35) and (8.36) yield accurate estimates of the waveform aberration for all values of the source point translation. However, it is seen that Eqs. (7.28) and (8.25) are unable to estimate the wavefront aberration $W\left(\overline{X}'_0\right)$ given a translation of the source point $\overline{P}_0$ to a neighboring source point $\overline{P}'_0$. This is due to the fact that Eqs. (7.28) and (8.25) do not consider the effects of the translation of source point, $\Delta\overline{P}_0 = [0 \quad \Delta P_{0y} \quad 0 \quad 0]^T$.

**Table 8.3** Comparison of estimated wavefront aberration for various translations of source point

| | $\Delta P_{0y} = 6$ | | $\Delta P_{0y} = 12$ | | $\Delta P_{0y} = 18$ | |
|---|---|---|---|---|---|---|
| | $W(\overline{X}'_0)$ | Error (%) | $W(\overline{X}'_0)$ | Error % | $W(\overline{X}'_0)$ | Error (%) |
| Raytracing | 0.010252 | | 0.010865 | | 0.011465 | |
| Equation (7.28) | −5.990622 | $-\infty$ | −11.99015 | $-\infty$ | −17.98968 | $-\infty$ |
| Equation (8.25) | −5.989751 | $-\infty$ | −11.98920 | $-\infty$ | −17.98866 | $-\infty$ |
| Equation (8.35) | 0.009378 | −8.52 | 0.009852 | −9.33 | 0.010325 | −9.94 |
| Equation (8.36) | 0.010248 | −0.04 | 0.010797 | −0.63 | 0.011334 | −1.15 |

## 8.6 Wavefront Shape, Irradiance, and Caustic Surface along a Ray Path

Figure 8.2 illustrates a general ray $\overline{R}_i = \left[\overline{P}_i \quad \overline{\ell}_i\right]^T = [P_{ix} \quad P_{iy} \quad P_{iz} \quad \ell_{ix} \quad \ell_{iy} \quad \ell_{iz}]^T$ reflected/refracted at the ith boundary surface in an optical system. Any intermediate point $\overline{P}'_i$ along this ray as it travels from $\overline{P}_i$ to $\overline{P}_{i+1}$, where $\overline{P}_{i+1}$ is the incidence point on the (i + 1)th boundary surface, can be expressed as

$$\overline{P}'_i = \begin{bmatrix} P'_{ix} \\ P'_{iy} \\ P'_{iz} \\ 1 \end{bmatrix} = \begin{bmatrix} P_{ix} \\ P_{iy} \\ P_{iz} \\ 1 \end{bmatrix} + \lambda \begin{bmatrix} \ell_{ix} \\ \ell_{iy} \\ \ell_{iz} \\ 0 \end{bmatrix}, \tag{8.37}$$

where $\lambda$ is the geometrical length from $\overline{P}_i$ to $\overline{P}'_i$. According to optical theory, the wavefront $\overline{\Omega}$ is the loci of all the points having the same OPL from the source point $\overline{P}_0$. Let $OPL\left(\overline{P}_0, \overline{P}'_i\right)$ be the OPL of the ray measured from the source point $\overline{P}_0$ to the intermediate point $\overline{P}'_i$ of Fig. 8.2. The condition for the wavefront $\overline{\Omega}$ having $OPL = OPL_{constant}$ is given as

$$\begin{aligned} OPL\left(\overline{P}_0, \overline{P}'_i\right) &= OPL_{constant} = \xi_0\lambda_1 + \xi_1\lambda_2 + \cdots + \xi_{i-1}\lambda_i + \xi_i\lambda \\ &= OPL_1 + OPL_2 + \cdots + OPL_i + \xi_i\lambda. \end{aligned} \tag{8.38}$$

Substituting $\lambda$ in Eq. (8.38) into Eq. (8.37), the following expression is obtained for the wavefront $\overline{\Omega}$ having a constant optical path length $OPL\left(\overline{P}_0, \overline{P}'_i\right) OPL_{constant}$:

$$\overline{\Omega} = \begin{bmatrix} \Omega_x \\ \Omega_y \\ \Omega_z \\ 1 \end{bmatrix} = \begin{bmatrix} P_{ix} \\ P_{iy} \\ P_{iz} \\ 1 \end{bmatrix} + \frac{[OPL_{constant} - (OPL_1 + OPL_2 + \cdots + OPL_i)]}{\xi_i} \begin{bmatrix} \ell_{ix} \\ \ell_{iy} \\ \ell_{iz} \\ 0 \end{bmatrix}. \tag{8.39}$$

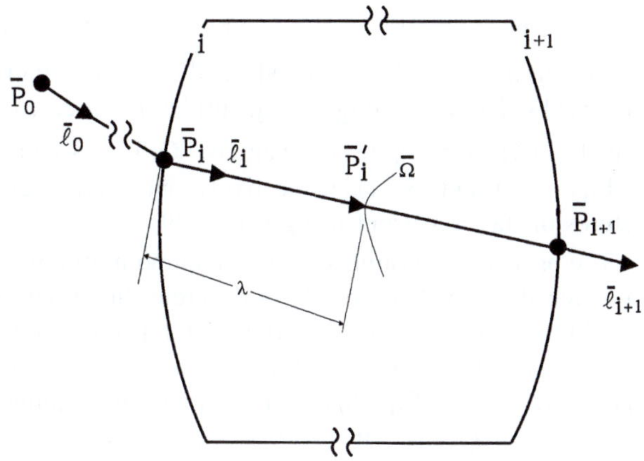


**Fig. 8.2** Wavefront $\overline{\Omega}$, defined as loci of all points having constant OPL from source point $\overline{P}_0$

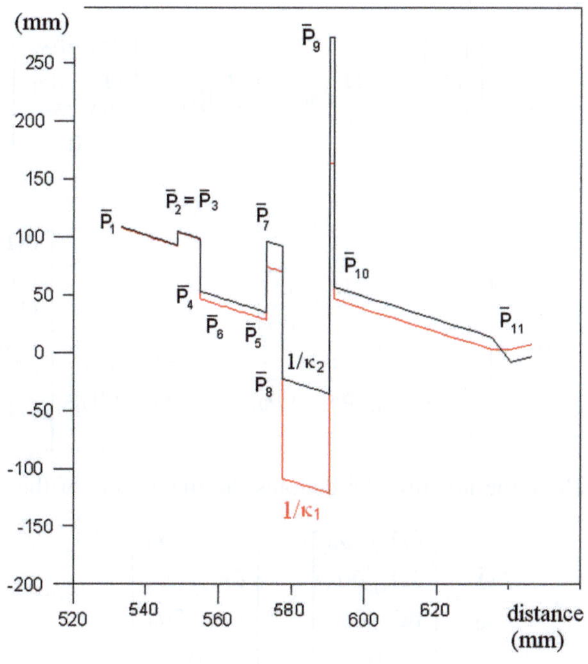


**Fig. 8.3** Variation of principal radii $1/\kappa_1$ and $1/\kappa_2$ with travel distance of chief ray originating from source point $\overline{P}_0 = [0 \quad -507 \quad 170 \; 1]^T$ in an axis-symmetrical system shown in Fig. 2.1

In this text, the first and second fundamental forms of the wavefront $\overline{\Omega}$ defined in Eq. (8.39) are used to investigate the wavefront principal curvatures and the irradiance. Since $\overline{R}_i = [P_i \quad \bar{\ell}_i]^T$ and $OPL_i$ (i = 1 to i = i) in Eq. (8.39) are both functions of $\alpha_0$ and $\beta_0$, the following quantities are required in order to investigate the shape of the wavefront $\overline{\Omega}$ from Eq. (8.39): (1) the Jacobian matrix $\partial\overline{R}_i/\partial\overline{X}_0$; (2) the Hessian matrix $\partial^2\overline{R}_i/\partial\overline{X}_0^2$; (3) the Jacobian matrix $\partial OPL_i/\partial\overline{X}_0$, and (4) the

Hessian matrix $\partial^2 \mathrm{OPL_i}/\partial \overline{X}_0^2$. In previous studies, the Jacobian matrix $\partial \overline{R}_i/\partial \overline{X}_0$ was given in Eq. (5.33). Meanwhile, the Hessian matrix $\partial^2 \overline{R}_i/\partial \overline{X}_0^2$ was derived in Eq. (8.11). The remaining two quantities required to investigate the shape of the wavefront $\overline{\Omega}$ from Eq. (8.39), namely $\partial \mathrm{OPL_i}/\partial \overline{X}_0$ and $\partial^2 \mathrm{OPL_i}/\partial \overline{X}_0^2$, were derived in Eqs. (7.4) and (8.21), respectively. Now one can compute the principal curvatures of the wavefront using Eq. (8.39).

The principal curvatures at any given point of a wavefront $\overline{\Omega}$ indicate the extent to which the wavefront bends in different directions at that point. In this section, the differential geometry-based method [3] is used to determine the principal curvatures and principal directions of the wavefront along the ray path through an optical system. In Eq. (8.39), the wavefront is defined in terms of the spherical coordinates $\alpha_0$ and $\beta_0$. When the wavefront is regular (i.e., $\partial \overline{\Omega}/\partial \alpha_0 \times \partial \overline{\Omega}/\partial \beta_0 \neq \overline{0}$), the tangent vector $\bar{t}_\Omega$ and unit normal vector $\bar{n}_\Omega$ can be determined respectively as

$$\bar{t}_\Omega = \begin{bmatrix} t_{\Omega x} \\ t_{\Omega y} \\ t_{\Omega z} \\ 0 \end{bmatrix} = \frac{\partial \overline{\Omega}}{\partial \alpha_0} d\alpha_0 + \frac{\partial \overline{\Omega}}{\partial \beta_0} d\beta_0 = \begin{bmatrix} \partial \Omega_x/\partial \alpha_0 \\ \partial \Omega_y/\partial \alpha_0 \\ \partial \Omega_z/\partial \alpha_0 \\ 0 \end{bmatrix} d\alpha_0 + \begin{bmatrix} \partial \Omega_x/\partial \beta_0 \\ \partial \Omega_y/\partial \beta_0 \\ \partial \Omega_z/\partial \beta_0 \\ 0 \end{bmatrix} d\beta_0, \tag{8.40}$$

$$\bar{n}_\Omega = \begin{bmatrix} n_{\Omega x} \\ n_{\Omega y} \\ n_{\Omega z} \\ 0 \end{bmatrix} = \frac{\dfrac{\partial \overline{\Omega}}{\partial \alpha_0} \times \dfrac{\partial \overline{\Omega}}{\partial \beta_0}}{\left\| \dfrac{\partial \overline{\Omega}}{\partial \alpha_0} \times \dfrac{\partial \overline{\Omega}}{\partial \beta_0} \right\|} = \frac{1}{\left\| \dfrac{\partial \overline{\Omega}}{\partial \alpha_0} \times \dfrac{\partial \overline{\Omega}}{\partial \beta_0} \right\|} \begin{bmatrix} \dfrac{\partial \Omega_y}{\partial \alpha_0}\dfrac{\partial \Omega_z}{\partial \beta_0} - \dfrac{\partial \Omega_z}{\partial \alpha_0}\dfrac{\partial \Omega_y}{\partial \beta_0} \\ \dfrac{\partial \Omega_z}{\partial \alpha_0}\dfrac{\partial \Omega_x}{\partial \beta_0} - \dfrac{\partial \Omega_x}{\partial \alpha_0}\dfrac{\partial \Omega_z}{\partial \beta_0} \\ \dfrac{\partial \Omega_x}{\partial \alpha_0}\dfrac{\partial \Omega_y}{\partial \beta_0} - \dfrac{\partial \Omega_y}{\partial \alpha_0}\dfrac{\partial \Omega_x}{\partial \beta_0} \\ 0 \end{bmatrix}, \tag{8.41}$$

where the notation $\| \, \|$ denotes the magnitude of the corresponding vector, and

$$\frac{\partial \overline{\Omega}}{\partial \alpha_0} = \begin{bmatrix} \partial \Omega_x/\partial \alpha_0 \\ \partial \Omega_y/\partial \alpha_0 \\ \partial \Omega_z/\partial \alpha_0 \\ 0 \end{bmatrix} = \begin{bmatrix} \partial P_{ix}/\partial \alpha_0 \\ \partial P_{iy}/\partial \alpha_0 \\ \partial P_{iz}/\partial \alpha_0 \\ 0 \end{bmatrix}$$
$$+ \frac{[\mathrm{OPL_{constant}} - (\mathrm{OPL_1} + \mathrm{OPL_2} + \cdots + \mathrm{OPL_i})]}{\xi_i} \begin{bmatrix} \partial \ell_{ix}/\partial \alpha_0 \\ \partial \ell_{iy}/\partial \alpha_0 \\ \partial \ell_{iz}/\partial \alpha_0 \\ 0 \end{bmatrix}$$

$$-\frac{1}{\xi_i}\left(\frac{\partial OPL_1}{\partial\alpha_0}+\frac{\partial OPL_2}{\partial\alpha_0}+\cdots+\frac{\partial OPL_i}{\partial\alpha_0}\right)\begin{bmatrix}\ell_{ix}\\ \ell_{iy}\\ \ell_{iz}\\ 0\end{bmatrix}, \tag{8.42}$$

$$\frac{\partial\overline{\Omega}}{\partial\beta_0}=\begin{bmatrix}\partial\Omega_x/\partial\beta_0\\ \partial\Omega_y/\partial\beta_0\\ \partial\Omega_z/\partial\beta_0\\ 0\end{bmatrix}=\begin{bmatrix}\partial P_{ix}/\partial\beta_0\\ \partial P_{iy}/\partial\beta_0\\ \partial P_{iz}/\partial\beta_0\\ 0\end{bmatrix}$$

$$+\frac{[OPL_{constant}-(OPL_1+OPL_2+\cdots+OPL_i)]}{\xi_i}\begin{bmatrix}\partial\ell_{ix}/\partial\beta_0\\ \partial\ell_{iy}/\partial\beta_0\\ \partial\ell_{iz}/\partial\beta_0\\ 0\end{bmatrix} \tag{8.43}$$

$$-\frac{1}{\xi_i}\left(\frac{\partial OPL_1}{\partial\beta_0}+\frac{\partial OPL_2}{\partial\beta_0}+\cdots+\frac{\partial OPL_i}{\partial\beta_0}\right)\begin{bmatrix}\ell_{ix}\\ \ell_{iy}\\ \ell_{iz}\\ 0\end{bmatrix}.$$

It can be noted from Fig. 2.2 that when the numerator of $\overline{n}_\Omega$ shown in Eq. (8.41) is $\partial\overline{\Omega}/\partial\alpha_0 \times \partial\overline{\Omega}/\partial\beta_0$, the component $n_{\Omega y}$ of $\overline{n}_\Omega$ is always positive. This indicates that the angle between $\overline{n}_\Omega$ and $y_o$ axis is an acute angle (see Figs. 2.1 and 3.12), since $\alpha_0$ is always confined in the range of $0° < \alpha_0 + 90° < 180°$. This finding is important since the principal curvature is taken to be positive if the wavefront turns in the same direction as the chosen unit normal vector $\overline{n}_\Omega$, and negative otherwise. The magnitude of vector $\partial\overline{\Omega}/\partial\alpha_0 \times \partial\overline{\Omega}/\partial\beta_0$in Eq. (8.41) is calculate as

$$\left\|\frac{\partial\overline{\Omega}}{\partial\alpha_0}\times\frac{\partial\overline{\Omega}}{\partial\beta_0}\right\|=\sqrt{\left(\frac{\partial\Omega_y}{\partial\alpha_0}\frac{\partial\Omega_z}{\partial\beta_0}-\frac{\partial\Omega_z}{\partial\alpha_0}\frac{\partial\Omega_y}{\partial\beta_0}\right)^2+\left(\frac{\partial\Omega_z}{\partial\alpha_0}\frac{\partial\Omega_x}{\partial\beta_0}-\frac{\partial\Omega_x}{\partial\alpha_0}\frac{\partial\Omega_z}{\partial\beta_0}\right)^2+\left(\frac{\partial\Omega_x}{\partial\alpha_0}\frac{\partial\Omega_y}{\partial\beta_0}-\frac{\partial\Omega_y}{\partial\alpha_0}\frac{\partial\Omega_x}{\partial\beta_0}\right)^2}. \tag{8.44}$$

In differential geometry, the first fundamental form is the inner product on the tangent space of a surface. As such, it permits the curvature and metric properties of the wavefront (e.g., the length and area) to be calculated in a manner consistent with the ambient space. The first fundamental form of the wavefront given in Eq. (8.39) is expressed as

$$I=\overline{t}_\Omega\bullet\overline{t}_\Omega=Ed\alpha_0^2+2Fd\alpha_0d\beta_0+Gd\beta_0^2, \tag{8.45}$$

in which the coefficients are given by

$$E=\frac{\partial\Omega_x}{\partial\alpha_0}\frac{\partial\Omega_x}{\partial\alpha_0}+\frac{\partial\Omega_y}{\partial\alpha_0}\frac{\partial\Omega_y}{\partial\alpha_0}+\frac{\partial\Omega_z}{\partial\alpha_0}\frac{\partial\Omega_z}{\partial\alpha_0}, \tag{8.45a}$$

$$F = \frac{\partial \Omega_x}{\partial \alpha_0}\frac{\partial \Omega_x}{\partial \beta_0} + \frac{\partial \Omega_y}{\partial \alpha_0}\frac{\partial \Omega_y}{\partial \beta_0} + \frac{\partial \Omega_z}{\partial \alpha_0}\frac{\partial \Omega_z}{\partial \beta_0}, \tag{8.45b}$$

$$G = \frac{\partial \Omega_x}{\partial \beta_0}\frac{\partial \Omega_x}{\partial \beta_0} + \frac{\partial \Omega_y}{\partial \beta_0}\frac{\partial \Omega_y}{\partial \beta_0} + \frac{\partial \Omega_z}{\partial \beta_0}\frac{\partial \Omega_z}{\partial \beta_0}. \tag{8.45c}$$

The second fundamental form of a surface is used to measure the change of the unit normal direction from point to point on the wavefront. Together with the first fundamental form, it serves to determine the principal curvatures of the wavefront. In this text, the second fundamental form of the wavefront is defined as the projections of the second-order derivatives of $\overline{\Omega}$ onto its unit normal vector $\bar{n}_\Omega$, i.e.,

$$II = Ld\alpha_0^2 + 2Md\alpha_0 d\beta_0 + Nd\beta_0^2, \tag{8.46}$$

where the coefficients are given by

$$\begin{aligned} L = \frac{\partial^2 \overline{\Omega}}{\partial \alpha_0 \partial \alpha_0} \bullet \bar{n}_\Omega = \frac{1}{\|\partial \overline{\Omega}/\partial \alpha_0 \times \partial \overline{\Omega}/\partial \beta_0\|} \Bigg[ \frac{\partial^2 \Omega_x}{\partial \alpha_0 \partial \alpha_0} \left( \frac{\partial \Omega_y}{\partial \alpha_0}\frac{\partial \Omega_z}{\partial \beta_0} - \frac{\partial \Omega_z}{\partial \alpha_0}\frac{\partial \Omega_y}{\partial \beta_0} \right) \\ + \frac{\partial^2 \Omega_y}{\partial \alpha_0 \partial \alpha_0} \left( \frac{\partial \Omega_z}{\partial \alpha_0}\frac{\partial \Omega_x}{\partial \beta_0} - \frac{\partial \Omega_x}{\partial \alpha_0}\frac{\partial \Omega_z}{\partial \beta_0} \right) + \frac{\partial^2 \Omega_z}{\partial \alpha_0 \partial \alpha_0} \left( \frac{\partial \Omega_x}{\partial \alpha_0}\frac{\partial \Omega_y}{\partial \beta_0} - \frac{\partial \Omega_y}{\partial \alpha_0}\frac{\partial \Omega_x}{\partial \beta_0} \right) \Bigg], \end{aligned} \tag{8.46a}$$

$$\begin{aligned} M = \frac{\partial^2 \overline{\Omega}}{\partial \alpha_0 d\beta_0} \bullet \bar{n}_\Omega = \frac{1}{\|\partial \overline{\Omega}/\partial \alpha_0 \times \partial \overline{\Omega}/\partial \beta_0\|} \Bigg[ \frac{\partial^2 \Omega_x}{\partial \alpha_0 \partial \beta_0} \left( \frac{\partial \Omega_y}{\partial \alpha_0}\frac{\partial \Omega_z}{\partial \beta_0} - \frac{\partial \Omega_z}{\partial \alpha_0}\frac{\partial \Omega_y}{\partial \beta_0} \right) \\ + \frac{\partial^2 \Omega_y}{\partial \alpha_0 \partial \beta_0} \left( \frac{\partial \Omega_z}{\partial \alpha_0}\frac{\partial \Omega_x}{\partial \beta_0} - \frac{\partial \Omega_x}{\partial \alpha_0}\frac{\partial \Omega_z}{\partial \beta_0} \right) + \frac{\partial^2 \Omega_z}{\partial \alpha_0 \partial \beta_0} \left( \frac{\partial \Omega_x}{\partial \alpha_0}\frac{\partial \Omega_y}{\partial \beta_0} - \frac{\partial \Omega_y}{\partial \alpha_0}\frac{\partial \Omega_x}{\partial \beta_0} \right) \Bigg], \end{aligned} \tag{8.46b}$$

$$\begin{aligned} N = \frac{\partial^2 \overline{\Omega}}{\partial \beta_0 \partial \beta_0} \bullet \bar{n}_\Omega = \frac{1}{\|\partial \overline{\Omega}/\partial \alpha_0 \times \partial \overline{\Omega}/\partial \beta_0\|} \Bigg[ \frac{\partial^2 \Omega_x}{\partial \beta_0 \partial \beta_0} \left( \frac{\partial \Omega_y}{\partial \alpha_0}\frac{\partial \Omega_z}{\partial \beta_0} - \frac{\partial \Omega_z}{\partial \alpha_0}\frac{\partial \Omega_y}{\partial \beta_0} \right), \\ + \frac{\partial^2 \Omega_y}{\partial \beta_0 \partial \beta_0} \left( \frac{\partial \Omega_z}{\partial \alpha_0}\frac{\partial \Omega_x}{\partial \beta_0} - \frac{\partial \Omega_x}{\partial \alpha_0}\frac{\partial \Omega_z}{\partial \beta_0} \right) + \frac{\partial^2 \Omega_z}{\partial \beta_0 \partial \beta_0} \left( \frac{\partial \Omega_x}{\partial \alpha_0}\frac{\partial \Omega_y}{\partial \beta_0} - \frac{\partial \Omega_y}{\partial \alpha_0}\frac{\partial \Omega_x}{\partial \beta_0} \right) \Bigg] \end{aligned} \tag{8.46c}$$

in which,

$$\frac{\partial^2\overline{\Omega}}{\partial\alpha_0^2}=\begin{bmatrix}\partial^2\Omega_x/\partial\alpha_0^2\\ \partial^2\Omega_y/\partial\alpha_0^2\\ \partial^2\Omega_z/\partial\alpha_0^2\\ 0\end{bmatrix}=\begin{bmatrix}\partial^2P_{ix}/\partial\alpha_0^2\\ \partial^2P_{iy}/\partial\alpha_0^2\\ \partial^2P_{iz}/\partial\alpha_0^2\\ 0\end{bmatrix}$$

$$+\frac{[OPL_{constant}-(OPL_1+OPL_2+\cdots+OPL_i)]}{\xi_i}\begin{bmatrix}\partial^2\ell_{ix}/\partial\alpha_0^2\\ \partial^2\ell_{iy}/\partial\alpha_0^2\\ \partial^2\ell_{iz}/\partial\alpha_0^2\\ 0\end{bmatrix}$$

$$-\frac{1}{\xi_i}\left(\frac{\partial^2OPL_1}{\partial\alpha_0^2}+\frac{\partial^2OPL_2}{\partial\alpha_0^2}+\cdots+\frac{\partial^2OPL_i}{\partial\alpha_0^2}\right)\begin{bmatrix}\ell_{ix}\\ \ell_{iy}\\ \ell_{iz}\\ 0\end{bmatrix}. \tag{8.47}$$

The components of $\partial^2\overline{\Omega}/\partial\alpha_0\partial\beta_0=\left[\partial^2\Omega_x/\partial\alpha_0\partial\beta_0 \quad \partial^2\Omega_y/\partial\alpha_0\partial\beta_0 \quad \partial^2\Omega_z/\partial\alpha_0\partial\beta_0 \quad 0\right]^T$ and $\partial^2\overline{\Omega}/\partial\beta_0^2=\left[\partial^2\Omega_x/\partial\beta_0^2 \quad \partial^2\Omega_y/\partial\beta_0^2 \quad \partial^2\Omega_z/\partial\beta_0^2 \quad 0\right]^T$ in Eqs. (8.46a), (8.46b) and (8.46c) can be obtained simply by replacing $\alpha_0^2$ in Eq. (8.47) with $\alpha_0\beta_0$ and $\beta_0^2$, respectively.

The principal curvatures of the wavefront at $\overline{P}'_i$ shown in Fig. 8.2 can be determined in several but equivalent ways. One elegant way involves solving the roots of the equation

$$\det\left(\begin{bmatrix}L-\kappa E & M-\kappa F\\ M-\kappa F & N-\kappa G\end{bmatrix}\right)=0. \tag{8.48}$$

If $\kappa$ is one of the principal curvatures, Eq. (8.48) is not invertible. Thus, there exists a non-zero unit vector column vector $\overline{T}=\left[h_1 \quad h_2\right]^T$ (denoted as the *principal direction* in differential geometry) with real number components such that

$$\begin{bmatrix}L-\kappa E & M-\kappa F\\ M-\kappa F & N-\kappa G\end{bmatrix}\begin{bmatrix}h_1\\ h_2\end{bmatrix}=0. \tag{8.49}$$

Let $\kappa_1$ and $\kappa_2$ be the principal curvatures at any designated point of the wavefront $\overline{\Omega}$, then: (1) $\kappa_1$ and $\kappa_2$ are real numbers; (2) if $\kappa_1=\kappa_2=\kappa$, then every tangent vector to the wavefront is a principal vector; (3) if $\kappa_1\neq\kappa_2$, then any two non-zero principal tangent vectors $\overline{T}_1$ and $\overline{T}_2$ corresponding to $\kappa_1$ and $\kappa_2$, respectively, are perpendicular (note that orthogonality of the principal tangent vectors is not general, but true if the second fundamental form is symmetric.); (4) the Gaussian curvature $\kappa_1\kappa_2$ is given by $\kappa_1\kappa_2=(LN-M^2)/(EG-F^2)$; and (5) the principal directions are unchanged when the ray travels through a medium with

**Table 8.4** Principal radii of curvature of wavefront at each boundary surface encountered by the chief ray originating from $\overline{P}_0 = [0 \quad -507 \quad 170 \quad 1]^T$ in the axis-symmetrical system shown in Fig. 2.1

| Incidence point $\overline{P}_i$ | $\overline{P}_1$ | $\overline{P}_2$ | $\overline{P}_3$ | $\overline{P}_4$ | $\overline{P}_5$ |
|---|---|---|---|---|---|
| Principal | 107.59722 | 103.94575 | 103.94575 | 46.33882 | 43.00113 |
| Radii | 108.45233 | 104.57090 | 104.57090 | 53.02069 | 49.68300 |
| | $\overline{P}_6$ | $\overline{P}_7$ | $\overline{P}_8$ | $\overline{P}_9$ | $\overline{P}_{10}$ |
| Principal | 34.59148 | 74.06272 | −108.87190 | 164.39263 | 46.36329 |
| Radii | 41.27335 | 96.17959 | −22.70970 | 273.53115 | 57.16198 |

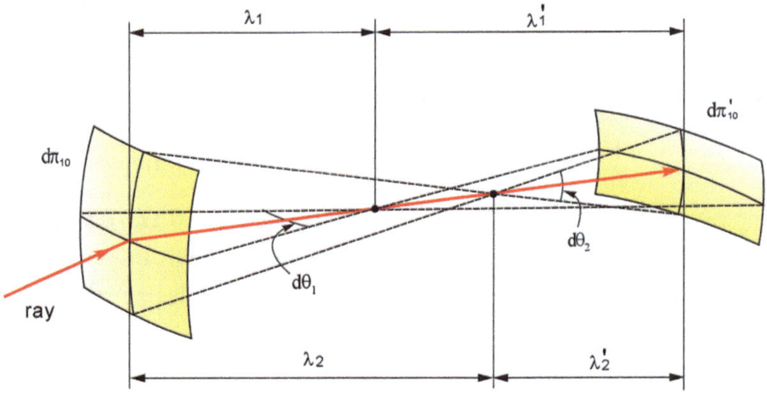


**Fig. 8.4** Element of wavefront after emerging from 10th boundary surface (modified from Fig. 2.1 of [12])

a constant refractive index. Furthermore, as described above, the curvature is taken to be positive if the wavefront turns in the same direction as the unit normal vector $\overline{n}_\Omega$. Thus, when looking in the direction of $\overline{n}_\Omega$, all concave wavefronts (i.e., diverging fronts) have negative curvatures, while all convex wavefronts (i.e., converging fronts) have positive ones.

*Example 8.3* Consider the wavefront along the path of the chief ray originating from a source point $\overline{P}_0 = [0 \quad -507 \quad 170 \quad 1]^T$in the axis-symmetrical optical system shown in Fig. 2.1. The principal radii of curvature ($1/\kappa_1$ and$1/\kappa_2$) of the wavefront at incidence points $\overline{P}_i$ (i = 1 to i = 10) are listed in Table 8.4. The variations of $1/\kappa_1$ and $1/\kappa_2$ with the travel distance of the chief ray are illustrated in Fig. 8.3. The wavefront from the source point $\overline{P}_0$ to point $\overline{P}_1$ is spherical. As a result, both principal radii of curvature are equal to the radius of the sphere and there are no distinguished principal directions. As discussed above, diverging wavefronts have a negative curvature while converging wavefronts have a positive curvature. It is noted from Fig. 8.3 that the wavefront along the chief ray diverges from $\overline{P}_8$ to $\overline{P}_9$.

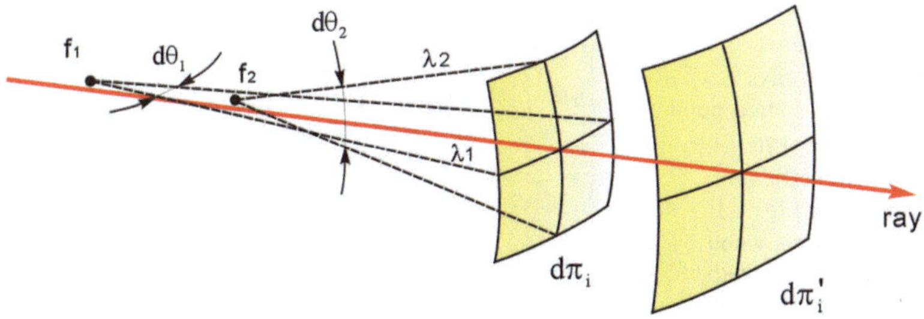


**Fig. 8.5** Wavefront and principal radii (modified from Fig. 3 of [10])

It is also noted that the principal radii of curvature approach zero as the ray approaches the 11th boundary surface (i.e., the image plane); indicating rays originating from $\overline{P}_0$ are focusing. Note that this finding is confirmed by the fact that the unit normal vector $\overline{n}_\Omega$ changes its sign as the ray approaches $\overline{P}_{11}$ (see Fig. 8.4).

*Example 8.4* Figure 8.5 shows a typical wavefront $\overline{\Omega}$ and the corresponding principal radii for a ray traveling through a medium with a constant refractive index. Let $d\pi_i$ denote an element of the area of this wavefront at incidence point $\overline{P}_i$. All of the rays passing through $d\pi_i$ intersect some subsequent wavefront $\overline{\Omega}$ within an elemental area $d\pi_i'$ at point $\overline{P}_i'$ (as shown in Fig. 8.2). Let $d\theta_1$ and $d\theta_2$ be the angles subtended at the centers of principal curvatures $\kappa_1$ and$\kappa_2$, respectively, by area $d\pi_i$. If $B_i$ is the irradiance at incidence point $\overline{P}_i$ (as computed from Eq. (6.9)), the irradiance $B'$ at any intermediate point $\overline{P}_i'$ in Fig. 8.2 can be computed via the following energy conservation law:

$$\frac{B_i'}{B_i} = \frac{d\pi_i}{d\pi_i'} = \frac{d\theta_1 d\theta_2/(\kappa_1\kappa_2)}{d\theta_1 d\theta_2/(\kappa_1'\kappa_2')} = \frac{\kappa_1'\kappa_2'}{\kappa_1\kappa_2}. \tag{8.50}$$

Equation (8.50) shows that the intensity along the ray is proportional to the Gaussian curvature $\kappa_1\kappa_2$ of the wavefront. For example, the irradiance of any intermediate point along the chief ray between points $\overline{P}_8$ and $\overline{P}_9$ in Example 8.3 is equal to $B_8' = B_8\ \kappa_1'\kappa_2'/(\kappa_1\kappa_2) = 15.457746\kappa_1'\kappa_2'$, where $B_8 = 0.006252$ when $B_8$ is normalized such that the total flux from source point $\overline{P}_0$ is equal to one (as shown by Eq. (6.9)). This intensity law suggests the feasibility of deriving an algorithm to determine the irradiance by tracking the Gaussian curvature as the ray travels through the optical system.

*Example 8.5* As shown in Table 8.4, the principal curvatures at the point of incidence of the chief ray on the 10th boundary surface are $\kappa_1 = 1/46.36329$ and $\kappa_2 = 1/57.16198$, respectively for the system in Fig. 2.1. The corresponding principal unit tangent vectors are $\overline{T}_1 = [0\quad 0.268532\quad 0.963271\quad 0]^T$ and

**Fig. 8.6** Variation of irradiance with travel distance for three rays originating from source point $\overline{P}_0 = [0 \quad -507 \quad 170 \quad 1]^T$ in an axis-symmetrical system shown in Fig. 2.1. **a** Irradiance of rays when traveling from 1st to 10th boundary surface. **b** Irradiance of rays when traveling from 10th boundary surface to an image plane. It shows that the irradiance increases rapidly in the region between the 10th boundary surface and the image plane

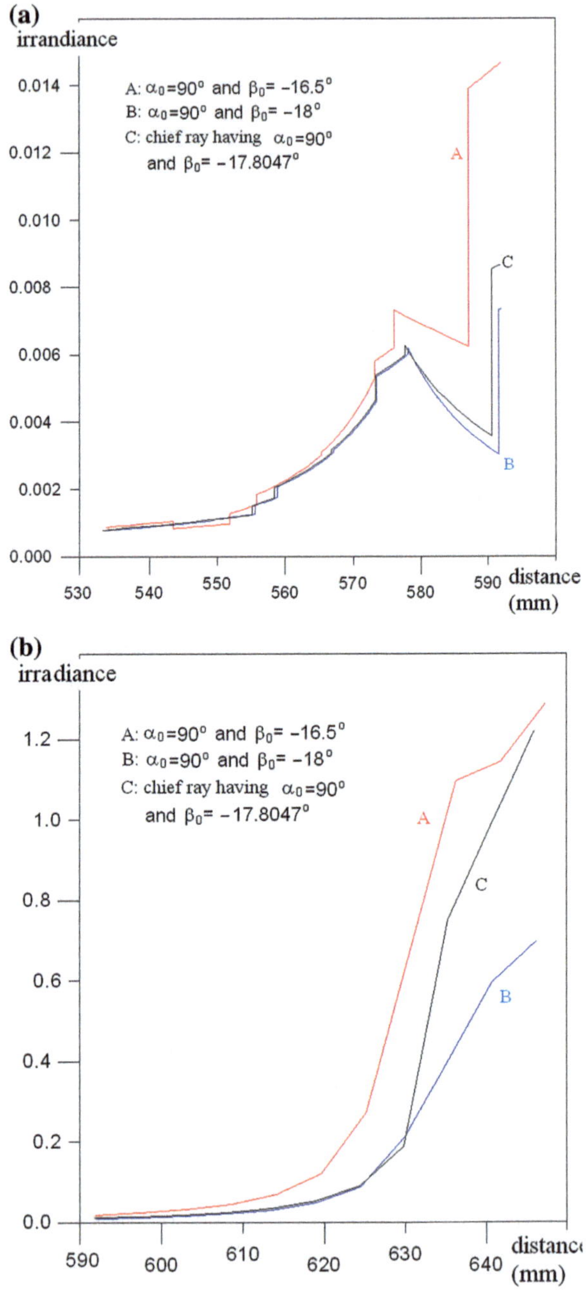

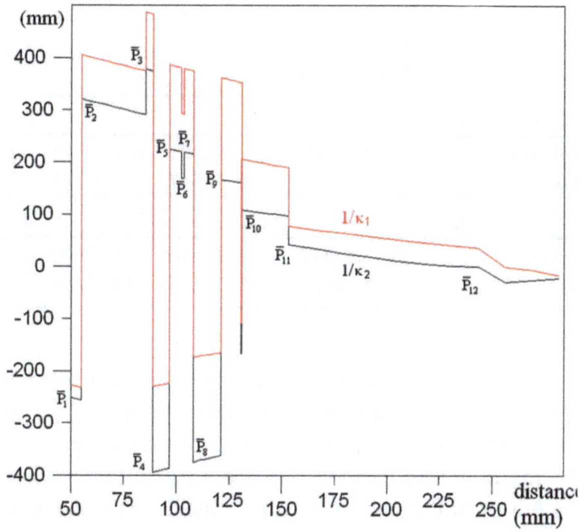


**Fig. 8.7** Variation of principal radii, $1/\kappa_1$ and $1/\kappa_2$, with travel distance for a ray originating from source point $\overline{P}_0 = [0 \quad -40 \quad 0 \quad 1]^T$ with $\alpha_0 = 4.5°$ and $\beta_0 = 1.5°$ in a non-axially symmetrical system shown in Fig. 3.12

$\overline{T}_2 = [-1 \quad 0 \quad 0 \quad 0]^T$, respectively. It was shown by Kneisly [9] that the principal directions remain unchanged for a ray traveling through a medium with a constant refractive index. It was also shown by Mitchell and Hanrahan [10] that when a ray travels from $\overline{P}_i$ (at which the principal curvatures of wavefront are $\kappa_1$ and $\kappa_2$) to an intermediate point $\overline{P}'_i$ (see Figs. 8.2 and 8.4), the principal curvatures $\kappa'_1$ and $\kappa'_2$ at $\overline{P}'_i$ have the forms

$$\kappa'_1 = \frac{\kappa_1}{1 - \lambda\kappa_1}, \tag{8.51}$$

$$\kappa'_2 = \frac{\kappa_2}{1 - \lambda\kappa_2}. \tag{8.52}$$

Inspecting the above equations, it is noted that after point $\overline{P}_{10}$, the denominator of either Eq. (8.51) or Eq. (8.52) becomes zero when the converging wavefront moves a distance $\lambda_1 = 1/\kappa_1 = 46.36329$ or $\lambda_2 = 1/\kappa_2 = 57.16198$, respectively; hence the irradiance becomes infinity (see Fig. 8.4). These positions of extremely high irradiance are the caustic surface of the wavefront. The caustic positions can also be determined by solving Eqs. (8.42) and (8.43) subject to $\partial\overline{\Omega}/\partial\alpha_0 = 0$ and $\partial\overline{\Omega}/\partial\beta_0 = 0$, respectively; resulting in values of $\lambda_1 = 46.36333$ and $\lambda_2 = 57.16197$, respectively.

**Table 8.5** Principal radii of curvature of wavefront at each boundary surface encountered by a ray originating from $\overline{P}_0 = [0 \quad -40 \quad 0 \quad 1]^T$ with $\alpha_0 = 4.5°$ and $\beta_0 = 1.5°$ in a non-axially symmetrical optical system shown in Fig. 3.12

| Incidence point $\overline{P}_i$ | $\overline{P}_1$ | $\overline{P}_2$ | $\overline{P}_3$ | $\overline{P}_4$ | $\overline{P}_5$ | $\overline{P}_6$ |
|---|---|---|---|---|---|---|
| Principal | −251.18806 | 320.94801 | 378.03749 | −394.22296 | 223.84250 | 169.00521 |
| Radii | −227.35411 | 404.74967 | 486.96250 | −229.40263 | 386.00136 | 292.72790 |

| Incidence point $\overline{P}_i$ | $\overline{P}_7$ | $\overline{P}_8$ | $\overline{P}_9$ | $\overline{P}_{10}$ | $\overline{P}_{11}$ | $\overline{P}_{12}$ |
|---|---|---|---|---|---|---|
| Principal | 218.44317 | −374.54701 | 165.92683 | −166.82421 | 108.66737 | 42.56389 |
| Radii | 378.88820 | −173.51977 | 361.77838 | −109.30423 | 204.95631 | 76.80608 |

*Example 8.6* The most convenient approach for determining the irradiance in an optical system is the ray-counting method [11], in which multiple rays are traced and a count is made of the number of rays hitting each grid of the mesh. However, the ray-counting method has a poor accuracy near the focal point of the optical system or in the vicinity of the caustic surfaces where the irradiance is very large. This problem arises since when the ray converges toward a focus point, the tube area becomes very small then the irradiance is significantly increased (see Fig. 8.4). However, deriving the wavefront shape using the method proposed herein involves the computation of local infinitesimal areas; therefore, the problem of poor accuracy is overcome. Figures 8.6a and b illustrate the variation of the irradiance along three different ray paths originating from the source point $\overline{P}_0 = [0 \quad -507 \quad 170 \quad 1]^T$ in the axis-symmetrical optical system in Fig. 2.1. It is noted from Fig. 8.6b that the irradiance increases rapidly between the 10th and 11th boundary surfaces. This is to be expected since all of the rays originating from $\overline{P}_0 = [0 \quad -507 \quad 170 \quad 1]^T$ are focused on the 11th boundary surface (i.e., the image plane).

*Example 8.7* Consider the wavefront along the path of the ray originating from source point $\overline{P}_0 = [0 \quad -40 \quad 0 \quad 1]^T$ with $\alpha_0 = 4.5°$ and $\beta_0 = 1.5°$ in the non-axially symmetrical system shown in Fig. 3.12. The principal radii of curvature of the wavefront at incidence points $\overline{P}_i$ (i = 1 to i = 12) are listed in Table 8.5. The variations of the principal radii of curvature with the travel distance of the ray are shown in Fig. 8.7. In general, the results presented in Table 8.5 and Fig. 8.7 indicate that the method proposed herein is applicable to both axis-symmetrical and non-axially symmetrical optical systems.

# Appendix A

$$
\begin{aligned}
\left[\frac{\partial^2 P_{ix}}{\partial x_w \partial x_v}\right] = &-\frac{1}{E_i}\begin{bmatrix} 0 & 0 & 0 & J_{ix} & 0 & 0 \\ & 0 & 0 & J_{iy} & 0 & 0 \\ & & 0 & J_{iz} & 0 & 0 \\ & & & 0 & 0 & 0 \\ & \text{symm.} & & & 0 & 0 \\ & & & & & 0 \end{bmatrix} + \frac{\ell_{i-1x}}{E_i^2}\begin{bmatrix} 0 & 0 & 0 & J_{ix}J_{ix} & J_{ix}J_{iy} & J_{ix}J_{iz} \\ & 0 & 0 & J_{iy}J_{ix} & J_{iy}J_{iy} & J_{iy}J_{iz} \\ & & 0 & J_{iz}J_{ix} & J_{iz}J_{iy} & J_{iz}J_{iz} \\ & & & 0 & 0 & 0 \\ & \text{symm.} & & & 0 & 0 \\ & & & & & 0 \end{bmatrix} \\
&+\frac{D_i}{E_i^2}\begin{bmatrix} 0 & 0 & 0 & 0 & 0 & 0 \\ & 0 & 0 & 0 & 0 & 0 \\ & & 0 & 0 & 0 & 0 \\ & & & 2J_{ix} & J_{iy} & J_{iz} \\ & \text{symm.} & & & 0 & 0 \\ & & & & & 0 \end{bmatrix} - \frac{2D_i\ell_{i-1x}}{E_i^3}\begin{bmatrix} 0 & 0 & 0 & 0 & 0 & 0 \\ & 0 & 0 & 0 & 0 & 0 \\ & & 0 & 0 & 0 & 0 \\ & & & J_{ix}J_{ix} & J_{ix}J_{iy} & J_{ix}J_{iz} \\ & \text{symm.} & & & J_{iy}J_{iy} & J_{iy}J_{iz} \\ & & & & & J_{iz}J_{iz} \end{bmatrix},
\end{aligned}
\tag{A.1}
$$

$$
\begin{aligned}
\left[\frac{\partial^2 P_{iy}}{\partial x_w \partial x_v}\right] = &-\frac{1}{E_i}\begin{bmatrix} 0 & 0 & 0 & 0 & J_{ix} & 0 \\ & 0 & 0 & 0 & J_{iy} & 0 \\ & & 0 & 0 & J_{iz} & 0 \\ & & & 0 & 0 & 0 \\ & \text{symm.} & & & 0 & 0 \\ & & & & & 0 \end{bmatrix} + \frac{\ell_{i-1y}}{E_i^2}\begin{bmatrix} 0 & 0 & 0 & J_{ix}J_{ix} & J_{ix}J_{iy} & J_{ix}J_{iz} \\ & 0 & 0 & J_{iy}J_{ix} & J_{iy}J_{iy} & J_{iy}J_{iz} \\ & & 0 & J_{iz}J_{ix} & J_{iz}J_{iy} & J_{iz}J_{iz} \\ & & & 0 & 0 & 0 \\ & \text{symm.} & & & 0 & 0 \\ & & & & & 0 \end{bmatrix} \\
&+\frac{D_i}{E_i^2}\begin{bmatrix} 0 & 0 & 0 & 0 & 0 & 0 \\ & 0 & 0 & 0 & 0 & 0 \\ & & 0 & 0 & 0 & 0 \\ & & & 0 & J_{ix} & 0 \\ & \text{symm.} & & & 2J_{iy} & J_{iz} \\ & & & & & 0 \end{bmatrix} - \frac{2D_i\ell_{i-1y}}{E_i^3}\begin{bmatrix} 0 & 0 & 0 & 0 & 0 & 0 \\ & 0 & 0 & 0 & 0 & 0 \\ & & 0 & 0 & 0 & 0 \\ & & & J_{ix}J_{ix} & J_{ix}J_{iy} & J_{ix}J_{iz} \\ & \text{symm.} & & & J_{iy}J_{iy} & J_{iy}J_{iz} \\ & & & & & J_{iz}J_{iz} \end{bmatrix},
\end{aligned}
\tag{A.2}
$$

$$
\begin{aligned}
\left[\frac{\partial^2 P_{iz}}{\partial x_w \partial x_v}\right] = &-\frac{1}{E_i}\begin{bmatrix} 0 & 0 & 0 & 0 & 0 & J_{ix} \\ & 0 & 0 & 0 & 0 & J_{iy} \\ & & 0 & 0 & 0 & J_{iz} \\ & & & 0 & 0 & 0 \\ & \text{symm.} & & & 0 & 0 \\ & & & & & 0 \end{bmatrix} + \frac{\ell_{i-1z}}{E_i^2}\begin{bmatrix} 0 & 0 & 0 & J_{ix}J_{ix} & J_{ix}J_{iy} & J_{ix}J_{iz} \\ & 0 & 0 & J_{iy}J_{ix} & J_{iy}J_{iy} & J_{iy}J_{iz} \\ & & 0 & J_{iz}J_{ix} & J_{iz}J_{iy} & J_{iz}J_{iz} \\ & & & 0 & 0 & 0 \\ & \text{symm.} & & & 0 & 0 \\ & & & & & 0 \end{bmatrix} \\
&+\frac{D_i}{E_i^2}\begin{bmatrix} 0 & 0 & 0 & 0 & 0 & 0 \\ & 0 & 0 & 0 & 0 & 0 \\ & & 0 & 0 & 0 & 0 \\ & & & 0 & 0 & J_{ix} \\ & \text{symm.} & & & 0 & J_{iy} \\ & & & & & 2J_{iz} \end{bmatrix} - \frac{2D_i\ell_{i-1z}}{E_i^3}\begin{bmatrix} 0 & 0 & 0 & 0 & 0 & 0 \\ & 0 & 0 & 0 & 0 & 0 \\ & & 0 & 0 & 0 & 0 \\ & & & J_{ix}J_{ix} & J_{ix}J_{iy} & J_{ix}J_{iz} \\ & \text{symm.} & & & J_{iy}J_{iy} & J_{iy}J_{iz} \\ & & & & & J_{iz}J_{iz} \end{bmatrix},
\end{aligned}
\tag{A.3}
$$

with $D_i$ and $E_i$ are given in Eqs. (2.39) and (2.40).

$$\left[\frac{\partial^2 \ell_{ix}}{\partial x_w \partial x_v}\right] = \frac{s_i N_i^2 (1 - N_i^2) J_{ix}}{\sqrt{\left(1 - N_i^2 + N_i^2 E_i^2\right)^3}} \begin{bmatrix} 0 & 0 & 0 & 0 & 0 & 0 \\ & 0 & 0 & 0 & 0 & 0 \\ & & 0 & 0 & 0 & 0 \\ & & & J_{ix}J_{ix} & J_{ix}J_{iy} & J_{ix}J_{iz} \\ & \text{symm.} & & & J_{iy}J_{iy} & J_{iy}J_{iz} \\ & & & & & J_{iz}J_{iz} \end{bmatrix}, \tag{A.4}$$

$$\left[\frac{\partial^2 \ell_{iy}}{\partial x_w \partial x_v}\right] = \frac{s_i N_i^2 (1 - N_i^2) J_{iy}}{\sqrt{\left(1 - N_i^2 + N_i^2 E_i^2\right)^3}} \begin{bmatrix} 0 & 0 & 0 & 0 & 0 & 0 \\ & 0 & 0 & 0 & 0 & 0 \\ & & 0 & 0 & 0 & 0 \\ & & & J_{ix}J_{ix} & J_{ix}J_{iy} & J_{ix}J_{iz} \\ & \text{symm.} & & & J_{iy}J_{iy} & J_{iy}J_{iz} \\ & & & & & J_{iz}J_{iz} \end{bmatrix}, \tag{A.5}$$

$$\left[\frac{\partial^2 \ell_{iz}}{\partial x_w \partial x_v}\right] = \frac{s_i N_i^2 (1 - N_i^2) J_{iz}}{\sqrt{\left(1 - N_i^2 + N_i^2 E_i^2\right)^3}} \begin{bmatrix} 0 & 0 & 0 & 0 & 0 & 0 \\ & 0 & 0 & 0 & 0 & 0 \\ & & 0 & 0 & 0 & 0 \\ & & & J_{ix}J_{ix} & J_{ix}J_{iy} & J_{ix}J_{iz} \\ & \text{symm.} & & & J_{iy}J_{iy} & J_{iy}J_{iz} \\ & & & & & J_{iz}J_{iz} \end{bmatrix}, \tag{A.6}$$

# Appendix B

$$\left[\frac{\partial^2 P_{ix}}{\partial x_w \partial x_v}\right] = \begin{bmatrix} 0 & 0 & 0 & \lambda_{i,1} & 0 & 0 \\ & 0 & 0 & \lambda_{i,2} & 0 & 0 \\ & & 0 & \lambda_{i,3} & 0 & 0 \\ & & & 2\lambda_{i,4} & \lambda_{i,5} & \lambda_{i,6} \\ & \text{symm.} & & & 0 & 0 \\ & & & & & 0 \end{bmatrix}$$

$$+\ell_{i-1x}\begin{bmatrix} \lambda_{i,11} & \lambda_{i,21} & \lambda_{i,31} & \lambda_{i,41} & \lambda_{i,51} & \lambda_{i,61} \\ & \lambda_{i,22} & \lambda_{i,32} & \lambda_{i,42} & \lambda_{i,52} & \lambda_{i,62} \\ & & \lambda_{i,33} & \lambda_{i,43} & \lambda_{i,53} & \lambda_{i,63} \\ & & & \lambda_{i,44} & \lambda_{i,54} & \lambda_{i,64} \\ & \text{symm.} & & & \lambda_{i,55} & \lambda_{i,65} \\ & & & & & \lambda_{i,66} \end{bmatrix}, \tag{B.1}$$

$$\left[\frac{\partial^2 P_{iy}}{\partial x_w \partial x_v}\right] = \begin{bmatrix} 0 & 0 & 0 & 0 & \lambda_{i,1} & 0 \\ & 0 & 0 & 0 & \lambda_{i,2} & 0 \\ & & 0 & 0 & \lambda_{i,3} & 0 \\ & & & 0 & \lambda_{i,4} & 0 \\ & \text{symm.} & & & 2\lambda_{i,5} & \lambda_{i,6} \\ & & & & & 0 \end{bmatrix} + \ell_{i-1y}\begin{bmatrix} \lambda_{i,11} & \lambda_{i,21} & \lambda_{i,31} & \lambda_{i,41} & \lambda_{i,51} & \lambda_{i,61} \\ & \lambda_{i,22} & \lambda_{i,32} & \lambda_{i,42} & \lambda_{i,52} & \lambda_{i,62} \\ & & \lambda_{i,33} & \lambda_{i,43} & \lambda_{i,53} & \lambda_{i,63} \\ & & & \lambda_{i,44} & \lambda_{i,54} & \lambda_{i,64} \\ & \text{symm.} & & & \lambda_{i,55} & \lambda_{i,65} \\ & & & & & \lambda_{i,66} \end{bmatrix}, \tag{B.2}$$

$$\left[\frac{\partial^2 P_{iz}}{\partial x_w \partial x_v}\right] = \begin{bmatrix} 0 & 0 & 0 & 0 & 0 & \lambda_{i,1} \\ & 0 & 0 & 0 & 0 & \lambda_{i,2} \\ & & 0 & 0 & 0 & \lambda_{i,3} \\ & & & 0 & 0 & \lambda_{i,4} \\ & \text{symm.} & & & 0 & \lambda_{i,5} \\ & & & & & 2\lambda_{i,6} \end{bmatrix} + \ell_{i-1z}\begin{bmatrix} \lambda_{i,11} & \lambda_{i,21} & \lambda_{i,31} & \lambda_{i,41} & \lambda_{i,51} & \lambda_{i,61} \\ & \lambda_{i,22} & \lambda_{i,32} & \lambda_{i,42} & \lambda_{i,52} & \lambda_{i,62} \\ & & \lambda_{i,33} & \lambda_{i,43} & \lambda_{i,53} & \lambda_{i,63} \\ & & & \lambda_{i,44} & \lambda_{i,54} & \lambda_{i,64} \\ & \text{symm.} & & & \lambda_{i,55} & \lambda_{i,65} \\ & & & & & \lambda_{i,66} \end{bmatrix}, \tag{B.3}$$

where $\partial^2\lambda_i/\partial x_w \partial x_v = \lambda_{i,wv}$ is given by

$$\lambda_{i,wv} = -D_{i,wv} \pm \frac{(2D_{i,w}D_{i,v} + 2D_i D_{i,wv} - E_{i,wv})}{2\sqrt{D_i^2 - E_i}} \pm \frac{(-2D_i D_{i,w} + E_{i,w})(2D_i D_{i,v} - E_{i,v})}{4(D_i^2 - E_i)\sqrt{D_i^2 - E_i}}, \tag{B.4}$$

with

$$\begin{aligned}D_{i,11} &= D_{i,12} = D_{i,21} = D_{i,13} = D_{i,31} = D_{i,15} = D_{i,51} = D_{i,16} = D_{i,61} = D_{i,22} = D_{i,23} = D_{i,32} = D_{i,24}\\ &= D_{i,42} = D_{i,26} = D_{i,62} = D_{i,33} = D_{i,34} = D_{i,43} = D_{i,35} = D_{i,53} = D_{i,44} = D_{i,45} = D_{i,54} = D_{i,46}\\ &= D_{i,64} = D_{i,55} = D_{i,56} = D_{i,65} = D_{i,66} = 0,\ D_{i,14} = D_{i,25} = D_{i,36} = D_{i,41} = D_{i,52} = D_{i,63} = 1,\end{aligned}$$

$$\begin{aligned}E_{i,12} &= E_{i,21} = E_{i,13} = E_{i,31} = E_{i,14} = E_{i,41} = E_{i,15} = E_{i,51} = E_{i,16} = E_{i,61} = E_{i,23} = E_{i,32} = E_{i,24} = E_{i,42}\\ &= E_{i,25} = E_{i,52} = E_{i,26} = E_{i,62} = E_{i,34} = E_{i,43} = E_{i,35} = E_{i,53} = E_{i,36} = E_{i,63} = E_{i,44} = E_{i,45} = E_{i,54}\\ &= E_{i,46} = E_{i,64} = E_{i,55} = E_{i,56} = E_{i,65} = E_{i,66} = 0,\ E_{i,11} = E_{i,22} = E_{i,33} = 2.\end{aligned}$$

$$\begin{aligned}\frac{\partial^2 \ell_{ix}}{\partial \overline{R}_{i-1}^2} &= \left[\frac{\partial^2 \ell_{ix}}{\partial x_w \partial x_v}\right] = \left(\frac{-n_{ix} N_i^2 C\theta_i}{\sqrt{1 - N_i^2 + (N_i C\theta_i)^2}} + N_i n_{ix}\right)\left[C\theta_{i,wv}\right]\\ &+ \left(-\sqrt{1 - N_i^2 + (N_i C\theta_i)^2} + N_i C\theta_i\right)\left[n_{ix,wv}\right]\\ &+ \frac{n_{ix} N_i^2 (N_i^2 - 1)}{\left(1 - N_i^2 + N_i^2 C\theta_i^2\right)\sqrt{1 - N_i^2 + (N_i C\theta_i)^2}}\left[C\theta_{i,w}\right]^T\left[C\theta_{i,v}\right]\\ &+ \left(\frac{-N_i^2 C\theta_i}{\sqrt{1 - N_i^2 + (N_i C\theta_i)^2}} + N_i\right)\left(\left[C\theta_{i,w}\right]^T\left[n_{ix,v}\right] + \left[n_{ix,w}\right]^T\left[C\theta_{i,v}\right]\right),\end{aligned} \tag{B.5}$$

$$\begin{aligned}\frac{\partial^2 \ell_{iy}}{\partial \overline{R}_{i-1}^2} &= \left[\frac{\partial^2 \ell_{iy}}{\partial x_w \partial x_v}\right] = \left(\frac{-n_{iy} N_i^2 C\theta_i}{\sqrt{1 - N_i^2 + (N_i C\theta_i)^2}} + N_i n_{iy}\right)\left[C\theta_{i,wv}\right]\\ &+ \left(-\sqrt{1 - N_i^2 + (N_i C\theta_i)^2} + N_i C\theta_i\right)\left[n_{iy,wv}\right]\\ &+ \frac{n_{iy} N_i^2 (N_i^2 - 1)}{\left(1 - N_{i^2} + (N_i C\theta_i)^2\right)\sqrt{1 - N_i^2 + (N_i C\theta_i)^2}}\left[C\theta_{i,w}\right]^T\left[C\theta_{i,v}\right]\\ &+ \left(\frac{-N_i^2 C\theta_i}{\sqrt{1 - N_i^2 + (N_i C\theta_i)^2}} + N_i\right)\left(\left[C\theta_{i,w}\right]^T\left[n_{iy,v}\right] + \left[n_{iy,w}\right]^T\left[C\theta_{i,v}\right]\right),\end{aligned} \tag{B.6}$$

$$
\begin{aligned}
\frac{\partial^2 \ell_{iz}}{\partial \bar{R}_{i-1}^2} &= \left[\frac{\partial^2 \ell_{iz}}{\partial x_w \partial x_v}\right] \\
&= \left(\frac{-n_{iz} N_i^2 C\,\theta_i}{\sqrt{1 - N_i^2 + (N_i C\,\theta_i)^2}} + N_i n_{iz}\right)\left[C\,\theta_{i,wv}\right] \\
&\quad + \left(-\sqrt{1 - N_i^2 + (N_i C\,\theta_i)^2} + N_i C\,\theta_i\right)\left[n_{iz,wv}\right] \\
&\quad + \frac{n_{iz} N_i^2 (N_i^2 - 1)}{\left(1 - N_i^2 + (N_i C\,\theta_i)^2\right)\sqrt{1 - N_i^2 + (N_i C\,\theta_i)^2}}\left[C\,\theta_{i,w}\right]^T\left[C\,\theta_{i,v}\right] \\
&\quad + \left(\frac{-N_i^2 C\,\theta_i}{\sqrt{1 - N_i^2 + (N_i C\,\theta_i)^2}} + N_i\right)\left(\left[C\,\theta_{i,w}\right]^T\left[n_{iz,v}\right] + \left[n_{iz,w}\right]^T\left[C\,\theta_{i,v}\right]\right),
\end{aligned}
\tag{B.7}
$$

where $C\theta_{i,wv} = \partial^2(C\theta_i)/\partial x_w \partial x_v$, $n_{ix,wv} = \partial^2 n_{ix}/\partial x_w \partial x_v$, $n_{iy,wv} = \partial^2 n_{iy}/\partial x_w \partial x_v$, $n_{iz,wv} = \partial^2 n_{iz}/\partial x_w \partial x_v$ are respectively given by:

$$
\begin{aligned}
C\theta_{i,11} &= -\left(\ell_{i-1x} n_{ix,11} + \ell_{i-1y} n_{iy,11} + \ell_{i-1z} n_{iz,11}\right), \\
C\theta_{i,12} &= -\left(\ell_{i-1x} n_{ix,12} + \ell_{i-1y} n_{iy,12} + \ell_{i-1z} n_{iz,12}\right) = C\theta_{i,21}, \\
C\theta_{i,13} &= -\left(\ell_{i-1x} n_{ix,13} + \ell_{i-1y} n_{iy,13} + \ell_{i-1z} n_{iz,13}\right) = C\theta_{i,31}, \\
C\theta_{i,14} &= -\left(\ell_{i-1x} n_{ix,14} + \ell_{i-1y} n_{iy,14} + \ell_{i-1z} n_{iz,14}\right) - n_{ix,1} = C\theta_{i,41}, \\
C\theta_{i,15} &= -\left(\ell_{i-1x} n_{ix,15} + \ell_{i-1y} n_{iy,15} + \ell_{i-1z} n_{iz,15}\right) - n_{iy,1} = C\theta_{i,51}, \\
C\theta_{i,16} &= -\left(\ell_{i-1x} n_{ix,16} + \ell_{i-1y} n_{iy,16} + \ell_{i-1z} n_{iz,16}\right) - n_{iz,1} = C\theta_{i,61},
\end{aligned}
$$

$$
\begin{aligned}
C\,\theta_{i,22} &= -\left(\ell_{i-1x} n_{ix,22} + \ell_{i-1y} n_{iy,22} + \ell_{i-1z} n_{iz,22}\right), \\
C\,\theta_{i,23} &= -\left(\ell_{i-1x} n_{ix,23} + \ell_{i-1y} n_{iy,23} + \ell_{i-1z} n_{iz,23}\right) = C\,\theta_{i,32}, \\
C\,\theta_{i,24} &= -\left(\ell_{i-1x} n_{ix,24} + \ell_{i-1y} n_{iy,24} + \ell_{i-1z} n_{iz,24}\right) - n_{ix,2} = C\,\theta_{i,42}, \\
C\,\theta_{i,25} &= -\left(\ell_{i-1x} n_{ix,25} + \ell_{i-1y} n_{iy,25} + \ell_{i-1z} n_{iz,25}\right) - n_{iy,2} = C\,\theta_{i,52}, \\
C\,\theta_{i,26} &= -\left(\ell_{i-1x} n_{ix,26} + \ell_{i-1y} n_{iy,26} + \ell_{i-1z} n_{iz,26}\right) - n_{iz,2} = C\,\theta_{i,62},
\end{aligned}
$$

$$
\begin{aligned}
C\theta_{i,33} &= -\left(\ell_{i-1x} n_{ix,33} + \ell_{i-1y} n_{iy,33} + \ell_{i-1z} n_{iz,33}\right), \\
C\theta_{i,34} &= -\left(\ell_{i-1x} n_{ix,34} + \ell_{i-1y} n_{iy,34} + \ell_{i-1z} n_{iz,34}\right) - n_{ix,3} = C\theta_{i,43}, \\
C\theta_{i,35} &= -\left(\ell_{i-1x} n_{ix,35} + \ell_{i-1y} n_{iy,35} + \ell_{i-1z} n_{iz,35}\right) - n_{iy,3} = C\theta_{i,53}, \\
C\theta_{i,36} &= -\left(\ell_{i-1x} n_{ix,36} + \ell_{i-1y} n_{iy,36} + \ell_{i-1z} n_{iz,36}\right) - n_{iz,3} = C\theta_{i,63},
\end{aligned}
$$

$$
\begin{aligned}
C\theta_{i,44} &= -\left(\ell_{i-1x} n_{ix,44} + \ell_{i-1y} n_{iy,44} + \ell_{i-1z} n_{iz,44}\right) - n_{ix,4} - n_{ix,4}, \\
C\theta_{i,45} &= -\left(\ell_{i-1x} n_{ix,45} + \ell_{i-1y} n_{iy,45} + \ell_{i-1z} n_{iz,45}\right) - n_{ix,5} - n_{iy,4} = C\theta_{i,54}, \\
C\theta_{i,46} &= -\left(\ell_{i-1x} n_{ix,46} + \ell_{i-1y} n_{iy,46} + \ell_{i-1z} n_{iz,46}\right) - n_{ix,6} - n_{iz,4} = C\theta_{i,64},
\end{aligned}
$$

$$C\theta_{i,55} = -(\ell_{i-1x}\, n_{ix,55} + \ell_{i-1y}\, n_{iy,55} + \ell_{i-1z}\, n_{iz,55}) - n_{iy,5} - n_{iy,5}\,,$$

$$C\theta_{i,56} = -(\ell_{i-1x}\, n_{ix,56} + \ell_{i-1y}\, n_{iy,56} + \ell_{i-1z}\, n_{iz,56}) - n_{iy,6} - n_{iz,5} = C\theta_{i,65},$$

$$C\theta_{i,66} = -(\ell_{i-1x}\, n_{ix,66} + \ell_{i-1y}\, n_{iy,66} + \ell_{i-1z}\, n_{iz,66}) - n_{iz,6} - n_{iz,6},$$

$$\begin{aligned} n_{ix,wv} =& s_i(-I_{ix}C\beta_i C\alpha_i - I_{iy}C\beta_i S\alpha_i - I_{iz}S\beta_i)\beta_{i,w}\beta_{i,v} + s_i(I_{ix}S\beta_i S\alpha_0 - I_{iy}S\beta_i C\alpha_i)\beta_{i,w}\alpha_{i,v} \\ &+ s_i(I_{ix}S\beta_i S\alpha_i - I_{iy}S\beta_i C\alpha_i)\alpha_{i,w}\beta_{i,v} + s_i(-I_{ix}C\beta_i C\alpha_i - I_{iy}C\beta_i S\alpha_i)\alpha_{i,w}\alpha_{i,v} \\ &+ s_i(-I_{ix}S\beta_i C\alpha_i - I_{iy}S\beta_i S\alpha_i + I_{iz}C\beta_i)\beta_{i,wv} + s_i(-I_{ix}C\beta_i S\alpha_i + I_{iy}C\beta_i C\alpha_i)\alpha_{i,wv}\,, \end{aligned}$$

$$\begin{aligned} n_{iy,wv} =& s_i(-J_{ix}C\beta_i C\alpha_i - J_{iy}C\beta_i S\alpha_i - J_{iz}S\beta_i)\beta_{i,w}\beta_{i,v} \\ &+ s_i(J_{ix}S\beta_i S\alpha_i - J_{iy}S\beta_i C\alpha_i)\beta_{i,w}\alpha_{i,v} \\ &+ s_i(J_{ix}S\beta_i S\alpha_i - J_{iy}S\beta_i C\alpha_i)\alpha_{i,w}\beta_{i,v} \\ &+ s_i(-J_{ix}C\beta_i C\alpha_i - J_{iy}C\beta_i S\alpha_i)\alpha_{i,w}\alpha_{i,v} \\ &+ s_i(-J_{ix}S\beta_i C\alpha_i - J_{iy}S\beta_i S\alpha_i + J_{iz}C\beta_i)\beta_{i,wv} \\ &+ s_i(-J_{ix}C\beta_i S\alpha_i + J_{iy}C\beta_i C\alpha_i)\alpha_{i,wv}, \end{aligned}$$

$$\begin{aligned} n_{iz,wv} =& s_i(-K_{ix}C\beta_i C\alpha_i - K_{iy}C\beta_i S\alpha_i - K_{iz}S\beta_i)\beta_{i,w}\beta_{i,v} + s_i(K_{ix}S\beta_i S\alpha_i - K_{iy}S\beta_i C\alpha_i)\beta_{i,w}\alpha_{i,v} \\ &+ s_i(K_{ix}S\beta_i S\alpha_i - K_{iy}S\beta_i C\alpha_i)\alpha_{i,w}\beta_{i,v} + s_i(-K_{ix}C\beta_i C\alpha_i - K_{iy}C\beta_i S\alpha_i)\alpha_{i,w}\alpha_{i,v} \\ &+ s_i(-K_{ix}S\beta_i C\alpha_i - K_{iy}S\beta_i S\alpha_i + K_{iz}C\beta_i)\beta_{i,wv} + s_i(-K_{ix}C\beta_i S\alpha_i + K_{iy}C\beta_i C\alpha_i)\alpha_{i,wv}, \end{aligned}$$

$$\begin{aligned} \beta_{i,wv} =& \frac{\partial^2\beta_i}{\partial x_w \partial x_v} \\ =& \frac{(\sigma_i^2+\rho_i^2)\tau_{i,wv} + 2(\sigma_i\sigma_{i,v}+\rho_i\rho_{i,v})\tau_{i,w} - \tau_i(\sigma_{i,v}\sigma_{i,w}+\sigma_i\sigma_{i,wv}+\rho_{i,v}\rho_{i,w}+\rho_i\rho_{i,uv}) - (\sigma_i\sigma_{i,w}+\rho_i\rho_{i,w})\tau_{i,v}}{(\sigma_i^2+\rho_i^2+\tau_i^2)\sqrt{(\sigma_i^2+\rho_i^2)}} \\ &+ \left[(\sigma_i^2+\rho_i^2)\tau_{i,w} - \tau_i(\sigma_i\sigma_{i,w}+\rho_i\rho_{i,w})\right]\left[\frac{-2(\sigma_i\sigma_{i,v}+\rho_i\rho_{i,v}+\tau_i\tau_{i,v})}{(\sigma_i^2+\rho_i^2+\tau_i^2)^2\sqrt{(\sigma_i^2+\rho_i^2)}} - \frac{(\sigma_i\sigma_{i,v}+\rho_i\rho_{i,v})}{(\sigma_i^2+\rho_i^2+\tau_i^2)(\sigma_i^2+\rho_i^2)\sqrt{(\sigma_i^2+\rho_i^2)}}\right], \end{aligned}$$

$$\begin{aligned} \alpha_{i,wv} =& \frac{\partial^2\alpha_i}{\partial x_w \partial x_v} = \frac{\sigma_i\rho_{i,wv} + \sigma_{i,v}\rho_{i,w} - \rho_{i,v}\sigma_{i,w} - \rho_i\sigma_{i,wv}}{\sigma_i^2+\rho_i^2} \\ &- \frac{2(\sigma_i\rho_{i,w} - \rho_i\sigma_{i,w})(\sigma_i\sigma_{i,v}+\rho_i\rho_{i,v})}{(\sigma_i^2+\rho_i^2)^2}, \end{aligned}$$

$$\begin{aligned} \sigma_{i,11} =& (I_{ix}\ell_{i-1x} + I_{iy}\ell_{i-1y} + I_{iz}\ell_{i-1z})\lambda_{i,11}, \\ \sigma_{i,12} =& (I_{ix}\ell_{i-1x} + I_{iy}\ell_{i-1y} + I_{iz}\ell_{i-1z})\lambda_{i,12} = \sigma_{i,21}, \\ \sigma_{i,13} =& (I_{ix}\ell_{i-1x} + I_{iy}\ell_{i-1y} + I_{iz}\ell_{i-1z})\lambda_{i,13} = \sigma_{i,31}, \\ \sigma_{i,14} =& (I_{ix}\ell_{i-1x} + I_{iy}\ell_{i-1y} + I_{iz}\ell_{i-1z})\lambda_{i,14} + I_{ix}\lambda_{i,1} = \sigma_{i,41}, \\ \sigma_{i,15} =& (I_{ix}\ell_{i-1x} + I_{iy}\ell_{i-1y} + I_{iz}\ell_{i-1z})\lambda_{i,15} + I_{iy}\lambda_{i,1} = \sigma_{i,51}, \\ \sigma_{i,16} =& (I_{ix}\ell_{i-1x} + I_{iy}\ell_{i-1y} + I_{iz}\ell_{i-1z})\lambda_{i,16} + I_{iz}\lambda_{i,1} = \sigma_{i,61}, \end{aligned}$$

$$\sigma_{i,22} = (I_{ix}\ell_{i-1x} + I_{iy}\ell_{i-1y} + I_{iz}\ell_{i-1z})\lambda_{i,22},$$
$$\sigma_{i,23} = (I_{ix}\ell_{i-1x} + I_{iy}\ell_{i-1y} + I_{iz}\ell_{i-1z})\lambda_{i,23} = \sigma_{i,32},$$
$$\sigma_{i,24} = (I_{ix}\ell_{i-1x} + I_{iy}\ell_{i-1y} + I_{iz}\ell_{i-1z})\lambda_{i,24} + I_{ix}\lambda_{i,2} = \sigma_{i,42},$$
$$\sigma_{i,25} = (I_{ix}\ell_{i-1x} + I_{iy}\ell_{i-1y} + I_{iz}\ell_{i-1z})\lambda_{i,25} + I_{iy}\lambda_{i,2} = \sigma_{i,52},$$
$$\sigma_{i,26} = (I_{ix}\ell_{i-1x} + I_{iy}\ell_{i-1y} + I_{iz}\ell_{i-1z})\lambda_{i,26} + I_{iz}\lambda_{i,2} = \sigma_{i,62},$$

$$\sigma_{i,33} = (I_{ix}\ell_{i-1x} + I_{iy}\ell_{i-1y} + I_{iz}\ell_{i-1z})\lambda_{i,33},$$
$$\sigma_{i,34} = (I_{ix}\ell_{i-1x} + I_{iy}\ell_{i-1y} + I_{iz}\ell_{i-1z})\lambda_{i,34} + I_{ix}\lambda_{i,3} = \sigma_{i,43},$$
$$\sigma_{i,35} = (I_{ix}\ell_{i-1x} + I_{iy}\ell_{i-1y} + I_{iz}\ell_{i-1z})\lambda_{i,35} + I_{iy}\lambda_{i,3} = \sigma_{i,53},$$
$$\sigma_{i,36} = (I_{ix}\ell_{i-1x} + I_{iy}\ell_{i-1y} + I_{iz}\ell_{i-1z})\lambda_{i,36} + I_{iz}\lambda_{i,3} = \sigma_{i,63},$$

$$\sigma_{i,44} = (I_{ix}\ell_{i-1x} + I_{iy}\ell_{i-1y} + I_{iz}\ell_{i-1z})\lambda_{i,44} + I_{ix}\lambda_{i,4} + I_{ix}\lambda_{i,4},$$
$$\sigma_{i,45} = (I_{ix}\ell_{i-1x} + I_{iy}\ell_{i-1y} + I_{iz}\ell_{i-1z})\lambda_{i,45} + I_{ix}\lambda_{i,5} + I_{iy}\lambda_{i,4} = \sigma_{i,54},$$
$$\sigma_{i,46} = (I_{ix}\ell_{i-1x} + I_{iy}\ell_{i-1y} + I_{iz}\ell_{i-1z})\lambda_{i,46} + I_{ix}\lambda_{i,6} + I_{iz}\lambda_{i,4} = \sigma_{i,64},$$

$$\sigma_{i,55} = (I_{ix}\ell_{i-1x} + I_{iy}\ell_{i-1y} + I_{iz}\ell_{i-1z})\lambda_{i,55} + I_{iy}\lambda_{i,5} + I_{iy}\lambda_{i,5},$$
$$\sigma_{i,56} = (I_{ix}\ell_{i-1x} + I_{iy}\ell_{i-1y} + I_{iz}\ell_{i-1z})\lambda_{i,56} + I_{iy}\lambda_{i,6} + I_{iz}\lambda_{i,5} = \sigma_{i,65},$$

$$\sigma_{i,66} = (I_{ix}\ell_{i-1x} + I_{iy}\ell_{i-1y} + I_{iz}\ell_{i-1z})\lambda_{i,66} + I_{iz}\lambda_{i,6} + I_{iz}\lambda_{i,6},$$

$$\rho_{i,11} = (J_{ix}\ell_{i-1x} + J_{iy}\ell_{i-1y} + J_{iz}\ell_{i-1z})\lambda_{i,11},$$
$$\rho_{i,12} = (J_{ix}\ell_{i-1x} + J_{iy}\ell_{i-1y} + J_{iz}\ell_{i-1z})\lambda_{i,12} = \rho_{i,21},$$
$$\rho_{i,13} = (J_{ix}\ell_{i-1x} + J_{iy}\ell_{i-1y} + J_{iz}\ell_{i-1z})\lambda_{i,13} = \rho_{i,31},$$
$$\rho_{i,14} = (J_{ix}\ell_{i-1x} + J_{iy}\ell_{i-1y} + J_{iz}\ell_{i-1z})\lambda_{i,14} + J_{ix}\lambda_{i,1} = \rho_{i,41},$$
$$\rho_{i,15} = (J_{ix}\ell_{i-1x} + J_{iy}\ell_{i-1y} + J_{iz}\ell_{i-1z})\lambda_{i,15} + J_{iy}\lambda_{i,1} = \rho_{i,51},$$
$$\rho_{i,16} = (J_{ix}\ell_{i-1x} + J_{iy}\ell_{i-1y} + J_{iz}\ell_{i-1z})\lambda_{i,16} + J_{iz}\lambda_{i,1} = \rho_{i,61},$$

$$\rho_{i,22} = (J_{ix}\ell_{i-1x} + J_{iy}\ell_{i-1y} + J_{iz}\ell_{i-1z})\lambda_{i,22},$$
$$\rho_{i,23} = (J_{ix}\ell_{i-1x} + J_{iy}\ell_{i-1y} + J_{iz}\ell_{i-1z})\lambda_{i,23} = \rho_{i,32},$$
$$\rho_{i,24} = (J_{ix}\ell_{i-1x} + J_{iy}\ell_{i-1y} + J_{iz}\ell_{i-1z})\lambda_{i,24} + J_{ix}\lambda_{i,2} = \rho_{i,42},$$
$$\rho_{i,25} = (J_{ix}\ell_{i-1x} + J_{iy}\ell_{i-1y} + J_{iz}\ell_{i-1z})\lambda_{i,25} + J_{iy}\lambda_{i,2} = \rho_{i,52},$$
$$\rho_{i,26} = (J_{ix}\ell_{i-1x} + J_{iy}\ell_{i-1y} + J_{iz}\ell_{i-1z})\lambda_{i,26} + J_{iz}\lambda_{i,2} = \rho_{i,62},$$

$$\rho_{i,33} = (J_{ix}\ell_{i-1x} + J_{iy}\ell_{i-1y} + J_{iz}\ell_{i-1z})\lambda_{i,33},$$
$$\rho_{i,34} = (J_{ix}\ell_{i-1x} + J_{iy}\ell_{i-1y} + J_{iz}\ell_{i-1z})\lambda_{i,34} + J_{ix}\lambda_{i,3} = \rho_{i,43},$$
$$\rho_{i,35} = (J_{ix}\ell_{i-1x} + J_{iy}\ell_{i-1y} + J_{iz}\ell_{i-1z})\lambda_{i,35} + J_{iy}\lambda_{i,3} = \rho_{i,53},$$
$$\rho_{i,36} = (J_{ix}\ell_{i-1x} + J_{iy}\ell_{i-1y} + J_{iz}\ell_{i-1z})\lambda_{i,36} + J_{iz}\lambda_{i,3} = \rho_{i,63},$$

$$\rho_{i,44} = (J_{ix}\ell_{i-1x} + J_{iy}\ell_{i-1y} + J_{iz}\ell_{i-1z})\lambda_{i,44} + J_{ix}\lambda_{i,4} + J_{ix}\lambda_{i,4},$$
$$\rho_{i,45} = (J_{ix}\ell_{i-1x} + J_{iy}\ell_{i-1y} + J_{iz}\ell_{i-1z})\lambda_{i,45} + J_{ix}\lambda_{i,5} + J_{iy}\lambda_{i,4} = \rho_{i,54},$$
$$\rho_{i,46} = (J_{ix}\ell_{i-1x} + J_{iy}\ell_{i-1y} + J_{iz}\ell_{i-1z})\lambda_{i,46} + J_{ix}\lambda_{i,6} + J_{iz}\lambda_{i,4} = \rho_{i,64},$$

$$\rho_{i,55} = (J_{ix}\ell_{i-1x} + J_{iy}\ell_{i-1y} + J_{iz}\ell_{i-1z})\lambda_{i,55} + J_{iy}\lambda_{i,5} + J_{iy}\lambda_{i,5},$$
$$\rho_{i,56} = (J_{ix}\ell_{i-1x} + J_{iy}\ell_{i-1y} + J_{iz}\ell_{i-1z})\lambda_{i,56} + J_{iy}\lambda_{i,6} + J_{iz}\lambda_{i,5} = \rho_{i,65},$$

$$\rho_{i,66} = (J_{ix}\ell_{i-1x} + J_{iy}\ell_{i-1y} + J_{iz}\ell_{i-1z})\lambda_{i,66} + J_{iz}\lambda_{i,6} + J_{iz}\lambda_{i,6},$$

$$\tau_{i,11} = (K_{ix}\ell_{i-1x} + K_{iy}\ell_{i-1y} + K_{iz}\ell_{i-1z})\lambda_{i,11},$$
$$\tau_{i,12} = (K_{ix}\ell_{i-1x} + K_{iy}\ell_{i-1y} + K_{iz}\ell_{i-1z})\lambda_{i,12} = \tau_{i,21},$$
$$\tau_{i,13} = (K_{ix}\ell_{i-1x} + K_{iy}\ell_{i-1y} + K_{iz}\ell_{i-1z})\lambda_{i,13} = \tau_{i,31},$$
$$\tau_{i,14} = (K_{ix}\ell_{i-1x} + K_{iy}\ell_{i-1y} + K_{iz}\ell_{i-1z})\lambda_{i,14} + K_{ix}\lambda_{i,1} = \tau_{i,41},$$
$$\tau_{i,15} = (K_{ix}\ell_{i-1x} + K_{iy}\ell_{i-1y} + K_{iz}\ell_{i-1z})\lambda_{i,15} + K_{iy}\lambda_{i,1} = \tau_{i,51},$$
$$\tau_{i,16} = (K_{ix}\ell_{i-1x} + K_{iy}\ell_{i-1y} + K_{iz}\ell_{i-1z})\lambda_{i,16} + K_{iz}\lambda_{i,1} = \tau_{i,61},$$

$$\tau_{i,22} = (K_{ix}\ell_{i-1x} + K_{iy}\ell_{i-1y} + K_{iz}\ell_{i-1z})\lambda_{i,22},$$
$$\tau_{i,23} = (K_{ix}\ell_{i-1x} + K_{iy}\ell_{i-1y} + K_{iz}\ell_{i-1z})\lambda_{i,23} = \tau_{i,32},$$
$$\tau_{i,24} = (K_{ix}\ell_{i-1x} + K_{iy}\ell_{i-1y} + K_{iz}\ell_{i-1z})\lambda_{i,24} + K_{ix}\lambda_{i,2} = \tau_{i,42},$$
$$\tau_{i,25} = (K_{ix}\ell_{i-1x} + K_{iy}\ell_{i-1y} + K_{iz}\ell_{i-1z})\lambda_{i,25} + K_{iy}\lambda_{i,2} = \tau_{i,52},$$
$$\tau_{i,26} = (K_{ix}\ell_{i-1x} + K_{iy}\ell_{i-1y} + K_{iz}\ell_{i-1z})\lambda_{i,26} + K_{iz}\lambda_{i,2} = \tau_{i,62},$$

$$\tau_{i,33} = (K_{ix}\ell_{i-1x} + K_{iy}\ell_{i-1y} + K_{iz}\ell_{i-1z})\lambda_{i,33},$$
$$\tau_{i,34} = (K_{ix}\ell_{i-1x} + K_{iy}\ell_{i-1y} + K_{iz}\ell_{i-1z})\lambda_{i,34} + K_{ix}\lambda_{i,3} = \tau_{i,43},$$
$$\tau_{i,35} = (K_{ix}\ell_{i-1x} + K_{iy}\ell_{i-1y} + K_{iz}\ell_{i-1z})\lambda_{i,35} + K_{iy}\lambda_{i,3} = \tau_{i,53},$$
$$\tau_{i,36} = (K_{ix}\ell_{i-1x} + K_{iy}\ell_{i-1y} + K_{iz}\ell_{i-1z})\lambda_{i,36} + K_{iz}\lambda_{i,3} = \tau_{i,63},$$

$$\tau_{i,44} = (K_{ix}\ell_{i-1x} + K_{iy}\ell_{i-1y} + K_{iz}\ell_{i-1z})\lambda_{i,44} + K_{ix}\lambda_{i,4} + K_{ix}\lambda_{i,4},$$
$$\tau_{i,45} = (K_{ix}\ell_{i-1x} + K_{iy}\ell_{i-1y} + K_{iz}\ell_{i-1z})\lambda_{i,45} + K_{ix}\lambda_{i,5} + K_{iy}\lambda_{i,4} = \tau_{i,54},$$
$$\tau_{i,46} = (K_{ix}\ell_{i-1x} + K_{iy}\ell_{i-1y} + K_{iz}\ell_{i-1z})\lambda_{i,46} + K_{ix}\lambda_{i,6} + K_{iz}\lambda_{i,4} = \tau_{i,64},$$

$$\tau_{i,55} = (K_{ix}\ell_{i-1x} + K_{iy}\ell_{i-1y} + K_{iz}\ell_{i-1z})\lambda_{i,55} + K_{iy}\lambda_{i,5} + K_{iy}\lambda_{i,5},$$
$$\tau_{i,56} = (K_{ix}\ell_{i-1x} + K_{iy}\ell_{i-1y} + K_{iz}\ell_{i-1z})\lambda_{i,56} + K_{iy}\lambda_{i,6} + K_{iz}\lambda_{i,5} = \tau_{i,65},$$

$$\tau_{i,66} = (K_{ix}\ell_{i-1x} + K_{iy}\ell_{i-1y} + K_{iz}\ell_{i-1z})\lambda_{i,66} + K_{iz}\lambda_{i,6} + K_{iz}\lambda_{i,6},$$

$$\frac{\partial^2 \ell_{ix}}{\partial \overline{R}_{i-1}^2} = \left[\frac{\partial^2 \ell_{ix}}{\partial x_w \partial x_v}\right]$$
$$= 2C\theta_i\left[n_{ix,wv}\right] + 2\left[C\theta_{i,v}\right]^T\left[n_{ix,w}\right] + 2\left[n_{ix,v}\right]^T\left[C\theta_{i,w}\right] + 2n_{ix}\left[C\theta_{i,wv}\right], \tag{B.8}$$

$$\frac{\partial^2 \ell_{iy}}{\partial \overline{R}_{i-1}^2} = \left[\frac{\partial^2 \ell_{iy}}{\partial x_w \partial x_v}\right]$$
$$= 2C\theta_i\left[n_{iy,wv}\right] + 2\left[C\theta_{i,v}\right]^T\left[n_{iy,w}\right] + 2\left[n_{iy,v}\right]^T\left[C\theta_{i,w}\right] + 2n_{iy}\left[C\theta_{i,wv}\right], \tag{B.9}$$

$$\frac{\partial^2 \ell_{iz}}{\partial \overline{R}_{i-1}^2} = \left[\frac{\partial^2 \ell_{iz}}{\partial x_w \partial x_v}\right]$$
$$= 2C\theta_i\left[n_{iz,wv}\right] + 2\left[C\theta_{i,v}\right]^T\left[n_{iz,w}\right] + 2\left[n_{iz,v}\right]^T\left[C\theta_{i,w}\right] + 2n_{iz}\left[C\theta_{i,wv}\right] \tag{B.10}$$

# References


1. O.N. Stavroudis, *The Optics of Rays, Wavefronts, and Caustics*. (Academic press, Cambridge, 1972)
2. O.N. Stavroudis, *The mathematics of geometricl and physical optics*. (Wiley, 2006)
3. A Pressley, *Elementary Differential Geometry*, The Springer Undergraduate Mathematics Series, (2001) p. 123
4. J.A. Hoffnagle, D.L. Shealy, Refracting the k-function: Stavroudis's solution to the eikonal equation for multielement optical systems. J. Opt. Soc. Am. A **28**, 1312–1321 (2011)
5. D.L. Shealy, D.G. Burkhard, Caustic surfaces and irradiance for reflection and refraction from an ellipsoid, elliptic parabolid and elliptic cone. Appl. Opt. **12**, 2955–2959 (1973)
6. D.L. Shealy, D.G. Burkhard, Caustic surface merit functions in optical design. J. Opt. Soc. Am. **66**, 1122 (1976)
7. D.L. Shealy, Analytical illuminance and caustic surface calculations in geometrical optics. Appl. Opt. **15**, 2588–2596 (1976)
8. D.L. Shealy, J.A. Hoffnagle, Wavefront and caustics of a plane wave refracted by an arbitrary surace. J. Opt. Soc. Am. A **25**, 2370–2382 (2008)
9. J.A. Kneisly II, Local curvature of wavefronts in an optical system. J. Opt. Soc. of Am. **54**, 229–235 (1964)
10. D.P. Mitchell, P. Hanrahan, Illumination from curved reflectors. in Proceedings of SIGGRAPH, computer graphics, vol 26, no. 2. , pp. 283–291 (1992)
11. W.J. Smith, *Modern Optical Engineering*, 3rd edn. (Edmund Industrial Optics, Barrington, 2001)
12. D.G. Burkhard, D.L. Shealy, Simplified formula for the illuminance in an optical system. Appl. Opt. **20**, 897–909 (1981)

# Curriculum Vitae

Psang Dain Lin
Distinguished Professor
National Cheng Kung University
Department of Mechanical Engineering
Tainan, Taiwan 70101
Phone: +886-6-2080563
Website: http://140.116.31.121/
E-mail: pdlin@mail.ncku.edu.tw

Education:
July 1979, BS, National Cheng Kung University, Department of Mechanical Engineering, Taiwan.
July 1984, MS, National Cheng Kung University, Department of Mechanical Engineering, Taiwan.
July 1989, PhD, Northwestern University, Department of Mechanical Engineering, USA.

PD Lin, *New Computation Methods for Geometrical Optics*,
Springer Series in Optical Sciences 178, DOI: 10.1007/978-981-4451-79-6,

MIX
Papier aus verantwortungsvollen Quellen
Paper from responsible sources
FSC® C105338

If you have any concerns about our products,
you can contact us on
**ProductSafety@springernature.com**

In case Publisher is established outside the EU,
the EU authorized representative is:
**Springer Nature Customer Service Center GmbH**
**Europaplatz 3, 69115 Heidelberg, Germany**

Printed by Libri Plureos GmbH
in Hamburg, Germany